Nanotechnology in Functional Foods

Scrivener Publishing
100 Cummings Center, Suite 541J
Beverly, MA 01915-6106

Publishers at Scrivener
Martin Scrivener (martin@scrivenerpublishing.com)
Phillip Carmical (pcarmical@scrivenerpublishing.com)

Nanotechnology in Functional Foods

Edited by

Tanima Bhattacharya
Shakeel Ahmed

Scrivener Publishing

WILEY

This edition first published 2022 by John Wiley & Sons, Inc., 111 River Street, Hoboken, NJ 07030, USA and Scrivener Publishing LLC, 100 Cummings Center, Suite 541J, Beverly, MA 01915, USA

For more information about Scrivener publications please visit www.scrivenerpublishing.com.

Wiley Global Headquarters
111 River Street, Hoboken, NJ 07030, USA

For details of our global editorial offices, customer services, and more information about Wiley products visit us at www.wiley.com.

Library of Congress Cataloging-in-Publication Data

ISBN 978-1-119-90482-3

Cover image: Pixabay.Com
Cover design by Russell Richardson

Set in size of 11pt and Minion Pro by Manila Typesetting Company, Makati, Philippines

Printed in the USA

10 9 8 7 6 5 4 3 2 1

Contents

Preface

Functional foods are an emerging trend in the food industry, whose potential value is determined by whether they are safe with respect to consumer health. *Nanotechnology in Functional Foods* was written to help the reader better understand the benefits and concerns associated with these foods. In addition to giving an overview of the current state-of-the-art in functional foods, different aspects of the advanced research being conducted on their extraction, synthesis, analysis, and biological effects are presented. Besides focusing on several synthesis techniques, the handbook also discusses the application of nanoparticles in nutrient delivery and pharmaceuticals; along with the advanced chemistry, consumer acceptance, and diversification of these nutrients. Moreover, new trends are discussed concerning the application of artificial intelligence in screening various components of functional foods. Therefore, as can be seen from the synopsis of each chapter given below, this handbook will be of great value to clinicians, scientists, research and development experts, research scholars, students and food industry personnel.

In Chapter 1, Animesh *et al.* describe various nanoencapsulation techniques, viz., nanoprecipitation, coacervation, nanoliposomes, solid lipid nanoparticles, nanoemulsions, nanofibers, and nanotubes, for specified nutrient delivery applications. The chapter focuses on various types of coating materials like chitosan, pectin, gelatin, phospholipids, tannic acids, and whey protein concentrates, and their efficacy in bioaccumulation of bioactive compounds, enhancement and improvement of stability, bioavailability under extrinsic parameters, controlled release and sensory attributes. The authors also address several research gaps for a better understanding of nanoparticles' fate in the metabolism of the human body.

In Chapter 2, Anjana *et al.* discuss recent trends in nutraceuticals and highlight their various attributes in the fortification, preservation, and enhancement of the shelf life of food; and discuss the relevance of nanomaterials in the nutraceutical industry. A detailed approach to the delivery of nutraceuticals in order to retain flavor, odor, and sensory qualities is outlined, along with the active intelligent technique of using nanonutraceuticals as preservatives in preventing deterioration of foods.

In Chapter 3, Anindita Deb Pal discusses the advanced biochemistry of bioactive components of functional foods in detail. The chapter throws light on several biopolymer nanoencapsulation systems and techniques, many of which change the chemical and physical properties of food, nutrient delivery at specified site of action, and the mechanism of nanoencapsulated bioactive compounds in the gastrointestinal tract, i.e., their fate in ingestion, absorption, and transmission in the animal body.

In Chapter 4, Keskin *et al.* elaborate on the ancient functional food "honey," and its beneficial effects on human health. The authors streamlined green synthesis techniques of metal nanoparticles with the help of honey and conducted a detailed survey of its pharmaceutical applications as antioxidant and anti-cancer agents.

In Chapter 5, Tanima *et al.* focus on nondairy yogurt and its nutritional importance in advancements in curing health disorders. Milk products are important daily use consumer products. However, though they have enormous health benefits, for those who are lactose intolerant they cause gastrointestinal problems; and in some cases, may contribute to cardiac diseases. Thus, the worldwide acceptance of nondairy or plant-based milk and its products are at their peak. This chapter gives a detailed outline of the various plant-based yogurts claiming to be functional foods. Also highlighted is the implementation of nanotechnology to increase the nutritive value and stability of products in order to cure lifestyle-related diseases which are comorbidity factors in the recent COVID-19 pandemic.

In Chapter 6, Shashank *et al.* discuss the potency of phyto oils as antimicrobial and antioxidant agents against multidrug-resistant bacteria. The authors focus on nanoformulations of phyto oils which improve stability, solubility, bioavailability, bioactivity, and drug delivery efficacy.

In Chapter 7, Mohamed *et al.* discuss the challenges involved in crossing the blood–brain barrier to treat neurological disorders. The chapter covers various lipid nanoformulations which overcome this barrier by increasing bioavailability and stability, and improving the brain targeting specificity of nuerotherapeutics, thereby curing neurological disorders.

In Chapter 8, Ahsan *et al.* summarize the effects of functional foods on the immune system of COVID-19 patients through influencing nanotechnology. They highlight the efficacy and mechanisms of functional foods like vitamins, natural plants, probiotics, and flavonoids in boosting immunity when nanoformulations are synthesized as well as applied.

In Chapter 9, Anindita *et al.* give a vivid description of dietary diversification and its impact on human health. The authors also raise toxicity concerns and the need to better address the regulatory aspects of nanofoods for further acceptance.

In Chapter 10, Dr. Luxita Sharma explains various types of organic, inorganic, and natural sources-based nanomaterials and their application in food fortification, nutrient delivery, and active, smart packaging. Nanofood safety, toxicity issues, and regulatory aspects of nanomaterials used in the food industry are also discussed.

In Chapter 11, Hitesh Chopra describes various routes of delivery of functional foods using nanotechnology. Also discussed are the fate and metabolism of those nanoformulations after being administered.

In Chapter 12, Abhishek *et al.* highlight nanotechnology-based approaches for various applications of functional foods like processing, preserving, extending shelf life, and packaging. The authors also present a market research report on the global rise in consumer acceptance for such nanofoods.

In Chapter 13, Mutoffar *et al.* present a combinatorial overview of the application of nanotechnology in enhancing physicochemical properties of functional foods and the role of artificial intelligence in exploring new bioactive components of functional food. The efficacy of their mechanism of action against health hazards is also discussed.

Finally, Chapter 14 by Shiladitya *et al.* sheds light on the functionalization of fortified foods by nanofortification in relation to toxicity issues and level of human consumption. Also discussed are several challenges facing the existing nanofortification of functional foods that need to be addressed in order for advancements in the food sector to be realized.

The editors are thankful for the work submitted by authors from different parts of the globe (India, China, Pakistan, and Indonesia), whose contributions will capture the interest of those who wish to study and research trends in functional foods.

The Editors
Tanima Bhattacharya
Shakeel Ahmed
June 2022

1

Advancement of Nanotechnology in Developing Functional Foods: Nanotechnology in Functional Foods

Animesh Naskar[1], Sebak Ranjan Roy[1]*, Ivi Chakraborty[2] and Tanima Bhattacharya[3,4]

[1]*Department of Food Science and Technology, Maulana Abul Kalam Azad University of Technology, West Bengal, India*
[2]*Department of Post Harvest Technology, Faculty of Horticulture, Bidhan Chandra Krishi Viswavidyalay, West Bengal, India*
[3]*College of Chemistry and Chemical Engineering, Hubei University, Wuhan, China*
[4]*Department of Science and Engineering, Novel Global Community Educational Foundation, Hebersham, Australia*

Abstract
Exposure of nanotechnology deals with several aspects of the food technology domain in the recent decade to provide significant quantity, safe, and healthy foods for human consumption. In this consequence, the research activities have been directed toward the advancement of nanotechnology in the food sector. Nano-based materials are not only involved in functional food development also uses in food processing, food packaging, and rapid detection of foodborne pathogens of food products. Other applications include the improvement of taste, flavor, color, the texture of foodstuffs. Recent research depicts the potential utilization of nanoparticles showing their efficacy in the delivery of bioactive ingredients, which explores the benefits not just within food products but also around food products. Nanoencapsulation is an emerging field in this regard, elevating the protection performance to sensitive bioactive ingredients by preventing unnecessary interactions with other constituents in foods. Moreover, the current data emphasizing the attributes of bioavailability and the degree of solubility of nanofoods ensure the optimum discharge of ingredients of food at an action taking place, thereby improving human health. However, successful applications of nanotechnology to

**Corresponding author*: roy_sebak@rediffmail.com

Tanima Bhattacharya and Shakeel Ahmed (eds.) Nanotechnology in Functional Foods, (1–24)

foods are paving new paths day by day, there persist many challenges that need to be concerned in evaluating the safety of food.

Keywords: Functional foods, Nanotechnology, Nanoscience, Food Product Development, Food Processing, Food Technology

1.1 Introduction

As people become more conscious of the effects of nutrition on their health, there is a growing demand for food products that provide health advantages. As a result, functional foods containing ingredients for health promotion, such as essential fatty acids, kind of antioxidants, pigments, like carotenoids, vitamins, various minerals, phytosterol, and fibers become more popular [1]. The term "functional food" first appeared in Japan in the 1980s, when hospitals were facing rising healthcare costs as the senior population grew rapidly [2]. The Ministry of Education, Science, and Culture, Government of Japan launched the initial National Project regarding Functional Foods in 1984, intending to assist universities with basic and applied research. There has been increasing recognition of the critical relation between diet and human health since the turn of the twentieth century. This has resulted in the creation of a new food category known as functional foods. Foods and their constituents that could result in health advantages in addition to nourishment are referred to as functional foods [3]. To put it another way, functional foods are made up of a variety of nutrients, nonnutrients and other components altering several physiological functions related to a state of well-being and health, as well as minimize the risk of disease [4]. Current research shows that the food industry's functional foods category is projected to be costing around 168000 million dollars and is expanding at an annual pace of roughly 9% [5].

Nanotechnology and Nanoscience have the efficacy to revolutionize the way of novel approaches to the functional foods' production, such as the incorporation of bioactive substances without compromising consumer's sensory discernment and enhancing uptake of certain components [5]. This technology is primarily concerned with the atomic, molecular, and macromolecular scales of materials and their management. Nanosized particles have a higher surface area per mass than larger-sized particles of the same chemistry, making them more physiologically active [6]. Nanotechnology, as is widely known, has opened up new horizons in disciplines, such as clinical diagnosis in medical science, drug transport, packaging of various cosmetics, tissue scaffolding, and, hence, its application possibilities are limitless. Apart from these significant applications, the nanoscale material has novel features and, hence, has a lot of potential in food processing industry, particularly in development of functional foods [7]. Sometimes, in the milk system, several biomolecules, such as network of fat globules and micelles of casein, already occur in their nanoparticle forms. Food nanotechnology, on the other hand, refers to nanoparticles that have been manufactured and added to food. Various forms of nanostructured constituents having functionality may be used as building blocks for creating unique structures, as well as add new functions to food products [8]. In other words, food that has been grown, produced, processed, or packed utilizing nanotechnology techniques or instruments is referred to as nanofood. Nanotechnology may create new opportunities and improvements in relation to food texture, flavor, processability, as well as improved product stability over time, in addition to enhancing the delivery system's performance and effectiveness [9]. Apart from these attributes of nanomaterials, when it comes to food processing, production, and packaging, key uses of nanotechnology include changes in food quality, effective sensor incorporation, and superior additives encapsulation. Nanocapsules, nanoliposomes, solid lipid nanostructure, nanoemulsions, nanotubes, nanofibers, and other nanoscale or modified nanoscale materials can all be used in this regard [6]. Nanomaterials can be fabricated from both organic and inorganic resources in these food applications.

However, worries about the potential negative effects of nano-based particles on the physiological profile of human health and the ecosystem may act as a deterrent to their use in this industry. According to early observations, nanoparticles' improved activity and penetrability capabilities may have negative consequences when used in biological systems [10]. As a result, risk assessments, regulatory policy, and monitoring should pay special attention to nanoscale applications. Several reports, each focusing on a different risk or societal issue [11, 12]. With the use of modern technological methodologies, this chapter discusses the importance of nanotechnology in the realm of making functional foods and nutraceuticals. It has also emphasized the use of nanoformulations for probiotics administration into proper destination and successful delivery of bioactive substances and minerals or micronutrients. In addition, the detection of food contamination is discussed, as well as the potential negative consequences.

1.2 Nanoencapsulation

The bioactive component serves as the core material in nanoencapsulation, and it must be covered with a coating material. The interaction of the covering material with the core influences the impact of nanoparticles' functionality since core materials differ from each other with their physical condition, solubility based on polarity, and atomic, as well as molecular weight. As a result, each core material must be enclosed by an appropriate wall material. The carrier material for food applications must be food grade and biodegradable. The physicochemical properties of the bioactive component which is to be encapsulated, as well as the technique of encapsulating influence the food industry's choice of carriers (covering materials) to nanoencapsulate the bioactive compounds [13]. Carbohydrate, lipid, and protein are the most prevalent wall components. Biopolymers (like starch, alginate, cellulose, etc.), xanthan gum, and pectin are all typical carbohydrate components. On the other hand, short-chain triglycerides, complex lipids, such as phospholipid from egg yolk, phosphatidylcholine, derived lipid-like cholesterol, oils like soybean oil, olive oil, and so on, are all employed in the lipid category. Whey proteins, soy proteins, gelatine, sodium caseinate, and zein are utilized in the protein type [14]. Bioactive ingredients can be nanoencapsulated to improve their water solubility and dispersibility in foods and beverages. The techniques improve bioavailability by displaying good dose-dependent functions, masks unpleasant flavors/tastes, and diminishes the negative impact on the mouthfeel. Moreover, it extends the shelf life and compatibility of food goods during manufacture, storage, transit, and use, as well as controlling the discharge/release rate or explicit delivery environment for better execution of their functions [15]. Nanoliposomes, nanoemulsions, nanoprecipitation, biopolymeric nanoparticles, and other technologies are utilized to create nanoencapsulated bioactive ingredients, which are depicted in Figure 1.1, and Table 1.1 illustrates a summary of these nanoformulation methods with their purposes in the development of functional food products.

1.2.1 Nanoliposomes

The liposome has a spherical form with particles ranging in size from nanometers to micrometers. Food nanotechnology has been interested in nanoliposomes because they are a suitable nanocarrier system for delivering the molecules relating to food and medicinal aspects like antibacterial, anti-inflammatory, and some critical therapeutic components, as well as providing bioactive component stabilization against numerous physiological barriers in the body of the living organism. This technology has progressed and shown its efficacy to entrap the desired compounds for improving stability, managing the sustained release of food ingredients, achieving successful encapsulation, and increasing bioavailability. Nanoliposomes

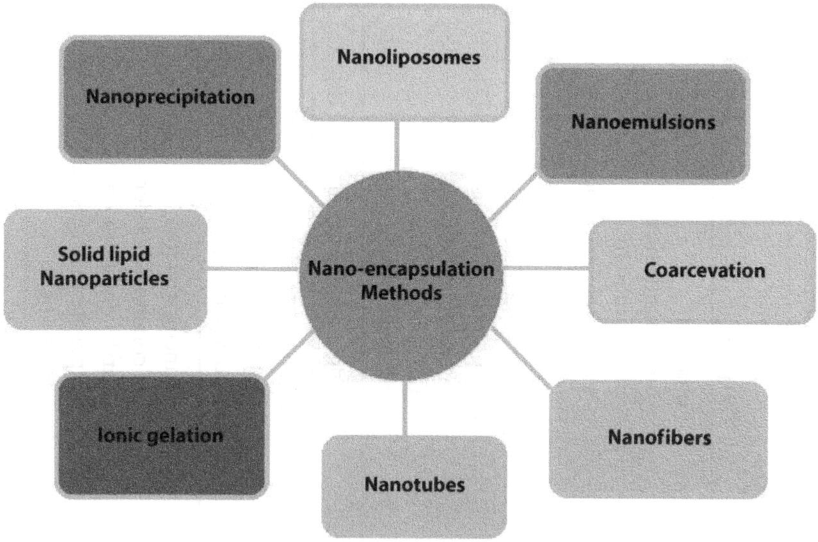

Figure 1.1 Nanoencapsulation techniques applied in the functionality of the final food products.

are mostly employed in the food industry to supply nutrients and flavors, but they have recently been explored, analyzed, and discovered to have the ability to incorporate antimicrobials that can protect food particles from microbial infection [26]. For applications in milk products, nanoliposomes having antimicrobial peptides were effectively used to prevent foodborne pathogens like gram-positive *Listeria monocytogenes* and gram-negative *Escherichia coli* [27]. Natural preservatives like essential oil occurring in Curry leaf that has been encapsulated in a liposome for use in functional foods. This liposome has antibacterial action against *Bacillus cereus*, one of the food poisoning organisms, and increased flour's shelf life obtained from rice [28]. To fabricate the nanoliposomes, some external and internal factors are considered, including acyl chain length of phospholipids constituents, polar head group's composition, pH, the saturation level of hydrocarbon chain, transition temperature, and strength of the suspension medium [29]. In addition, several characteristics influence the choice of a proper and suitable approach for the synthesis of liposomes, such as physicochemical characteristics of trapping substance, effective concentration of the chemical employed for entrapment, as well as its toxicological potential, the lipid vesicles' dispersion medium's intrinsic characteristics, and optimum and effective vesicle size and half life [30]. Nanoliposomes made from fish oil comprising phosphatidylcholine and sunflower oil

Table 1.1 Few nanoencapsulation methodologies developed for targeted delivery of bioactive components.

Raw materials (wall materials/emulsifiers)	Bioactive ingredients	Nanoencapsulation method	Carrier size (nm)	Purpose	Reference
Protein concentrate (Whey)/pectin	Curcumin	Nanoemulsions	145–305	Enhanced bioaccumulation of curcumin in transparent nematodes (*Caenorhabditis elegans*)	[16]
Chitosan and gelatin	Tocopherol, cinnamaldehyde, and garlic oil	Nanoemulsions	100–155	Boosting the physical occurrence, reducing the unfavorable sensorial outcome	[17]
Lecithin (marine)	Salmon oil	Nanoemulsions	150–210	Rising oxidative firmness	[18]
Phospholipids obtained from rice bran	Quercetin	Nanoliposomes	155	Develop enhanced solubility and bioavailability	[19]
Phosphatidylcholine from Soya	Vitamin C	Nanoliposomes	100–110	Advancing the stability of compound under harsh conditions on exposure to excessive heat or light	[20]

(*Continued*)

Table 1.1 Few nanoencapsulation methodologies developed for targeted delivery of bioactive components. (*Continued*)

Raw materials (wall materials/ emulsifiers)	Bioactive ingredients	Nanoencapsulation method	Carrier size (nm)	Purpose	Reference
Emulsifier (Tween-80)	α-Tocopherol, γ-Oryzanol	Nanoliposomes	200	Developing the stability of components and their controlled release	[21]
Polylactic-coglycolic acid (D/L), Polylactic acid (D/L) Tween 20, Gelatin	β-carotene	Nanoprecipitation	75–85	Developing bioavailability and stability under chemical as well as the physical stressed environment	[22]
Pectin/Zein	Resveratrol	Nanoprecipitation	235	Increasing chemical stability	[23]
Pectin and sodium caseinate	Rutin	Nanoprecipitation	215	Controlled release in developing intestinal circumstances	[24]
Gelatin, Tannins, Tween 60, glutaraldehyde	Capsaicin	Coacervation	110	Prohibit the release of foul odor and advance stability	[25]

were created using the sonication method. Using a probe sonicator with a 20-kHz optimal frequency, the generated liposomal dispersion was sonicated for 7 minutes at 25°C. Liposomes had an EE of 92.22%, with particle sizes ranging from 300 to 500 nm [31]. Risaliti *et al.* (2020) used nanoliposomal encapsulation to increase the biological and antifungal activities of essential oil obtained from sweet wormwood (*Artemisia annua*) [32]. Therefore, antimicrobial encapsulation has a significant impact on the use of nanoliposomes in food technology. Despite the fact that liposome technology has been used widely in milk and dairy products for a long time. Proteolytic enzymes including proteinases, lipolytic enzymes including lipases, bacteriocin like nisin, and types of flavors encapsulated with liposomes have been used in food fortification and processing of dairy products [33]. Nanoliposomes, on the other hand, got their start in the food technology industry by being used in the manufacturing of cheese, or dairy products. When compared with other enzyme encapsulation technologies for usage in the cheese industry, employing nanoliposomes has several advantages. Sometimes nanoliposomes can be made from ingredients found in cheese in their natural state, and these nanoliposomes prevent casein from early hydrolysis during cheese manufacture. Alongside, the manufacture of industry-based food-grade qualities is aided by excellent partitioning aptitude in the curd and modern techniques through tonal adjustments satisfying key standards, such as the Mozafari method [33, 34]. Apart from reducing processing time, liposomes are also used to fortify dairy products with vitamins, resulting in an increase in nutritional quality and a faster rate of digestion. The fact that washed curds perish due to the presence of spore-formers especially bacteria. Therefore, antimicrobial agent (nitrate) can be advised for production of these types of curds. However, As a result, nitrate usage is recommended in these types of curds. However, nitrate executes its toxicity, which is responsible for several health problems. Thus, antimicrobial lysozyme encapsulated with liposomes has been utilized, and it primarily aims the matrix area of cheese where typically the bacteria favors for their colonization. Nanoliposomes have been used to encapsulate a kind of antimicrobials, such as Nisin Z, fluorescein complex, bacteriocin, and so on [35]. Another application has been noticed where Skim milk was fortified with nanoliposomes made from shrimp oil. The effect of using fish oil nanoliposomes to make a functional food, namely fortified bread where technological, as well as sensory aspects, was studied. It was shown that the texture, nutritional value, and sensory attributes of the bread were unaffected by the addition of fish oil encapsulated nanoliposomes [36]. Because of its health benefits, nanoliposomes associated with vitamin using is appropriate for use as a food fortification agent.

A stable aqueous dispersion of vitamin D-encapsulated nanoliposomes were produced, and this formulation can be employed as a fortification ingredient in beverages [37]. In this manner, liposomal encapsulation helps to protect vitamins from degradation. The liposomal encapsulation of food preservatives is another significant involvement in the instance of processing dairy products. As a biopreservative alternative for milk shelf life extension, nisin and garlic extract were coencapsulated into phosphatidylcholinenanoliposomes. This nanoliposome technology can potentially be used as a biopreservative agent in other foods to extend shelf life and improve food quality [38]. Focusing on various bioactive compounds (BACs), carotenoids, such as lutein, carotene, lycopene, and canthaxanthin acting as antioxidants, were encapsulated utilizing liposomes in the form of nanoliposomes or microliposomes for use in functional foods [39]. In the recent decade, nanoliposome technology was used to make a stable green tea extract [40].

1.2.2 Nanoemulsions

Nanoemulsions are colloidal systems with minuscule droplet sizes that are kinetically stable. Nanoemulsions are emulsions with particle sizes ranging from 20 to 500 nanometers. Likewise nanoliposomes, In the food sector, nanoemulsions could be used to deliver nutraceuticals, antimicrobial compounds, coloring and flavoring agents, in addition to pharmaceuticals and cosmetics. Their increased surface area, ability to efficiently encapsulate, lesser sensitivity under physicochemical changes and tiny volume are all advantages that qualify them as possible candidates for use in the food sector. For the purpose of encapsulating and optimal distribution of bioactive lipophilic substances, the content and structure of nanoemulsions can be regulated. These nanoemulsions are created using two ways. One is high-energy mode and other is low-energy approach. As nanoemulsions contain many droplets, that may contribute to increasing the surface area. Although, creating more surface area necessitates a significant quantity of energy [41]. High-energy approaches consist of homogenization, microfluidization, ultrasonication with high-pressure valve, whereas low-energy approaches comprise phase inversion composition, phase inversion temperature, and spontaneous emulsification [42]. The delivery system based on nanoemulsion must be compatible with the food medium and have least impact lying on meal's organoleptic qualities, such as texture, flavor, and appearance [43]. Stabilizers, such as small molecule surfactants, are commonly utilized in the manufacture of nanoemulsions. Phospholipids like dairy lecithin, egg, or soy detergents like Twins/Spans, proteins having amphiphilic nature like caseinate, isolate from whey protein, and amphiphilic based polysaccharides (gum arabic and tailored starch) are examples of surfactants. Various stabilizers are being considered for further enhancement of the stability characteristics, such as modifying the textural properties, delaying ripening, and making it weighing agents to the liquid nanoemulsions to balance their densities [5]. Nanoemulsions derived from food-grade components are used in larger proportions in the process of encapsulating biologically active lipids, such as omega-3 fatty acids and polyunsaturated fatty acids in functional foods [44]. Several studies have focused on edible nanocoating derived from nanoemulsion enclosing flavor and coloring components, antimicrobials, enzymes, antioxidants, antibrowning substances that may be utilized as a coating material for several food products to extend their long-lasting expectancy. Furthermore, this process can inhibit gaseous exchange, as well as limit moisture loss and food deterioration. Another important feature of nanoemulsions in the food processing is their ability to encapsulate numerous bioactive substances, such as beta-carotene, resveratrol, alphalinolenic acid, curcumin, lycopene and alpha-tocopherol, and so provide an effective way of molecular delivery [42, 45].

1.2.3 Nanoprecipitation

The nanoprecipitation approach involves the organic internal phase spontaneously emulsifying into the aqueous external phase. The process includes organic solvent, bioactive compounds, and dissolved polymer. Encapsulation of lipophilic bioactive substances is the most common application of this easy and repeatable process. Solvent displacement, or desolvation, is another name for this process. On a magnetic stirrer, two miscible phases (water phase and oil phase) including necessary components that are miscible are mixed together. At the interphase of liquids, vortices of solvent occur because of the surface tension differential between the two phases. This results in polymer precipitation on the oil phase, as well as the creation of nanoparticles in the form of nanospheres and nanocapsules [46]. In this nanoprecipitation technique, biodegradable polymers, such as polylactic acid, polycaprolactone, poly-alkyl cyanoacrylate, and polylactide-coglycolide, are used [47]. In an aquatic environment, the polar particles dissolve fast during nanoprecipitation process. Nonpolar particles, on the other hand, tend to form nanoaggregates when they come into contact with other nonpolar particles. As a result, depending on the wall materials utilized and the environmental circumstances, nanoparticles might have varied charges (negative, positive, or neutral) [48]. The precipitation

process was also used to make zein–curcumin nanoparticles [49]. Lutein nanocrystals are employed in low-fat diets because they have better water solubility and bioavailability [50]. Curcumin, a weakly water soluble antioxidant bioactive molecule, was made into nanoparticles in another study [51]. Suwannateep *et al.* (2011) also created modified curcumin via nanoencapsulation to increase its sustainability and bioavailability when consumed orally [52]. Also employed is the inclusion complexation process to encapsulate lipophilic chemicals, such as volatile organic compounds (essential oils and vitamins), and is effective to hide odors and flavors while also preserving aromas. The ability to make nanocapsules as small as 100 nm is one of the key advantages of the nanoprecipitation process. As a result, this is very beneficial nanoencapsulation technique for the administration of biologically active substances with superior effectiveness, resistance to deterioration and enhanced cellular absorption [53, 54].

1.2.4 Nanofibers

Nanofibers have prospective uses not only in drug delivery, medical aspects, tissue scaffolds, but also in the food industry due to some essential physicochemical qualities, such as high porosity, extraordinarily high surface to mass ratio with excellent perfunctory capability. Nanofibers also execute outstanding thermal stability because they stabilize volatile or unstable food additives [55]. Electrospun nanofibers play a significant role in the variety of delivery systems because they behave as microemulsions containing solubilized lipophilic functional chemicals, such as pharmaceuticals, antimicrobials, a variety of bioactive chemicals, antioxidants, and flavors. Cellulose acetate nanofibers have also been used to immobilize vitamins. These nanofibers are immobilized with fat soluble Vitamin A and vitamin E creating a smooth texture and spherical cross-sectional shape (240–270 nm in size), resulting in sustained release of vitamins over the testing period [56]. In the sphere of growing food nanotechnology, another benefit of nanofibers is enzyme immobilization. The lipase enzyme is very useful among the most crucial enzymes in food processing industries, particularly in bakery, dairy, and oil products. The electrospun nanofibers that engage in the lipase process have been embedded with polyethene glycol diacyl chloride and get attached to it by using carbodiimide chemistry on the nanofibers' outer surface [57]. Nanofibers have been shown to have antibacterial as well as antimicrobial properties [58]. The incorporation of bacteriophage viruses into electrospun fibers having cellulose diacetate core, polyethylene oxide, and their combination aid in the prevention of bacterial contamination on the food's surface [59]. In food

nanotechnology, nanofibers can be employed as biosensors and antifouling agents. Cellulosic nanofibers, on the other hand, appear to be a potential candidate in the packaging industry because it helps to develop composite thin films that are utilized in the packaging of food products [60].

1.2.5 Solid Lipid Nanoparticles (SLNs)

Solidified lipid nanoparticles are one of the strategic ways for targeted delivery of bioactive substances that physically entrap and preserve them from the aqueous environment. These biologically active compounds are EGCG, carotene, vitamin B_2, vitamin B_{12}, peppermint essential oil, rosmarinic acid, resveratrol, tocopherol, and so on. The prepared SLNs are usually spheroid in shape and range in size from a few micrometers to a few millimeters. High-melting-point waxes (such as beeswax, hydrogenated palm oil, butterfat, carnauba and candelilla wax) or food-grade triglycerides are used to make these particles [54]. Disseminating the bioactive all through the liquefied fat juncture is a simple way to build delivery systems for hydrophilic bioactive. Hydrophilic bioactive compounds may be encapsulated within SLNs by deliquescing them in water and then creating a W/O suspension which finally produces solid fat units [61]. After a few weeks of storage, substantial proportions of bioactive embedded inside the solid wax spheres have been recorded [54]. High-energy procedures, such as homogenization or high shear procedures, as well as low-energy procedures, like spontaneous emulsification, and so on, can be used to fabricate nanosized particles within 100 nm [62]. However, there are several obstacles to developing feasible SLNs-based administration systems to be used in the food sectors, including unattractive amorphous "sandy" texture, unwanted sensory effects, susceptible to agglomeration because of feasible hydrophobic attraction, seep out of bioactive during storage, and restricted loading capacity [63].

1.3 Nanotechnology in Dietary Supplements and Probiotics Formulation

Dietary supplements and foods for particular medical purposes are designated as special medical items under the law [64]. Food fortification with nutrients or nonnutrient bioactive components can help to balance a diet's total nutritional profile and replenish elements lost in processing, thereby preventing nutritional deficiencies [65]. As nanoscience offers a versatile dimension to many current materials, with new or modified

properties it can be associated with various types of nutritional supplements. These nanoformulation-based supplements are specifically designed to protect active components from degradation, improve bioavailability, and reduce negative impacts. Nanofibers, nanoliposomes, nanoemulsions, core-shell nanoparticles, cyclodextrin complexes, dual encrusted hydroxides, nanotubes, mesoporous nanoparticles, SLNs, and nanocapsules are some of the nanocarriers used in the delivery of dietary supplements. Sometimes the absorption capacity of nutrients can be increased by using nano-sized powders. For patients suffering from osteoporosis, iron deficiency, inequity of motor activity, and many other diseases, these nutritional compositions are clinically useful in their treatment and prevention [66]. The most common commercialized nanotechnology-based products on the market are Neosino capsules (dietary supplements), Canola active oil, Nanogreen tea, Nutralease (fortifying nanomaterials carrying drugs and nutraceuticals), and Aquanova. Similarly, oat nutritional drinks, nanoceuticals slim shakes, fortified fruit juices, and tuna fish oil nanocapsules in pieces of bread, are a few commercially produced nano-processed foods that are extensively available in the US, China, Japan, and Australia [67].

Now impending to the term "probiotics," where the presently accepted meaning of probiotics can be explained by the addressing of live microbes that, while provided in sufficient proportions, can benefit the consumer's health and well-being. As a result, the advantages of probiotics extend beyond the gut microbiota's mediation through related processes. *Lactobacilli*, *Enterococci*, *Bifidobacteria*, and *Leuconostoc* spp. are the most frequently used probiotics found in functional food items and other fermented products. In addition to these lactic acid bacteria, few *Saccharomyces* spp. executing their beneficial functions are being considered as probiotics [68]. Cheese, yogurts, and yogurt-like fermented milk, puddings, and fruit-based drinks are all common probiotic foods. One of the issues that manufacturers of probiotic-based functional foods are the natural instability of living bacteria. Probiotic nanoencapsulation has also been reported in the earlier period. Encapsulation predominantly extends the product's shelf life and probiotic bacteria encased in nanoparticles is designed in such a way to release in a controlled manner at specific regions of the digestive tract, where they interact with specific receptors [69]. Food manufacturers facing one of the most difficult challenges is to keep bacteria alive until they reach the intestines. They may die during the manufacturing process, or in the stomach with high acidity after oral ingestion. Ebrahimnejad *et al.* made *Lactobacillus acidophilus* nanocapsules based on chitosan indicating enhanced viability and survival rates in the gastrointestinal tract [70]. Kalal *et al.* demonstrated the 4-week survivability of nano-encapsulated *Lactobacillus casei* in bitter gourd juice comprising maltodextrin [71]. Additionally, the most promising nano-encapsulated product is the milk protein-La-associated nanotube, which might have a significant impact on the administration of probiotics [72]. Noori *et al.* developed nanoliposomes consisting of *Lactobacillus casei* and *Bifidobacterium lactis* [73].

1.4 Nanotechnology in Food Packaging and Nanosensors

All of us are aware that freezing, drying, and canning, are some of the classic ways for food preservation. Nanotechnology has recently emerged as a potent interdisciplinary way of developing smart packaging systems. Targeted nutrition delivery systems could be made from a variety of functional nanomaterials that are employed as active packaging materials and have significant mechanical and barrier qualities [74]. According to research, between 2011 and 2021, the worldwide active and smart packaging market would increase by a factor of two, and still rising at a rate of 17% per year until 2021, attaining 24,650 million dollars. Consequently, electronic smart packaging will exceed 1.45 billion dollars for global demand in the next ten years [75]. The majority of nanoparticles carrying polypeptides used in food packaging have antimicrobial capabilities and protecting the deterioration caused by microorganisms. In some circumstances, nanoparticles are used to carry various beneficial components which in turn extends the long-lasting expectancy of a product after it has been exposed to air [7]. Nanoscale techniques are used in food packaging in the forms of nanocomposites, inorganic, metallic, and polymeric nanoparticles, nanosensors, and so on. Nanocomposites are frequently used in packaging and coating applications [76]. Nanocomposite films based on chitosan, mainly silver-containing nanocomposites, have been found to have antimicrobial properties [77]. For possible food packaging applications, nanoemulsions can be integrated into films and coatings. One of the benefits of adding nanoemulsified bioactives and flavors to beverages is that the product's appearance is unaffected. The continuous phase consists of biopolymer matrix sheets and coatings that provide nanoemulsion droplets with stability and monodispersity. The coalescence of droplets is reduced when the continuous phase's viscosity rises. Food packaging with silicate nanoparticles also functions as a barrier to gases and moisture, reducing food deterioration and drying [78].

Nanotechnology allows for the use of selective nanosensors in smart food packaging to examine the quality at different steps of the process ensuring that the ultimate consumers receive eminence products. In general, the introduction of nanosensors alerts customers to food spoilage or contamination by detecting pesticides, toxins, and microbiological contamination in foodstuffs based on color formation and favor creation [79]. The smart nanosensors are also capable of detecting various gases associated with spoilage, microbial metabolism's byproducts, and so on [80]. Metals for instance platinum, palladium, and gold are commonly used in gas-detecting sensors. Fertilizers, pesticides, and other harmful compounds have all been detected using nanoparticles. The gold nanoparticles have been employed as fluorometric and colorimetric pesticide sensors, in particular for organophosphorus and carbamate pesticide detection [81]. Palytoxin, a type of bacterial toxin, alpha-toxin generated by *Aspergillus flavus* and *Aspergillus parasiticus*, can be detected by various nano-based sensors, such as gold nanoparticle-based immunochromatographic strips, carbon nanotubes-based electrochemiluminescence, and so on [82]. To identify the folic acid's level in wheat flour, milk samples, and juices, single-walled carbon nanotube-ionic liquid nanocomposites and multi-walled carbon nanotubes have been reported. Likewise, nickel oxide nanoparticles have also been employed to detect ascorbic acid, a vitamin found in foods [83]. Table 1.2 represents the type of nanosensors and their use in food products.

1.5 Obstacles Coupled with Nanoencapsulation of Bioactive Components

One of the primary goals of food law is to assure food safety. The legislation also governs the communication role, and how a product is displayed has an impact on consumer approval [75]. Although it has been shown that integrating bioactive components through nanoencapsulation into functional foods is an advanced way for delivering such physiological endorsing compounds into the human body, there are still some significant obstacles to overcome prior to fully developed functional products are made accessible to general public. Apart from the wall material, the procedure of manufacturing nanoparticles containing bioactive ingredients is extremely important and challenging. The fact is that while the manufactured nanostructure containing bioactive components may execute the pure stable form in a regulated environment, their stability becomes a serious concern when they are introduced into food [93]. Catalytic and kinetic characteristics,

Table 1.2 Nanosensors and their applications in food industries.

Nanosensor	Sample	Analyte	Reference
Silver nanoparticles	Food preservation and packaging	Antibacterial agent absorbs and decomposes ethylene in vegetables and fruit	[84]
Silicon dioxide	Food preservation and packaging	Food colorant, anticaking, and drying agent.	[85]
Gold nanoparticles	Food storage in meat and dairy industries, glucose	Incorporation of enzymes or antibodies or DNA with Au NP	[86]
Au and Cu nanoparticles	Surface water	Pathogens	[87]
Quantum dots	Chicken carcass wash water	*Salmonella typhi*	[88]
Graphene	Spots contaminants in food	Nanoplate-based nanocomposites	[89]
Carbon nanotubes	vacuum proof food packaging and food inspection	Electrical, optical, thermal, and mechanical conductivity	[90]
Multi-walled carbon nanotubes (MWCNT)	Food industry	Integration biomolecules Fructose Paraoxon	[91]
Chitosan	Anti-fungicide	Covering/coating agent for strawberries, fresh fruits, mandarin	[92]

as well as functionalization, agglomeration, critical particle reactivity, and the practical environment, all contribute to nanoparticle-mediated toxicity [7]. The toxicity of some bioactive compounds beyond a given dosage and the fabricated nanoparticles' toxicity are currently being debated [54]. Another major stumbling block is the difficulties of pure absorption of nanoformulation and their proper dispersion in human physiology, with a particular focus on genotoxicity and cytotoxicity [10]. In point of fact, the kind of enzymes, high acidity in stomach, mucus membrane, and semi permeability of intestinal cells must all be overcome in order for nanoparticles to be absorbed [94]. Furthermore, the size and hydrophobicity of fabricated nanocomposites all are responsible for significantly influencing the consumer's health. On the basis of the absorption criterion, nanoparticles with a diameter as small as 100 nm have high absorption potential, those with a diameter of 100–200 nm have shown moderate absorption, those with a diameter of 300–500 nm have low absorption, and nanoparticles with a diameter of more than 500 nm have no absorption through the mucus membrane of the intestine [95]. Some nanomaterials attach to enzymes, and proteins in the cell causing oxidative damage by the action of reactive oxygen species (ROS). The buildup of reactive oxygen species causes mitochondrial degeneration and death [96]. Another essential consideration is the nanoparticle preparation process, which should be feasible technically as well as economically along with up-scalable potential from laboratory to pilot scale. As a result, more scientific research is needed to set criteria for encapsulated bioactive component safety limits. For lawful nanotechnological applications, comprehensive government rules and laws, as well as thorough toxicological screening procedures, are mandatory. Hence, several governments have required a regulatory structure to deal with the risks linked with nanofood as a result of these rising regulatory issues. Alongside, for the control of nanoparticles in the food sector, a broadly approved international regulatory mechanism is urgently needed [7].

1.6 Conclusion

Nanotechnology is a new discipline of research that has emerged in the last few decades. Many countries have developed nanotechnology applications in the preparation of functional food items, and developments in new technologies introduce to food production have also aided this rapidly expanding field of study. Nanotechnology allows for superior food processing, smart packaging, and long-term food preservation. Aside from that, it gives huge growth in the food industry by enhancing food quality by boosting flavor and texture. The majority of food bioactive used to treat various ailments are usually hydrophobic, with poor stability and bioavailability. As a result, methods based on nanotechnology improved food bioactive compound bioavailability and targeted delivery. Nanoformulated probiotics, for example, are the future of health-promoting foods. As a result, a greater understanding of the mechanisms involved in nano-based probiotics is required, as well as significant human trials to substantiate their high potential. Regardless of these facts, another potential application of nanomaterials is in the domain of sensors, which aid customers by providing information about the condition of the food on the inside. Nanosensors ensure the nutritional status of food items with improved security through the identification of chemical pollutants, colors, and pathogenic microorganisms, among other things. However, numerous significant hurdles must be overcome before functional foods containing nanoencapsulated bioactive may be made available to consumers. Only a few nations have clear regulatory guidelines in place for using nanotechnology approaches in food. For the effective and widespread future implementation of nanotechnology in the field of functional foods, it is also necessary to look up public acceptance, economics, and pragmatic directive. If properly supervised and regulated, it has the potential to improve the processing part of food and the quality of food product, that will promote better health and mankind.

References

1. Tarabella, A., Varese, E., Buffagni, S., Functional Foods, in: *Food Products Evolution: Innovation Drivers and Market Trends*, pp. 117–142, Springer, Cham, 2019.
2. Henry, C., Functional foods. *Eur. J. Clin. Nutr.*, 64, 657–659, 2010.
3. Bigliardi, B. and Galati, F., Innovation trends in the food industry: The case of functional foods. *Trends Food Sci. Technol.*, 31, 118–129, 2013.
4. Cencic, A. and Chingwaru, W., The role of functional foods, nutraceuticals, and food supplements in intestinal health. *Nutrients*, 2, 611–625, 2010.
5. Singh, H., Nanotechnology applications in functional foods; opportunities and challenges. *Prev. Nutr. Food Sci.*, 21, 1, 1–8, 2016.
6. Sastry, R.K., Anshul, S., Rao, N., Nanotechnology in food processing sector-An assessment of emerging trends. *J. Food Sci. Technol.*, 50, 831–841, 2013.
7. Nile, S.H., Baskar, V., Selvaraj, D., Nile, A., Xiao, J., Kai, G., Nanotechnologies in food science: applications, recent trends, and future perspectives. *Nanomicro Lett.*, 12, 1–34, 2020.

8. Momin, J.K., Jayakumar, C., Prajapati, J.B., Potential of nanotechnology in functional foods. *Emir. J. Food Agric.*, 25, 10–19, 2013.
9. Sekhon, B.S., Food nanotechnology–an overview. *Nanotechnol. Sci. Appl.*, 3, 1, 2010.
10. McClements, D.J. and Xiao, H., Is nano safe in foods? Establishing the factors impacting the gastrointestinal fate and toxicity of organic and inorganic food-grade nanoparticles. *NPJ Sci. Food*, 1, 1–13, 2017.
11. ETC, Down on the Farm: The Impact of Nanoscale Technologies on Food and Agriculture, *ETC Group publications*, 1–64, 2003.
12. Miller, G. and Senjen, R., *Out of the Laboratory and onto our Plates: Nanotechnology in Food & Agriculture. A report prepared for Friends of the Earth Australia*, Friends of the Earth Europe and Friends of the Earth United States and supported by Friends of the Earth Germany Friends of the Earth Australia Nanotechnology Project, Australia, 2008.
13. Taheri, A. and Jafari, S.M., Gum-based nanocarriers for the protection and delivery of food bioactive compounds. *Adv. Colloid Interface Sci.*, 269, 277–295, 2019.
14. Singh, R., Mann, B., Sharma, R., Singh, S., Application of Nanotechnology in Functional Foods, in: *Nanoscience for Sustainable Agriculture*, pp. 547–579, Springer, Cham, 2019.
15. Zhu, J. and Huang, Q., Nanoencapsulation of functional food ingredients. *Adv. Food Nutr. Res.*, 88, 129–165, 2019.
16. Shen, P., Zhang, R., McClements, D.J., Park, Y., Nanoemulsion-based delivery systems for testing nutraceutical efficacy using Caenorhabditis elegans: Demonstration of curcumin bioaccumulation and body-fat reduction. *Food Res. Int.*, 120, 157–166, 2019.
17. Pérez-Córdoba, L.J., Norton, I.T., Batchelor, H.K., Gkatzionis, K., Spyropoulos, F., Sobral, P.J., Physico-chemical, antimicrobial and antioxidant properties of gelatin-chitosan based films loaded with nanoemulsions encapsulating active compounds. *Food Hydrocolloids*, 79, 544–559, 2018.
18. Belhaj, N., Arab-Tehrany, E., Linder, M., Oxidative kinetics of salmon oil in bulk and in nanoemulsion stabilized by marine lecithin. *Process Biochem.*, 45, 187–195, 2010.
19. Rodriguez, E.B., Almeda, R.A., Vidallon, M.L.P., Reyes, C.T., Enhanced bioactivity and efficient delivery of quercetin through nanoliposomal encapsulation using rice bran phospholipids. *J. Sci. Food Agric.*, 99, 1980–1989, 2019.
20. Yang, S., Liu, W., Liu, C., Liu, W., Tong, G., Zheng, H., Zhou, W., Characterization and bioavailability of vitamin C nanoliposomes prepared by film evaporation-dynamic high pressure microfluidization. *J. Dispers. Sci. Technol.*, 33, 1608–1614, 2012.
21. Huynh Mai, C., Thanh Diep, T., Le, T.T., Nguyen, V., Advances in colloidal dispersions: A review. *J. Dispers. Sci. Technol.*, 41, 479–494, 2020.

22. Ribeiro, H.S., Chu, B.-S., Ichikawa, S., Nakajima, M., Preparation of nanodispersions containing β-carotene by solvent displacement method. *Food Hydrocolloids*, 22, 12–17, 2008.
23. Huang, X., Dai, Y., Cai, J., Zhong, N., Xiao, H., McClements, D.J., Hu, K., Resveratrol encapsulation in core-shell biopolymer nanoparticles: Impact on antioxidant and anticancer activities. *Food Hydrocolloids*, 64, 157–165, 2017.
24. Luo, Y., Pan, K., Zhong, Q., Casein/pectin nanocomplexes as potential oral delivery vehicles. *Int. J. Pharm.*, 486, 59–68, 2015.
25. Jincheng, W., Xiaoyu, Z., Sihao, C., Preparation and properties of nanocapsulated capsaicin by complex coacervation method. *Chem. Eng. Commun.*, 197, 919–933, 2010.
26. Reza Mozafari, M., Johnson, C., Hatziantoniou, S., Demetzos, C., Nanoliposomes and their applications in food nanotechnology. *J. Liposome Res.*, 18, 309–327, 2008.
27. Cantor, S., Vargas, L., Rojas A, O.E., Yarce, C.J., Salamanca, C.H., Oñate-Garzón, J., Evaluation of the antimicrobial activity of cationic peptides loaded in surface-modified nanoliposomes against foodborne bacteria. *Int. J. Mol. Sci.*, 20, 680, 2019.
28. Cui, H., Li, W., Lin, L., Antibacterial activity of liposome containing curry plant essential oil against Bacillus cereusin rice. *J. Food Saf.*, 37, e12302, 2017.
29. Khosravi-Darani, K. and Mozafari, M., Nanoliposome potentials in nanotherapy: A concise overview. *Int. J. Nanosci. Nanotechnol.*, 6, 3–13, 2010.
30. Mozafari, M.R., Liposomes: an overview of manufacturing techniques. *Cell. Mol. Biol. Lett.*, 10, 711, 2005.
31. Ghorbanzade, T., Jafari, S.M., Akhavan, S., Hadavi, R., Nano-encapsulation of fish oil in nano-liposomes and its application in fortification of yogurt. *Food Chem.*, 216, 146–152, 2017.
32. Risaliti, L., Pini, G., Ascrizzi, R., Donato, R., Sacco, C., Bergonzi, M.C., Salvatici, M.C., Bilia, A.R., Artemisia annua essential oil extraction, characterization, and incorporation in nanoliposomes, smart drug delivery systems against Candida species. *J. Drug Deliv. Sci. Technol.*, 59, 101849, 2020.
33. Khanniri, E., Bagheripoor-Fallah, N., Sohrabvandi, S., Mortazavian, A., Khosravi-Darani, K., Mohammad, R., Application of liposomes in some dairy products. *Crit. Rev. Food Sci. Nutr.*, 56, 484–493, 2016.
34. Thompson, A.K., Mozafari, M.R., Singh, H., The properties of liposomes produced from milk fat globule membrane material using different techniques. *Le Lait, INRA Editions, EDP Sciences*, 87, 349–360, 2007.
35. Ajeeshkumar, K.K., Aneesh, P.A., Raju, N., Suseela, M., Ravishankar, C.N., Benjakul, S., Advancements in liposome technology: Preparation techniques and applications in food, functional foods, and bioactive delivery: A review. *Compr. Rev. Food Sci. Food Saf.*, 20, 1280–1306, 2021.
36. Ojagh, S.M. and Hasani, S., Characteristics and oxidative stability of fish oil nano-liposomes and its application in functional bread. *J. Food Meas. Charact.*, 12, 1084–1092, 2018.

37. Mohammadi, M., Ghanbarzadeh, B., Hamishehkar, H., Formulation of nanoliposomal vitamin D3 for potential application in beverage fortification. *Adv. Pharm. Bull.*, 4, 569, 2014.
38. Pinilla, C.M.B. and Brandelli, A., Antimicrobial activity of nanoliposomes co-encapsulating nisin and garlic extract against Gram-positive and Gram-negative bacteria in milk. *Innov. Food Sci. Emerg. Technol.*, 36, 287–293, 2016.
39. Subramani, T. and Ganapathyswamy, H., An overview of liposomal nano-encapsulation techniques and its applications in food and nutraceutical. *J. Food Sci. Technol.*, 1–11, 57, 10, 2020.
40. Dag, D., Guner, S., Oztop, M.H., Physicochemical mechanisms of different biopolymers'(lysozyme, gum arabic, whey protein, chitosan) adsorption on green tea extract loaded liposomes. *Int. J. Biol. Macromol.*, 138, 473–482, 2019.
41. Tadros, T., Izquierdo, P., Esquena, J., Solans, C., Formation and stability of nano-emulsions. *Adv. Colloid Interface Sci.*, 108, 303–318, 2004.
42. Aswathanarayan, J.B. and Vittal, R.R., Nanoemulsions and their potential applications in food industry. *Front. Sustain. Food Syst.*, 3, 95, 2019.
43. Pathak, M., Nanoemulsions and their stability for enhancing functional properties of food ingredients, in: *Nanotechnology applications in food*, pp. 87–106, Elsevier, USA, 2017.
44. Mohd-Setapar, S.H., Nian-Yian, L., Kamarudin, W.N.W., Idham, Z., Norfahana, A.-T., Omega-3 emulsion of rubber (Hevea brasiliensis) seed oil. *Agric. Sci.*, 4, 84, 2013.
45. Silva, H.D., Poejo, J., Pinheiro, A.C., Donsì, F., Serra, A.T., Duarte, C.M., Ferrari, G., Cerqueira, M.A., Vicente, A.A., Evaluating the behaviour of curcumin nanoemulsions and multilayer nanoemulsions during dynamic *in vitro* digestion. *J. Funct. Foods*, 48, 605–613, 2018.
46. Galindo-Rodriguez, S., Allemann, E., Fessi, H., Doelker, E., Physicochemical parameters associated with nanoparticle formation in the salting-out, emulsification-diffusion, and nanoprecipitation methods. *Pharm. Res.*, 21, 1428–1439, 2004.
47. Reis, C.P., Neufeld, R.J., Veiga, F., Ribeiro, A.J., Preparation of drug-loaded polymeric nanoparticles. In: *Nanomedicine in Cancer*, L. Balogh (ed.), Jenny Stanford Publishing, Singapore, 171-214, 2017.
48. Joye, I.J., Davidov-Pardo, G., McClements, D.J., Nanotechnology for increased micronutrient bioavailability. *Trends Food Sci. Technol.*, 40, 168–182, 2014.
49. Patel, A., Hu, Y., Tiwari, J.K., Velikov, K.P., Synthesis and characterisation of zein–curcumin colloidal particles. *Soft Matter*, 6, 6192–6199, 2010.
50. Mitri, K., Shegokar, R., Gohla, S., Anselmi, C., Müller, R.H., Lutein nanocrystals as antioxidant formulation for oral and dermal delivery. *Int. J. Pharm.*, 420, 141–146, 2011.
51. Kakran, M., Sahoo, N.G., Tan, I.-L., Li, L., Preparation of nanoparticles of poorly water-soluble antioxidant curcumin by antisolvent precipitation methods. *J. Nanopart. Res.*, 14, 1–11, 2012.

52. Suwannateep, N., Banlunara, W., Wanichwecharungruang, S.P., Chiablaem, K., Lirdprapamongkol, K., Svasti, J., Mucoadhesive curcumin nanospheres: biological activity, adhesion to stomach mucosa and release of curcumin into the circulation. *J. Controlled Release*, 151, 176–182, 2011.
53. Ezhilarasi, P., Karthik, P., Chhanwal, N., Anandharamakrishnan, C., Nanoencapsulation techniques for food bioactive components: a review. *Food Bioprocess Technol.*, 6, 628–647, 2013.
54. Rashidinejad, A. and Mahdi, S., Nanoencapsulation of bioactive food ingredients, in: *Handbook of Food Nanotechnology. Applications and Approaches*, pp. 279–291, 2017.
55. Xue, J., Wu, T., Dai, Y., Xia, Y., Electrospinning and electrospun nanofibers: Methods, materials, and applications. *Chem. Rev.*, 119, 5298–5415, 2019.
56. Taepaiboon, P., Rungsardthong, U., Supaphol, P., Vitamin-loaded electrospun cellulose acetate nanofiber mats as transdermal and dermal therapeutic agents of vitamin A acid and vitamin E. *Eur. J. Pharm. Biopharm.*, 67, 387–397, 2007.
57. Wang, Z., Ma, H., Hsiao, B.S., Chu, B., Nanofibrous ultrafiltration membranes containing cross-linked poly (ethylene glycol) and cellulose nanofiber composite barrier layer. *Polymer*, 55, 10, 366–372, 2014.
58. Song, K., Wu, Q., Qi, Y., Kärki, T., Electrospun nanofibers with antimicrobial properties, in: *Electrospun nanofibers*, pp. 551–569, 2017.
59. Korehei, R. and Kadla, J.F., Encapsulation of T4 bacteriophage in electrospun poly (ethylene oxide)/cellulose diacetate fibers. *Carbohydr. Polym.*, 100, 150–157, 2014.
60. Rezaei, A., Nasirpour, A., Fathi, M., Application of cellulosic nanofibers in food science using electrospinning and its potential risk. *Compr. Rev. Food Sci. Food Saf.*, 14, 269–284, 2015.
61. Katouzian, I., Esfanjani, A.F., Jafari, S.M., Akhavan, S., Formulation and application of a new generation of lipid nano-carriers for the food bioactive ingredients. *Trends Food Sci. Technol.*, 68, 14–25, 2017.
62. Oxley, J., Spray cooling and spray chilling for food ingredient and nutraceutical encapsulation, in: *Encapsulation technologies and delivery systems for food ingredients and nutraceuticals*, pp. 110–130, Elsevier, USA, 2012.
63. Mellema, M., Scope and limitations of using wax to encapsulate water-soluble compounds, in: *Food Colloids*, pp. 103–115, 2007.
64. Jampilek, J., Kos, J., Kralova, K., Potential of nanomaterial applications in dietary supplements and foods for special medical purposes. *Nanomaterials*, 9, 296, 2019.
65. Dwyer, J.T., Wiemer, K.L., Dary, O., Keen, C.L., King, J.C., Miller, K.B., Philbert, M.A., Tarasuk, V., Taylor, C.L., Gaine, P.C., Fortification and health: challenges and opportunities. *Adv. Nutr.*, 6, 124–131, 2015.
66. Hill, C.J., Lynch, D.B., Murphy, K., Ulaszewska, M., Jeffery, I.B., O'Shea, C.A., Watkins, C., Dempsey, E., Mattivi, F., Tuohy, K., Evolution of gut

microbiota composition from birth to 24 weeks in the INFANTMET Cohort. *Microbiome*, 5, 1–18, 2017.
67. Poças, M.F., Delgado, T., Oliveira, F.A., Kerry, J., Butler, P., Smart packaging technologies for fruits and vegetables, in: *Smart Packaging Technologies for Fast Moving Consumer Goods*, pp. 151–166, 2008.
68. Ashaolu, T.J., Immune boosting functional foods and their mechanisms: A critical evaluation of probiotics and prebiotics. *Biomed. Pharmacother.*, 130, 110625, 2020.
69. Vidhyalakshmi, R., Bhakyaraj, R., Subhasree, R., Encapsulation "the future of probiotics"-a review. *Adv. Biol. Res.*, 3, 96–103, 2009.
70. Ebrahimnezhad, P., Khavarpour, M., Khalili, S., Survival of Lactobacillus Acidophilus as probiotic bacteria using chitosan nanoparticles. *Int. J. Eng.*, 30, 456–463, 2017.
71. Kalal, A.Y., Hiregoudar, D.S., Nidoni, U., *Nanoencapsulation of Probiotic Bitter Gourd Juice Powder*, LAP LAMBERT Academic Publishing, India, 2017.
72. Mandal, A. and Banerjee, E.R., Nanotechnology and Functional Food, in: *Nanomaterials and Biomedicine*, pp. 85–112, Springer, Singapore, 2020.
73. Noori, N., Noudoost, B., Nasrabadi, H., Basti, A., Effects of green tea extract nanoencapsulation on the survival of Lactobacillus casei and Bifidobacterium lactis in symbiotic ice cream. *J. Vet. Res.*, 72, 195–205, 2017.
74. Sharma, C., Dhiman, R., Rokana, N., Panwar, H., Nanotechnology: an untapped resource for food packaging. *Front. Microbiol.*, 8, 1735, 2017.
75. Fuertes, G., Soto, I., Carrasco, R., Vargas, M., Sabattin, J., Lagos, C., Intelligent packaging systems: sensors and nanosensors to monitor food quality and safety. *J. Sens.*, 2016, 1–8, 2016.
76. Pereda, M., Marcovich, N.E., Ansorena, M.R., Nanotechnology in food packaging applications: Barrier materials, antimicrobial agents, sensors, and safety assessment, in: *Handbook of Ecomaterials*, L. Martínez, O. Kharissova, B. Kharisov (Eds), pp. 2035–2056, Springer, Cham., 2019.
77. Rhim, J.-W., Park, H.-M., Ha, C.-S., Bio-nanocomposites for food packaging applications. *Prog. Polym. Sci.*, 38, 1629–1652, 2013.
78. Neethirajan, S. and Jayas, D.S., Nanotechnology for the food and bioprocessing industries. *Food Bioprocess Technol.*, 4, 39–47, 2011.
79. He, X., Deng, H., Hwang, H.-m., The current application of nanotechnology in food and agriculture. *J. Food Drug Anal.*, 27, 1–21, 2019.
80. King, T., Osmond-McLeod, M.J., Duffy, L.L., Nanotechnology in the food sector and potential applications for the poultry industry. *Trends Food Sci. Technol.*, 72, 62–73, 2018.
81. Zamora-Sequeira, R., Starbird-Pérez, R., Rojas-Carillo, O., Vargas-Villalobos, S., What are the main sensor methods for quantifying pesticides in agricultural activities? *A review. Molecules*, 24, 2659, 2019.
82. Kumar, V., Guleria, P., Mehta, S.K., Nanosensors for food quality and safety assessment. *Environ. Chem. Lett.*, 15, 165–177, 2017.

83. Shafiq, M., Anjum, S., Hano, C., Anjum, I., Abbasi, B.H., An overview of the applications of nanomaterials and nanodevices in the food industry. *Foods*, 9, 148, 2020.
84. Acosta, E., Bioavailability of nanoparticles in nutrient and nutraceutical delivery. *Curr. Opin. Colloid Interface Sci.*, 14, 3–15, 2009.
85. Jones, N., Ray, B., Ranjit, K.T., Manna, A.C., Antibacterial activity of ZnO nanoparticle suspensions on a broad spectrum of microorganisms. *FEMS Microbiol. Lett.*, 279, 71–76, 2008.
86. Ozdemir, C., Yeni, F., Odaci, D., Timur, S., Electrochemical glucose biosensing by pyranose oxidase immobilized in gold nanoparticle-polyaniline/AgCl/gelatin nanocomposite matrix. *Food Chem.*, 119, 380–385, 2010.
87. Zhang, H., Wang, J., Ye, S., Predictions of acidity, soluble solids and firmness of pear using electronic nose technique. *J. Food Eng.*, 86, 370–378, 2008.
88. Yang, L. and Li, Y., Quantum dots as fluorescent labels for quantitative detection of Salmonella typhimurium in chicken carcass wash water. *J. Food Prot.*, 68, 1241–1245, 2005.
89. Sundramoorthy, A.K. and Gunasekaran, S., Applications of graphene in quality assurance and safety of food. *TrAC - Trends Anal. Chem.*, 60, 36–53, 2014.
90. Yadav, S., Tissue science & engineering realizing the potential of nanotechnology for agriculture and food technology. *J. Tissue Sci. Eng.*, 8, 8–11, 2017.
91. Tominaga, M., Nomura, S., Taniguchi, I., d-Fructose detection based on the direct heterogeneous electron transfer reaction of fructose dehydrogenase adsorbed onto multi-walled carbon nanotubes synthesized on platinum electrode. *Biosens. Bioelectron.*, 24, 1184–1188, 2009.
92. Xing, Y., Xu, Q., Li, X., Chen, C., Ma, L., Li, S., Che, Z., Lin, H., Chitosan-based coating with antimicrobial agents: preparation, property, mechanism, and application effectiveness on fruits and vegetables. *Int. J. Polym. Sci.*, 2016, 1–24, 2016.
93. Bajpai, V.K., Kamle, M., Shukla, S., Mahato, D.K., Chandra, P., Hwang, S.K., Kumar, P., Huh, Y.S., Han, Y.-K., Prospects of using nanotechnology for food preservation, safety, and security. *J. Food Drug Anal.*, 26, 1201–1214, 2018.
94. Ensign, L.M., Cone, R., Hanes, J., Oral drug delivery with polymeric nanoparticles: the gastrointestinal mucus barriers. *Adv. Drug Deliv. Rev.*, 64, 557–570, 2012.
95. Lai, S.K., O'Hanlon, D.E., Harrold, S., Man, S.T., Wang, Y.-Y., Cone, R., Hanes, J., Rapid transport of large polymeric nanoparticles in fresh undiluted human mucus. *Proc. Natl. Acad. Sci.*, 104, 1482–1487, 2007.
96. Abdal Dayem, A., Hossain, M.K., Lee, S.B., Kim, K., Saha, S.K., Yang, G.-M., Choi, H.Y., Cho, S.-G., The role of reactive oxygen species (ROS) in the biological activities of metallic nanoparticles. *Int. J. Mol. Sci.*, 18, 120, 2017.

2

Role of Nanotechnology in Fortifying Nutraceuticals

Anjana Singh[1], Samira Chugh[2], Dwaipayan Sinha[3*] and Arpita De[4]

[1]Department of Botany, Deshbandhu College, University of Delhi, Kalkaji, New Delhi, India
[2]Department of Botany, Gargi College, University of Delhi, Sirifort Road, New Delhi, India
[3]Department of Botany, Government General Degree College, Mohanpur, Paschim Medinipur, West Bengal, India
[4]Department of Botany, Government Science College (Autonomous) Bangalore, Karnataka, India

Abstract

Nanotechnology has provided new insights in the arena of food science, especially in packaging and storage. Various nanostructured materials are involved largely in the preservation and packaging of food items. Among various types of functional food, nutraceuticals have been in limelight for a decade or so for their positive effect on health. Nanotechniques are instrumental in the nutraceutical food industry. Nanoencapsulation results in enhanced bioavailability and solubility of nutritional ingredients especially minerals and vitamins. This results in controlled release and protection of essential molecules and micronutrients during processing and enhancing its beneficial action in the living system. In addition, nanoparticles also provide an efficient antimicrobial activity thereby preventing the foodstuffs from decay or unwanted contamination. The chapter is composed to highlight the various actions of nanomaterials in the handling and safeguarding of food nutraceuticals. Efforts have been made to highlight the various beneficial activities of nutraceuticals and how they are involved in their fortification and preservation of targeted delivery. The mechanistic aspects of nanomaterials in bringing about the desirable actions have also been highlighted.

**Corresponding author*: dwaipayansinha@hotmail.com

Tanima Bhattacharya and Shakeel Ahmed (eds.) Nanotechnology in Functional Foods, (25–60)

***Keywords*:** Nanotechnology, nutraceuticals, nanomaterials, nanosensors, food industry

2.1 Abbreviations

•OH	Hyroxyl radical
Ag^+	Silver ion
AgNPs	Silver nanoparticles
Au^+	Gold ion
AuNPs	Gold nanoparticles
CuO	Cupric oxide
Fe^{2+}	Ferrous ion
Fe^{3+}	Ferric ion
H_2O_2	Hydrogen peroxide
HO–	Hyroxide ion
nm	Nanometer
NMs	Nanomaterials
NPs	Nanoparticles
PEF	Pulse electric field
ROS	Reactive oxygen species
SiO_2	Silicon dioxide
TiO_2	Titanium dioxide
ZnO	Zinc oxide

2.2 Introduction

Nanotechnology is a revolutionary science that manipulates matter at its molecular and atomic scale rendering them exceptionally reactive. It has a colossal contribution to the multidimensional development of mankind. The very distinctive properties of nanoparticles (NPs) are being broadly studied and implemented in commercial, industrial, therapeutic, domestic, agricultural, and environmental applications. Nanoparticles are tiny particulate matter having a diameter stretching from 1 to 100 nanometers (nm) and are have three layers namely (a) surface (b) shell and (c) core [1]. Large surface area to volume ratio augments their physicochemical properties including reactivity and absorptivity [2]. These unparalleled properties are primarily attributed to their structural uniqueness like shape and size [3] and largely govern their reactivity. Based on their dimension (D), NPs can be categorized into 0 dimension (NPs, quantum dots), 1 dimension (nanowires and nanofibers), 2 dimensions (thin films and nanosheets), or 3 dimensions (bulk materials) [2, 4]. Depending on their chemical properties, nanomaterials (NMs) can further be categorized into metal NPs namely nanoparticles of gold (AuNPs) and silver (AgNPs), metal oxide NPs (Cupric Oxide-CuO, Zinc oxide-ZnO, etc.), and carbon-based nanomaterials (fullerenes, nanotubes, etc.) [5]. Metal NPs have crucial applications in medical science. The oxide NPs have proven their efficacy in degradation reactions which serve as an indispensable tool in environmental monitoring. Investigations reveal respective bulk counterparts of the NPs are considerably inert and thus less toxic [6]. Nanoparticles are manufactured following both breakdown (top-down) and buildup (bottom-up) methods. The former involves the application of external forces to break larger particles into smaller ones while the latter method produces NPs from atoms (from liquids or gases) [7].

Nanoparticles can get transferred and consequently biomagnify in the food chain [8]. A detailed analysis of the nature of interaction, competence, behavior, and fate of NPs is thus a prerequisite before its application in any ecological or industrial platform. Though the NPs are an embodiment of a plethora of benefits, a subtle alteration in their doses can convert these otherwise virtuous particles to extremely toxic and dangerous ones. Thus, implementation of NPs should be accompanied by thorough toxicity profiling.

Different European, Asian and American countries have sought to put nanotechnology into practice to deliver the desired quality of life to the masses [9]. The manufacture and supply of food and nutraceuticals fortified with adequate nutrients will break the cycle of poverty and malnutrition that perpetuates in our society. Integration of nano and food technologies cater to address the above cause. Implementation of nanotechnology has catapulted food science to a noticeably advanced level altogether. This chapter attempts to present a succinct description of the role of nanotechnology in fortifying nutraceuticals and its incredible contributions in the industrial food production, emphasizing on the packaging of nutraceuticals and preservation of food. The mode of action of these nanomaterials (NMs) has also been discussed highlighting appropriate examples.

2.3 What Are Nutraceuticals?

Nutraceuticals are nutritional products consisting of compounds of largely natural origin, such as carotenoids, vitamins, peptides, polyphenols, phytosterols, fibers, essential minerals, lipids, and so on, which offer physiological benefits in addition to nutritional requirements [10]. The nutrient and pharmaceutical forms two major components of the nutraceutical domain and was used by Stephen DeFelice in late eighties, who established the Foundation for Innovation in Medicine, Cranford, New Jersey [11]. Nutraceuticals work by preventing the occurrence of the untoward physiological condition in the body and Hippocrates, emphasized food as a source of medicine [12]. Nutraceuticals are categorized based on source (plants, animals, minerals, or microbes), pharmacological conditions, and chemical constitution of the products [13]. Nutraceuticals can also be of potential and established type. Established nutraceuticals are those potential nutraceuticals that are fortified with sufficient clinical data of their health and medical benefits [14, 15]. All-natural food sources which act as nutraceuticals are constituted by dietary fiber, antioxidant vitamins, polyphenols, probiotics, polyunsaturated fatty acids, prebiotics, spices, and so on [16]. They are used for the improvement of health, prevent chronic diseases, support bodily functions, delay aging, and increase life expectancy [17]. In comparison to pharmaceuticals that have a governmental sanction, nutraceuticals usually are not patent protected [18]. Thus, the product domain is largely overlapping and includes nutraceuticals, herbals, dietary supplements, functional foods due to minimal international regulation in the area.

Nutraceuticals may have restrictions like absorption and availability in biological systems, reduced solubility, and chemical instability. The use of nanotechnology can help overcome these limitations [19]. Modification of properties of active constituents of nutraceuticals using nanoformulations boosts their availability as NPs have better penetration into cells because of their small size ensuring rapid delivery of high concentrations of active ingredients. This allows consumption of lesser amounts of the nutraceuticals decreasing dose-dependent toxicity or side effects if any [20]. The low oral bioavailability of some natural bioactive compounds like curcumin restricts their usage in functional foods [21]. The oral bioavailability of such sparsely soluble nutraceuticals is improved using novel NPs like organogel-based nanoemulsions which can significantly affect functional food and pharmaceutical industries [22].

Other major potential roles reported are protection from oxidation, masking of taste, controlled release of nutrients, and safe packaging [23, 24]. Modification of many general features of nutraceuticals, namely texture, taste, and stability during shelf life with this enabling technology can bring a vast change in industries dealing with functional foods [25].

2.3.1 Nutraceuticals and Their Beneficial Roles

Nutraceuticals, at present have been a focus of research for immense nutritional therapeutic properties [26]. It acts potent antioxidant [27] and used for relief from Alzheimer's disease [28], autoimmune disorders [29], obesity [30], cardiac diseases [31], eye disease [32], cancer [33], Parkinson's disease [34], blood sugar related disorders [35] and stroke [36] (Figure 2.1). According to the theory of nutraceutical and nutritional therapy, foods provide nutrition and help to avoid nutritional deficiencies. They also offer medicinal benefits and help in body detoxification, restoration of healthy digestion, and dietary habits [37].

In India, the concept of nutraceuticals is still in its early stages, the annual growth of the nutraceutical market in India is 21% [37]. According to the report from ITA's Country Commercial Guide, the nutraceuticals market in India is likely to contribute around at least 3.5% of the worldwide market stake by 2023 and progress to $18 billion 2025. At present, the Indian market imports worth US$ 2.7 billion of nutraceuticals. The government of India has opened 100% foreign direct investment under automatic route in this manufacturing domain, and such entities can trade their merchandises through retail, wholesale, or e-commerce platform [38], thereby, resulting in the increase of the foreign direct investment from US$ 131.4 million in 2012 to US$584.7 million in 2019 [38]. The Food Safety and Standards Act, 2006 (FSSA), India for the first time has developed a new category called foods for special dietary purposes in addition to the normal ones. . They are included as a "special category" within the "general category" of foods and attract regulatory requirements as earlier nutraceuticals were not defined or recognized in India [39].

Nutraceuticals have pioneered a new episode in the domain of medicine and healthcare, where food industries are initiating research activities to

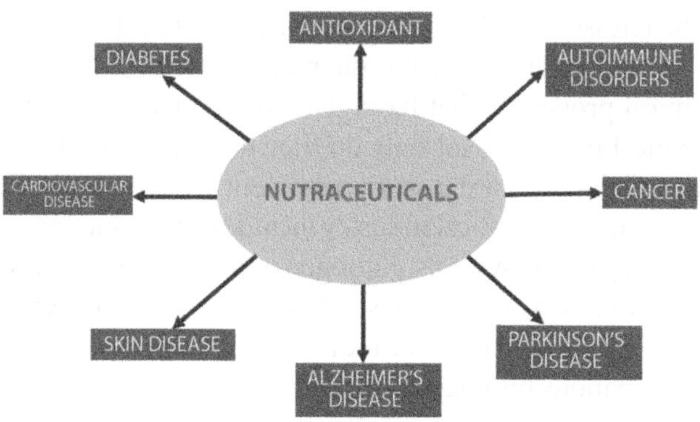

Figure 2.1 Diseases that can be cured or managed by nutraceuticals.

explore the medicinal potential within food [40]. To make the research and development activity of the nutraceutical producing food industry foolproof, healthcare professionals and toxicologists need a synchronized action to formulate appropriate regulations for the benefit and safety of mankind from a health and therapeutic point of view. This requires the formation of regulatory bodies to standardize the safety parameters of food industries focussing on nutraceuticals [16]. While earlier nutraceuticals came in medicinal formats in prescribed doses, modern nutraceuticals have evolved themselves in the form of whole food or as probiotic drinks [41]. The main reason for the worldwide popularity of the nutraceutical market is the increasing population, changing lifestyle, a varied biological activities, cost-effectiveness, and easy availability with no harmful effects. Coral calcium, Rox®, Chaser™, Mushroom optimizer™, Proplus®, WelLife®, Olivenol™, PNer plus™, Appetite Intercept™ are neutraceuticals that are marketed by the USA; Omega woman manufactured by the UK; GRD®, Proteinex®, Calcirol D-3®, and Threptin® Diskettes manufactured by India [42, 43]. Table 2.1 depicts examples of nutraceuticals, sources from where we can obtain them, and their influence on human health while Figure 2.2 depicts the chemical structure of selected nutraceutical molecules.

2.3.2 Relevance of NMs in the Nutraceutical Industry

Nanotechnology is at present of great use and operates in the dimension of one billionth of a meter for the synthesis of several products that are beneficial to humans. Quantum effects are observed in nanomaterials and their amplified relative surface area attributes to unique properties related to strength, electrical conductivity, and reactivity that demarcates them from conventional materials [71]. They are used for the improvement of stability, solubilization of active ingredients, and increasing bioavailability of nutraceuticals products [72]. They are now being designed for increasing shelf life, preventing the growth of microbes, the release of antioxidants enzymes, and flavors into the nutraceuticals [73]. Nanomaterials like cellulosic nanomaterials (CNMs) have unique physicochemical properties that have given them great potential in the food and nutraceutical industry. Not only do they have great mechanical strength and biocompatibility, but are also environment friendly [74]. Their various applications in the nutraceutical industry include the production of intelligent, active, and green packing systems, pathogen or pH detectors, and so on. In this chapter, we discuss these applications of NMs related to nutraceuticals.

Several government bodies and industries have channelized their investment towards nanotechnological application in the fields of agricultural production, packaging processing, and the safety of food [75]. In this case, Nano-sensors can operate as signaling mediators in this scenario, alerting

Table 2.1 Some examples of nutraceuticals, their food sources, and associated health benefits.

S. no.	Nutraceuticals	Food sources	Health benefits	References
1.	Vitamin C (Ascorbic Acid)	Tomatoes, pink grapefruits, watermelon, apricots.	1. Antioxidant, required to make bones, teeth gum, and skin healthy. 2. Prevention of common cold and influenza.	[44, 45]
2.	Lycopene	Apricots, tomatoes, watermelon, pink grapefruits.	1. Confers protection of skin from ultraviolet rays. 2. Beneficial for cardiovascular health. 3. Protection from prostate cancer.	[45, 46]
3.	Flavonoids	Most fruits and vegetables, wine.	1. Prevents atherosclerosis, aggregation of platelets, thrombosis, ulcer formation, inflammation, and diarrhea. 2. Inhibits arthritis and osteoporosis.	[45, 47]
4.	Vitamin E (Tocopherol)	Almonds, oils of sunflower, various nuts and canola, bran.	1. Counters oxidative stress. 2. Initiates blood cell formation, fortifies muscle, lungs, and nerves. 3. Immunity boosting capacity.	[44, 45]
5.	Calcium and Magnesium	Milk products, vegetables, seeds.	1. Reduces risk of high blood pressure during pregnancy. 2. Fortifies maternal and Infant Health.	[41, 45, 48]

(Continued)

Table 2.1 Some examples of nutraceuticals, their food sources, and associated health benefits. (*Continued*)

S. no.	Nutraceuticals	Food sources	Health benefits	References
6.	Milk's dipeptides	Milk and dairy products.	1. Confers protection from cancer. 2. Increase cellular glutathione. 3. Induction of hormonal and cell-mediated immune system.	[45, 49, 50]
7.	Coenzyme Q10 (CoQ10)	Heart and liver of animals, olive, corn, various nuts. Animal liver, soybean, nuts, corn, olive.	1. Confers protection from cardiac and vascular diseases. 2. Confers protection from gum disease, radiation-related injury, mitochondrial disorders, obesity. 3. Inhibits gastric ulceration, allergic responses, migraine, renal failure, and muscular dystrophy.	[45,51]
8.	Polyunsaturated Fatty Acids	Various vegetable and seed oils, meat, and milk of ruminants.	1. Confers protection from cardiovascular and inflammatory diseases. 2. Improves brain function and mental health.	[45, 52]
9.	*Zingiber officinale* - Ginger (Gingerol, Shagaol, Zingiberene, Curcumene)	Rhizome.	1. Inhibits the progression of diabetes, obesity, and cancer. 2. Inhibits enzymes related to the progression of hyperglycemia and modulates metabolic pathways.	[45, 53, 54]

(*Continued*)

Table 2.1 Some examples of nutraceuticals, their food sources, and associated health benefits. (*Continued*)

S. no.	Nutraceuticals	Food sources	Health benefits	References
10.	*Curcuma longa*- Turmeric	Turmeric.	1. Inhibition of oxidative stress and inflammatory condition. 2. Management of arthritis, anxiety, and hyperlipidemia.	[45, 55]
11.	Cinnamon	Bark.	1. Anti-inflammatory, antimicrobial, antitumor, 2. Confers protection against cardiovascular disease, cholesterol-lowering, and immunomodulatory effects.	[45, 56]
12.	Saffron (crocin, crocetin, picrocrocin and safranal)	*Crocus sativus* stigma.	1.Antidepressant and anticonvulsant, analgesic, anti-cancer. 2. Confers protection from cardiac and vascular disease, digestive, immune, respiratory, genital-urinary, and central nervous system.	[45, 57]
13.	Zinc	Beef, legumes, poultry, hot cereals, and pork.	1. Reduces preeclampsia in pregnant women. 2. Confers protection from diabetes, Alzheimer's disease depression, Wilson's disease. 3. Inhibits proliferation and induces apoptosis.	[58–60]
14.	Vitamin D (Cholecalciferol)	Oil-rich fish and eggs, UVB sunlight exposure.	1. Induces the development of bones and teeth 2. Helps in calcium metabolism.	[44, 61]

(*Continued*)

Table 2.1 Some examples of nutraceuticals, their food sources, and associated health benefits. (*Continued*)

S. no.	Nutraceuticals	Food sources	Health benefits	References
15.	Melatonin	Nuts, eggs, fish, mushrooms, cereals, and legumes.	1. Counteracts oxidative stress and inhibits inflammation and fortifies immunity. 2. Confers protection against cancer, cardiac and vascular diseases, and diabetes, nerve disease, and obesity.	[62, 63]
16.	Vitamin A (Retinol)	Fruits like mangoes, roots, tubers, and leafy vegetables, yellow or orange sweet potatoes, red palm oil, etc.	1. Counteracts oxidative stress. 2. Prevents cancer and skin diseases. 3. Required for proper growth and development.	[44, 64]
17.	Ellagic acid	Bilberry, grape, walnut, and pomegranate.	1. Prevents inflammatory disorders and cancer.	[65]
18.	Phytosterols	Wheat, corn, soybean, Centella, and pomegranate.	1. Inhibits hyperlipoproteinemia and inflammation.	[66]
19.	Saponins	Most fruits and vegetables.	1. Inhibits tumor formation and mutagens.	[67]
20.	Tannins and Proanthocyanidins	Various berries, grapes, and wine.	1. Potent antioxidant 2. Helps in detoxification of carcinogens.	[67, 68]
21.	Carotenoids	Various fruits and vegetables.	1. Counteracts oxidative stress. 2. Protects against cardiac and vascular diseases, inflammatory responses.	[69,70]

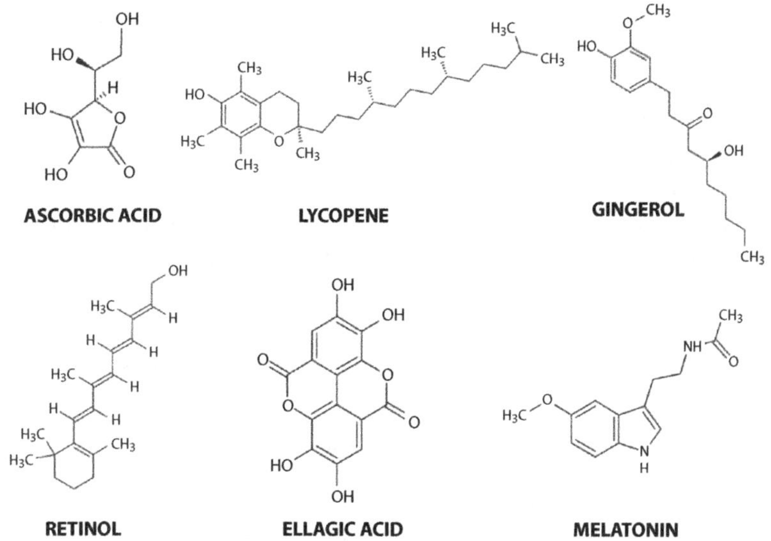

Figure 2.2 Structures of some selected nutraceutical molecules.

consumers to food quality [76]. Nanotechnology offers the potential to create packaging with improved mechanical and thermal properties [77]. Delivery systems could be used in nanotechnology to make functional foods by delivering active ingredients to desired action [78] keeping the food products fresh for a prolonged storage, secure during transportation, and safe from microbial pathogens [79]. The AgNPs used for the packaging permit longer product shelf life but possibilities of the transfer of silver particles into the eatable cannot be ruled out along with deleterious effect associated with the disposal process [80]. Irradiation of food with various types of penetrating rays kills microorganisms and is used for preservation without having much influence on the food quality [81]. High-intensity pulsed electric fields (PEF) are for the incapacitation of microbes and enzymes by breaking microbial cell membranes with high voltage thus, facilitating small molecules to enter the cell, resulting in swelling and rupture of microorganisms. They can be used to extend the shelf life of liquid and semi-liquid foods [82]. Coating with NPs confers protection of tomatoes improves shelf life and maintains nutritional quality [83]. Nanocapsules containing nutrients and minerals can be integrated into bread and juice for improving their nutritional value [80]. Nanoformulated nutrients and supplements possess better uptake, absorption [84], and bioavailability [85] than their bulk equivalents and also lead to less usage of preservatives and salt [86] in foods. The nanostructured foods also improve tastes and textures throughout the food [79]. Nano-coatings with antibacterial properties are used in food processing to maintain hygiene and also as markers of the food safety and genuineness in the production and distribution series [79]. Water can be decontaminated using nano-porous substances, and also to filter to remove improper taste, flavors, and allergens from foods [79].

However, strict regulations are required to reassure that safe, secure nanotechnology-based products are in the market. Consumers should be safeguarded from any potential risks, which is only possible by prior risk assessment and analysis of the product in detail before bringing them to the general people for use. Figure 2.3 shows different types of food packaging and nano-based devices associated with it. Figure 2.4 depicts the role of nanotechnology in food packaging from a broad perspective.

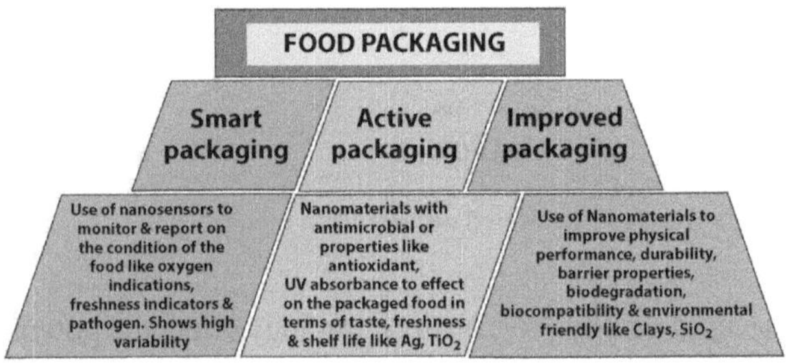

Figure 2.3 Types of food packaging with associated nano-based techniques.

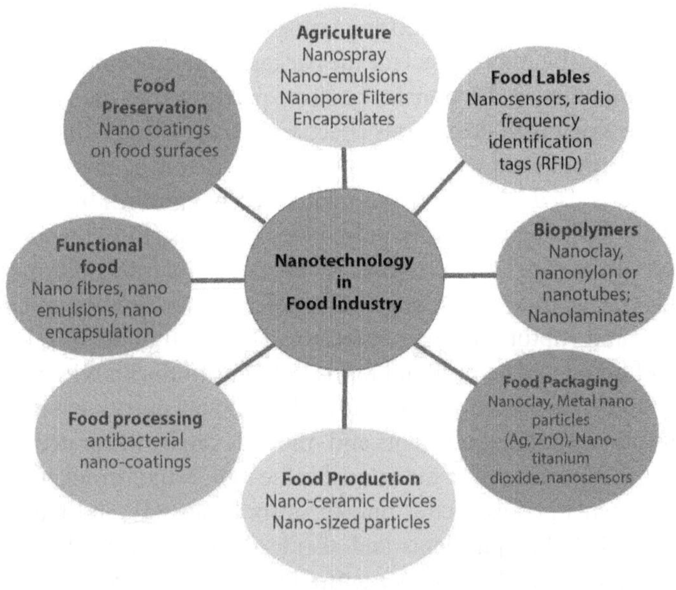

Figure 2.4 Diagrammatic illustration of relevant nanotechnology in the food industry.

2.3.3 Nanomaterials in the Packaging of Nutraceuticals

The consumption of nutraceuticals has significantly increased worldwide as nutraceuticals are thought to be advantageous over modern medicines because of probable fewer side effects, easy and economical availability [19]. Nutraceuticals require good protective packaging systems to avoid contamination and spoilage during distribution and transport. The type of packaging depends on mechanical, thermal, and barrier characteristics with which the material is composed with the primary motive of packaging and preservation [87]. Currently, most of the packaging of these nutraceuticals is done utilizing nonbiodegradable materials especially petroleum-based plastic polymer materials. This adds to the already existing grave global environmental problem of pollution [88]. Permeability of food packaging materials which are also used for nutraceutical packaging to gas and other small molecules is a big problem [89]. The application of nanotechnology in functional food packaging can not only provide a solution to these problems but also provide other benefits like prolonging the shelf life and increasing safety of the product. Gradually, food monitoring and packaging are becoming the primary focus in the domain of nanomaterial research [87].

Nanotechnology offers improved packaging options like better gas and moisture barrier (improved packaging properties) and tensile strength, antimicrobial, and self-cleaning properties. Other advantages reported are intelligent packaging and active packaging [24, 90].

2.3.3.1 Green Packaging

Biodegradable products produced from biopolymers from biomass such as agropolymers from agro-resources may be the answer to the ever-increasing menace of plastic packaging. The use of bio-based materials for packaging is currently limited due to natural polymers' poor barrier and mechanical qualities. The products are yet to attain sufficient popularity owing to their brittle nature, reduced heat distortion, enhanced permeability for gases, and unsatisfactory resistance during processing. Nanotechnology can aid the improvement of these properties and make them more cost-efficient [77, 91] and the limitations can be overcome by the application of nanocomposites in packaging [92]. Three main types of composites: micro-composites, intercalated nano-composite, and exfoliated nanocomposites are made using layered clay incorporated with a polymer [93]. These bio-nanocomposites outperform the basic polymers in terms of mechanical, barrier, and thermal properties. Presently, bio-nano composites made from natural polymers such as polyhydroxybutyrate, polybutylene succinate polylactic acid, aliphatic polyester are being improved by nanotechnological approaches [94].

2.3.3.2 Improved Packaging

The addition of NMs like nanoclays and Silicon dioxide (SiO_2) to NPs can increase the resilience, strength, barrier, and thermal characteristics of packaging [87]. The incorporation of fine NPs (100 nm or less) into plastics has improved properties as compared to conventional plastic packaging material [94]. The potential of thermoplastic starch (TPS) as a packaging material can be improved when nano clay is used as a filler [95].

2.3.3.3 Smart Packaging/Intelligent

Intelligent packaging offers a certain extra function outside the regular role of providing a physical barrier between the contents and its surrounding environment [96]. This type of packaging has the quality to sense and signal microbial growth or biochemical changes in its contents and extend shelf life [90]. Nutraceuticals and other packaged foods are now in high demand worldwide which leads to the requirement of new approaches for guaranteed safety and maintenance of quality at the time of harvesting and further processing, transport, and ultimately storage. In the last few years, sensitive biosensors have been designed using NMs for quality analysis and assurance in packaged foods and nutraceuticals [24, 97]. Sundramoorthy *et al.* (2018) have reviewed graphene-based NMs having properties like high conductivity, biocompatibility, mechanical flexibility, ultrahigh surface area, and so on suitable for use in developing biosensors used in food quality analysis [97].

2.3.3.4 Active Packaging

This innovative technology is capable of interacting with the food product and confer antimicrobial and oxygen scavenging properties [24, 90]. Nanomaterials of silver, copper oxide, magnesium oxide, titanium dioxide (TiO_2), and so on, are being suggested as active packaging materials [98]. An active packaging may be used to inhibit microbial growth once the functional food package has been opened [99]. Al-Naamani *et al.*, 2016 reported that coatings of chitosan-ZnO nanocomposite on PE films were very effective as active packaging material as the use of nanomaterial reduced the viability of cultures of *Escherichia coli*, *Salmonella enterica*, and *Staphylococcus aureus* by 99.9% [100]. Figure 2.5 represents the schematic outline of active packaging.

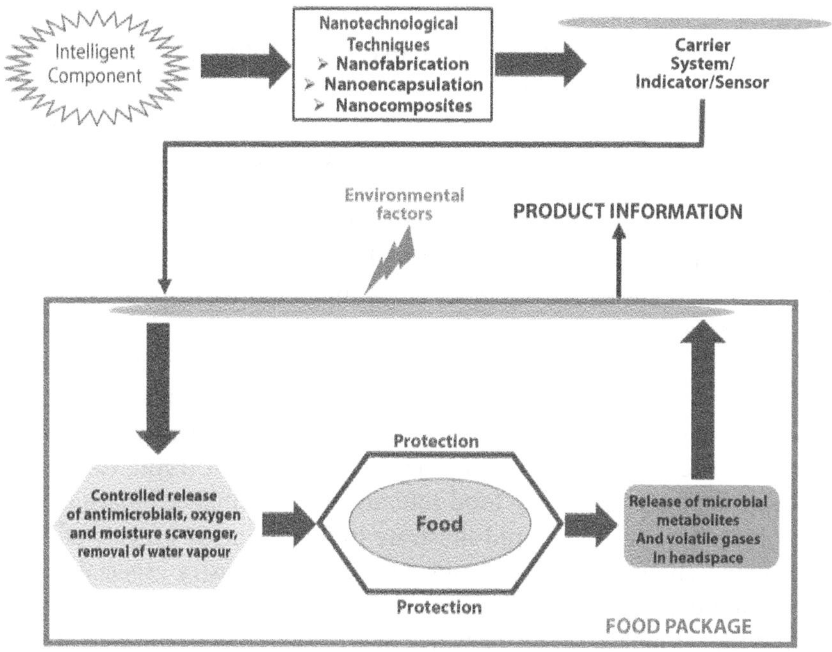

Figure 2.5 Schematic representation of concept Smart/Active/Intelligent packaging. Intelligent active components either in nanoparticulate form or tagged to NPs protect packaged food from spoilage and can also act as sensors thereby indicating the quality and state of the food through the interaction of the external environment or internal packaging atmosphere (headspace).

2.4 Mechanism of Action of Nanomaterials

There has been an inherent tendency of humans to preserve food right from their evolutionary origin. As time proceeded, humans became more civilized, rational, advanced, and devised simple methods of food preservation. Records state that the consumption of fermented food material existed 10,000 years ago which almost parallels the age of introduction of agriculture and animal husbandry [101]. It is also reported that ancient Egyptians used sundried food to prevent spoilage while Romans resorted to pickling for preventing food from microbial infestation [102]. The first technological breakthrough in the preservation of food came through the hands of William Cullen in 1784 who was credited with the development of a crude form of present-day refrigerator [103]. In the early 1800s, the process of canning and salting was adopted to preserve foods [87]. Louis Pasteur devised the pasteurization procedure in 1862, which allowed milk, wine, and beer to be preserved for longer periods of time [105]. With the invention of refrigerators in the modern age, there have been rapid progress and advancement in the preservative process of food. This also led to the greater purchase of perishable foods like milk, eggs, cheese, and meat [106]. However, none of the materials were effective in preserving food for a prolonged period. This lead to a search for more effective strategies for the preservation of food. The rapid development of nanoscience has revolutionized many domains of the food industry especially related to processing and preservation. At present food industry finds the use of various NMs and each has its role but with an ultimate motive of preservation of food and also fortification of the same for the nutritive benefit of mankind and animals. NMs and nanotechnological approaches have been steadily making their strides in the food industry. At present keeping food safe and nutritious is of topmost priority and a matter of concern for the overall benefit of society. For this reason, scientists have been trying to develop innovative strategies to keep food safe, stable and enhance bioavailability [107, 108]. This section would illustrate the mechanistic aspects of NMs in bringing about the desired result in the food industry especially with respect to decontamination, preservation, fortification, and bioavailability with special emphasis on its relevance in the domain of nutraceuticals.

2.4.1 Nanomaterials in Preservative Process

The potential contribution of nanotechnology has propelled the food industry to a completely new dimension. It is perceived in various sectors of this industry like food packaging, its preservation, safety, and migration [109]. Wastage of edible products is one of the strongest impetus that urges researchers to devise cost-effective, environment-friendly strategies that help in the storage and sustenance of food. From time immemorial man has been putting in undaunting efforts to preserve food. This age-old practice has a significant role in feeding and sustaining the blooming population. Nanomaterials used in the food industry generally furnish to fulfill certain benchmarks like enhancement of the nutritive value of food (without tampering or compromising with its taste and appearance), increasing its shelf life along with better packaging and transportation of food materials. The primary norms that govern the procedure of preservation are retardation or the prevention of microbial activities.

Nanocapsules are widely being used in food processing. Nanoparticles are integrated into food materials to buttress them with desired qualities [110]. The fundamental principle of food processing and preservation involving nano-sized materials pivots around techniques like the addition of nutrients, vitamins, minerals, nanoencapsulation of flavors, and so on

[25]. Nanoencapsulation is becoming obligatory in food technology as it not only aids in trapping the aroma but also assists in delivering the nutrients to the target. Nanocapsules are stable and easy to handle and serve as a high-class delivery tool. Organic NPs like lipid and carbohydrate NPs are extensively being used in nutraceuticals and encapsulating oil respectively [111, 112]. Nanoemulsions have anti-bacterial properties that are reported to deter spoilage of food [113].

Nanoparticles should be processed and rendered nontoxic, biocompatible, and fit for consumption before incorporation into food substances. Apart from extending the shelf life, this technology has greatly contributed towards combining anti-microbial agents in food and screen for contaminants in it [114]. Starch nanospheres are generally used in the packaging of food [115]. They are used as adhesives that require lesser time to dry since moisture requirement is very less. Silica NPs in plastic does not allow gases to penetrate and interact with the food, thus helping in preserving the food for a longer period [116]. Foiling the food from rotting faster indirectly paves way for better management of food scarcity at different corners of the globe. Silver NPs (AgNPs) are used to inhibit microbial infection and keep pathogenic microbes from contaminating the food [117]. Titanium dioxide (TiO_2) and SiO_2 NPs helps in preserving the freshness of lettuce and promote reduction in stages of development in grapes respectively [118, 119]. Nanomaterials have demonstrated their competence in serving as reliable sensors (electrical as well as optical sensors) in food industries (soft drinks, sauce, milk, etc.) assisting in the detection of various toxins, pathogens, and adulterants [120]. Apart from these food-grade coatings (on vegetables and fruits) made of nanomaterials help in the proper storage and maintenance of food value. Nanocomposites are bestowed with excellent thermal properties and are capable of precluding moisture entry in food [121].

This section, thus. gives a vivid description of how the intertwining of nano and food science can minimize wastage of food by improvised preservation strategies and thus help in addressing the global food crisis to a great extent.

2.4.2 Nanomaterials as Antimicrobial Agents

The antimicrobial properties of NMs have been exploited in the food industry and are being closely associated with the active packaging procedures. Active packaging is a recent technology that is developed to prolong shelf life along with maintaining excellence, security, and reliability of food [122]. As defined by the European regulation (EC) No 450/2009 active packaging consists of packaging schemes that interact with food in such a way to include components that release or absorb substances into or from the packed food or the environment around the food on purpose [123]. Active packaging systems are further categorized into active scavenging systems and active releasing systems. Unwanted components are removed from the food via an active scavenging system from the surrounding milieu such as moisture, oxygen, ethylene, carbon dioxide, while the emitters add up desired compounds into the food such as antioxidants, flavors, ethylene, an antimicrobial compound, and so on [122].

At present, antimicrobial NPs have become a vital component of active food packing. Its goal is to stop germs from growing in food and boost shelf life and microbiological safety for consumers while maintaining quality and other desirable features[124]. In this case, Antimicrobial NPs are integrated into a polymer matrix to extend shelf life. In addition, because of the high surface-to-volume ratio and higher surface reactivity of NMs, bacteria in food are inhibited [125]. The various antimicrobial packaging system and their associated NPs are tabulated in Table 2.2.

Nutraceuticals are extremely popular among people for a wide array of health benefits. It is preferred because it is less likely to exhibit a side effect and is, thus, perceived as safe for consumption [137]. Presently, plant nutraceuticals are made to associate with various NPs for fortifying their antimicrobial activity. A report stated that two polyphenols namely quercetin and afzelin extracted from *Crotalaria tetragona* when fabricated to AgNPs exhibited potent inhibitory activity against *Pseudomonas aeruginosa* and *Salmonella typhi*. It was further witnessed that the antimicrobial action was due to reduction in cell surface hydrophobicity, decrease in adherence to the surface, loss of membrane potential, all of which resulted in inhibition of biofilm production [138]. In another study, it was reported that zero valent silver-containing extracts of *Capsicum annuum* exhibited inhibitory activity against *Staphylococcus aureus* which is probably due to its active principle capsaicin [139]. It was further reported that citrus flavonoids hesperidin and naringin loaded green synthesized gold NPs and stabilized by gum tragacanth reduced viability of *Acanthamoeba castellanii* and *Naegleria fowleri*. In addition to it, the nanocomposite also exhibited inhibitory activity against *Escherichia coli* [140].

The mechanism by which the metal-based NPs acts as antimicrobial agents can be illustrated with respect to AgNPs. Four main mechanisms are involved in the antimicrobial activity of these NPs namely (i) attraction of NPs to the cell wall, (ii) Alteration of bacterial membrane permeability resulting in destabilization of a cell, (iii) generation of oxidative stress, (iv) initiation of signal transduction pathways [141]. The adhesion of AgNPs is governed by both the size and zeta potential of the NPs [142, 143]. It is further stated that the positive charges of AgNPs are electrostatically attracted to negative charges of the bacterial membrane and this facilitates

Table 2.2 Antimicrobial packaging system and associated NPs.

Name of NPs	Name of polymer matrix/ medium	Microorganism inhibited	Reference
Silver and Chitosan	Polylactide	*Staphylococcus aureus* (ATCC 6538) *Escherichia coli* (DSMZ 30083)	[126]
Silver	Binary blend films of agar and banana powder	*Escherichia coli* and *Listeria monocytogenes*	[127]
TiO_2	low-density polyethylene (LDPE) film	*Escherichia coli*	[128]
TiO_2	chitosan matrix	*Escherichia coli* and *Staphylococcus aureus*	[129]
ZnO/Nanocopper Composite	Silica Gel	*Xanthomonas alfalfae* subsp. *citrumelonis*, *Pseudomonas syringae* pv. *syringae*, *Clavibacter michiganensis* subsp. *michiganensis*	[130]
Copper	Chitosan	*Bacillus cereus*, *Escherichia coli*	[131]
Silicon dioxide	poly(butylene adipate-co-terephthalate)	*Staphylococcus aureus*, and *Escherichia coli*	[132]

(*Continued*)

Table 2.2 Antimicrobial packaging system and associated NPs. (*Continued*)

Name of NPs	Name of polymer matrix/ medium	Microorganism inhibited	Reference
Nano ZnO and nano-kaolin	Semolina	*Staphylococcus aureus*, and *Escherichia coli,* *Candida albicans* *Aspergillus niger*	[133]
ZnO	Chitosan-Carboxymethyl cellulose	*Staphylococcus aureus* *Pseudomonas aeruginosa*, *Escherichia coli* *Candida albicans*	[134]
ZnO prepared from Roselle calyx extract	Chitosan/guar gum	*Bacillus cereus* *Staphylococcus aureus* *Pseudomonas aeruginosa* *Listeria monoytogenes* *Escherichia coli* *Yersinia enterocolitica* *Salmonella typhiurium* *Aspergillus* sp.	[135]
TiO_2 NPs	chitosan, polyvinyl alcohol, glycerol	*Bacillus cereus* *Bacillus subtilis* *Staphylococcus aureus* *Pseudomonas aeruginosa* *Listeria monoytogenes* *Escherichia coli* *Yersinia enterocolitica* *Salmonella typhimurium*	[136]

the attachment of the NPs [144]. After adhesion to the bacterial cell surface, AgNP interacts with the cell in two different ways. It is found that the small AgNPs directly enter the bacteria [145] while large particles remain outside. However, in both cases, AgNPs uninterruptedly discharge silver ions (Ag^+) which consequently attach to bacterial cell membrane thereby altering its potential and ultimately leads to leakage of contents [146]. This results in destabilization of cell wall and membrane thereby providing access to larger AgNPs within the cell [141]. The modes of action of AgNPs in the bacterial cell membrane is depicted schematically in Figure 2.6.

Once the AgNPs get inside the bacterial cell, the Ag^+ gets a chance to interact with several biomolecules and alters their function. The AgNPs attach to bacterial DNA thereby preventing the replication process [147]. It also binds to the ribosomal proteins leading to ribosomal denaturation native structure and ultimately attenuation of translation [148]. It is also reported that AgNPs firmly bind to proteins with a negative charge and nucleic acids and also block the respiratory chain by binding to the enzymes [149]. AgNP is also reported to disrupt the production of

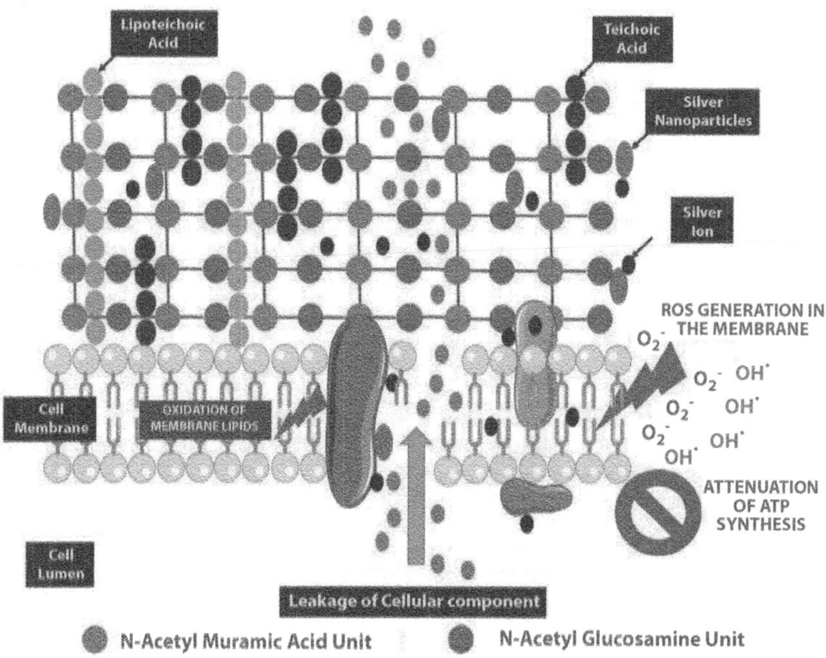

Figure 2.6 Pictorial representation of anchorage of AgNP and released Ag^+ in the peptidoglycan layer of cell wall resulting in oxidation of membrane, attenuation of ATP production of leakage of intracellular contents.

adenosine triphosphate due to the deactivation of enzymes in the cytoplasmic membrane [150].

Another mechanism through which AgNPs exhibit the antimicrobial effect is the onset of oxidative stress. A study reported the production of reactive oxygen species (ROS) in *Candida albicans* upon treatment with AgNPs. This was also accompanied by a delay in the cell cycle [151]. In most cases, the metal ions released from the NPs amalgamates into redox cycling process of bacterial cell and chemocatalysis through Fenton's reaction [$H_2O_2 + Fe^{2+} \rightarrow Fe^{3+} + HO^- + \bullet OH$] or reactions similar to Fenton's reaction [$Ag^+ \; H_2O_2 + H^+ = Ag+ + \bullet OH + H_2O$] [152, 153]. Moreover, the mechanism of ROS can also be explained with the help of the Ti-AgNP system in which the electron from the titanium ion may go to the intracellular part of the bacteria and results in abnormal production of ROS within the cell [154]. All these processes result in the accumulation of ROS within the bacterial cell which further proceeds towards oxidation of biomolecules thereby altering their normal function and ultimately death of bacteria (Figure 2.7).

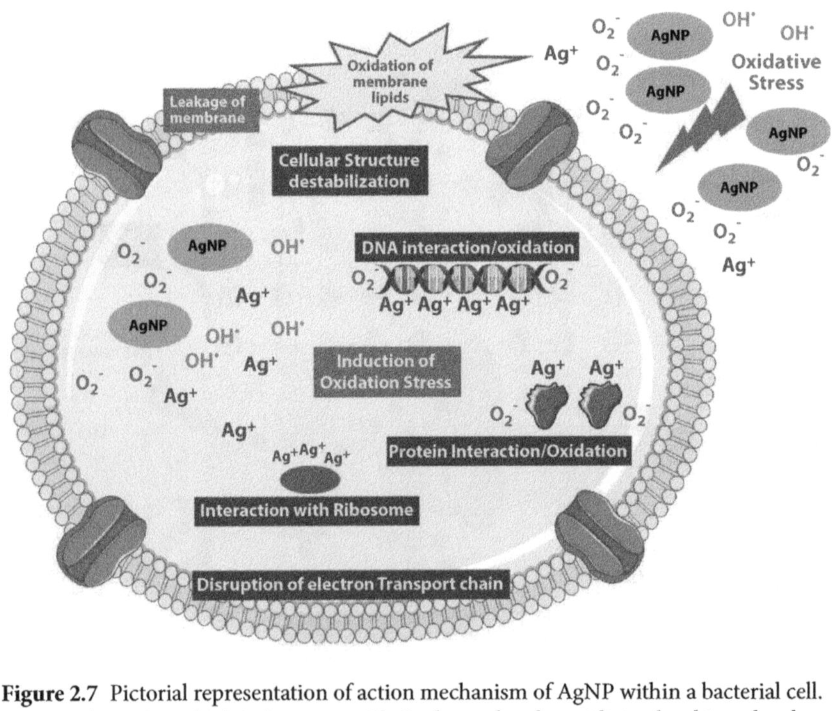

Figure 2.7 Pictorial representation of action mechanism of AgNP within a bacterial cell. AgNPs release Ag^+ which induces an oxidative burst thereby oxidizing key biomolecules and leakage of the membrane, The Ag^+ also interacts with DNA, protein, and ribosome causing an altered function.

2.4.3 Nanomaterials in Increasing Bioavailability of Nutraceuticals

Nanomaterials are frequently used as delivery systems for the controlled discharge of their cargo which enhances the bioavailability. Nanoscale emulsions improve lipophilic chemical bioavailability by increasing solubilization, increasing passive diffusion rate, and promoting direct uptake of the compounds by the intestine's lymphatic system [155]. It is reported from a study that puerarin, an isoflavonoid present in *Pueraria lobata* roots exhibited greater oral bioavailability when loaded in polybutylcyanoacrylate NPs than crude compound [156]. Another study reported that the oral bioavailability of curcumin was amplified by nine folds in novel organogel-based nanoemulsions [157]. It is further reported that a self nano emulsifying drug delivery system of resveratrol increased the bioavailability by 3.2 times in comparison to the native compound [158]. Lipid NPs have to capacity to enhance the bioavailability of orally administered compounds through several mechanisms. Firstly, they possess enhanced solubilization and dissolute rapidly due to the inclusion of lipid within lipid NPs. Moreover, lipid administration results in results in activation of biliary and pancreatic secretion which helps in the digestion process [159]. Moreover, the adhesive property of lipid NPs results in presence of bioactive compounds for an elongated period within the gastrointestinal tract thereby increasing the absorption [160]. Drugs or nutraceuticals carried by lipid-based NPs which have high solubility for triglycerides undergo transportation through lymph ducts which increases the bioavailability as they do not undergo first-pass metabolism [161]. The transportation of lipid NPs into the epithelium cells involves the process of lipid raft-dependent endocytosis, clathrin-mediated endocytosis and macropinocytosis [162].

2.4.4 Nanomaterials in Protecting From Decomposition

Deterioration of food quality can be attributed to the chemical degeneration in which various compounds present within the food undergo spontaneous reaction in normal conditions thereby distorting the quality and taste of the food. Nanoparticles can prove to be a very effective solution in stopping these unwanted chemical reactions. In this aspect, the polymeric NPs are the most suitable candidates for encapsulation or nutraceuticals or other natural products and discharge in the acidic location particularly the stomach [163]. Nanoencapsulation is a useful method for improving the delivery of biomolecules and antibiotics in food in the food industry. Most of the nutraceuticals are sensitive and may easily degenerate, the primary aim of nanoencapsulation is a reduction of damage and protecting the biomolecules from untoward circumstances [164]. Nanoparticles functionalized with natural antioxidants are found to enhance the antioxidant activity and also provide targeted delivery of those compounds which have poor membrane permeability and cellular internalization [165]. The protection of food from drying out, dampness, and oxidation is another goal of the food packaging sector. This purpose is often achieved by a hybrid system containing silicate NPs. They reduce the entry of gases and also prevent the exit of moisture, thereby slowing down the process of decay [107].

2.5 Prospects and Concluding Remarks

Humans need balanced nutrition for their normal physiological functioning and healthy growth. These nutrients are obtained from various types of food. Also, a large number of deleterious compounds and free radicals are produced in the body because of numerous biochemical reactions. The body is equipped with mechanisms that can nullify the deleterious effects of these harmful biochemical byproducts. The antioxidant mechanism in this regard plays a very vital role in counteracting ROS. However, with age, the capacity of the body to counter the ROS as well as the detoxification mechanism weakens. Under those conditions consumption of dietary supplements becomes a necessity which then fortifies the physiological state of the body. Nutraceuticals can provide medical and health benefits. They have become very popular throughout the globe and people are becoming increasingly dependent on the same primarily due to their natural source and also perceived to be safe for consumption. Thus, there is an increasing trend of fortification of food with nutraceuticals. The priority of the food industry is largely concentrated on packaging whose primary aim is maintaining the food quality and prevent degradation. Hence, they rely on various nanoparticulate matters which in several ways prevent degradation or acts as sensors to indicate the onset of degradation or contamination. Therefore, they provide the consumers vital information regarding the state of food.

Nanoparticles are being widely synthesized for application in the various domain including the food industry. However, only a few nations have established uniform regulatory guidelines for the use of NPs in food [166]. Moreover, scientific validation about the efficacy of NPs in the food industry is scarcely new. In addition, there is always a possibility that the NMs used in the packaging may enter the food chain and bioaccumulate in various trophic levels. This may in turn create damage to the living system at a molecular level which can have a far-fetched effect on the living system. In addition, proper disposal of nanoparticle-based packaging systems is also required as there is also a chance of getting accumulated in the soil. This may harm the microbial flora of the soil. Thus, proper investigation about the safety of NPs in the packaging of food is very relevant. In addition, *in-vivo* studies require to be performed to check for any adverse effects of nanoparticle-based food and food supplements. In countries where contamination of food is frequent, these products are instrumental in offering safe food to the population.

Since many nanotechnology-based products are still in the research and development stage, there are several substantial knowledge gaps in our present understanding of the behaviour, properties, and impacts of NMs that need to be addressed. People should be made aware of nanoparticle-based food and efforts should be undertaken from governmental levels to popularise the same and increase the acceptability. In third world countries, where there is often an imbalance of nutrient uptake, nanoparticle-based nutraceuticals can be instrumental in nourishing the population. Care should be taken to make these food products in an affordable range. Thus, there are both positivities and challenges of NMs in the food industry. However, judicious use of the NPs can make the situation better and it should be accompanied by proper safety measures backed by sufficient scientific data. Also, there is an urgent need for the producers not only to guarantee the security, quality of the product but also to provide the associated risks, benefits, and regulations, then only acceptance of the products by the consumers is made successful. This can be achieved by utilizing the methods for finding, characterization of NMs in multifaceted food matrices, adopting best practices and self-regulation, establishing international research collaborations and networks, developing nano food technologies through exploration of political, societal, psychological, and historical topics so as, to address different facets of the prevailing and novel nanotechnology applications in food and agriculture areas which is a necessary element of commercialization. Thus, a universal regulatory system is needed to enable premarket review of nano-food items, as well as to establish obligations and restrictions for any nano-additives used in feeding and related applications.

References

1. Shin, W.K., Cho, J., Kannan, A.G., Lee, Y.S., Kim, D.W., Cross-linked composite gel polymer electrolyte using mesoporous methacrylate-functionalized SiO2 nanoparticles for lithium-ion polymer batteries. *Sci. Rep.*, 6, 1, 1–10, 2016.

2. Singh, A.K., *Engineered Nanoparticles*, pp. 1–18, Academic Press, Cambridge, MA, USA, 2016.
3. Khan, I., Saeed, K., Khan, I., Nanoparticles: Properties, applications and toxicities. *Arab. J. Chem.*, 12, 7, 908–931, 2019.
4. Tiwari, J.N., Tiwari, R.N., Kim, K.S., Zero-dimensional, one-dimensional, two-dimensional and three-dimensional nanostructured materials for advanced electrochemical energy devices. *Prog. Mater. Sci.*, 57, 4, 724–803, 2012.
5. Paramo, L.A., Feregrino-Pérez, A.A., Guevara, R., Mendoza, S., Esquivel, K., Nanoparticles in agroindustry: Applications, toxicity, challenges, and trends. *Nanomaterials*, 10, 9, 1654, 2020.
6. Saleh, T.A., Nanomaterials: Classification, properties, and environmental toxicities. *Environ. Technol. Innov.*, 20, 101067, 2020.
7. Horikoshi, S.A. and Serpone, N.I., Introduction to nanoparticles, in: *Microwaves in nanoparticle synthesis: Fundamentals and applications*, S.A. Horikoshi and N.I. Serpone (Eds.), pp. 1–24, John Wiley & Sons, Inc., Weinheim, Germany, 2013.
8. Judy, J.D., Unrine, J.M., Bertsch, P.M., Evidence for biomagnification of gold NPs within a terrestrial food chain. *Environ. Sci. Technol.*, 45, 2, 776–781, 2011.
9. Baeumner, A., Nanosensors identify pathogens in food. *Food Technol. (Chicago)*, 58, 8, 51–55, 2004.
10. Ting, Y., Jiang, Y., Ho, C.T., Huang, Q., Common delivery systems for enhancing in vivo bioavailability and biological efficacy of nutraceuticals. *J. Funct. Foods*, 7, 112–128, 2014.
11. Andlauer, W. and Fürst, P., Nutraceuticals: A piece of history, present status and outlook. *Food Res. Int.*, 35, 2-3, 171–176, 2002.
12. Yeung, A.W.K., Mocan, A., Atanasov, A.G., Let food be thy medicine and medicine be thy food: A bibliometric analysis of the most cited papers focusing on nutraceuticals and functional foods. *Food Chem.*, 269, 455–465, 2018.
13. Mandala, F., Guida, R., Guida, G.F., Nutraceuticals: What they are and how they work. *E-Journal of the ESC Council for Cardiology Practice.*, 9, 11, 2010.
14. Kokate, C.K., Purohit, A.P., Gokhale, D.S., *Pharmacognosy*, pp. 81–94, Nirali Prakashan, Pune, 2008.
15. Pandey, M., Verma, R.K., Saraf, S.A., Nutraceuticals: New era of medicine and health. *Asian J. Pharm. Clin. Res.*, 3, 1, 11–15, 2010.
16. Das, L., Bhaumik, E., Raychaudhuri, U., Chakraborty, R., Role of nutraceuticals in human health. *J. Food Sci. Technol.*, 49, 2, 173–183, 2012.
17. Kalra, E.K., Nutraceutical-definition and introduction. *AAPS PharmSci.*, 5, 3, 27–28, 2003.
18. Palthur, M.P., Palthur, S.S., Chitta, S.K., Emerging product categories in India: A regulatory view. *Food Drug Law J.*, 64, 4, 677–692, 2009.
19. Surve, D.H., Paul, A.T., Jindal, A.B., Dasgupta, N., Ranjan, S., Lichtfouse, E., Nanotechnology based delivery of nutraceuticals, in: *Environmental Nanotechnology Environmental Chemistry for a Sustainable World*,

N. Dasgupta, S. Ranjan, E. Lichtfouse (Eds.), pp. 63–107, Springer, Gewerbestrasse, Switzerland, 2019.

20. Jampilek, J., Kos, J., Kralova, K., Potential of nanomaterial applications in dietary supplements and foods for special medical purposes. *Nanomaterials (Basel)*, 9, 2, 296, 2019.
21. Lopresti, A.L., The problem of curcumin and its bioavailability: Could its gastrointestinal influence contribute to its overall health-enhancing effects? *Adv. Nutr.*, 9, 1, 41–50, 2018.
22. Yu, H. and Huang, Q., Improving the oral bioavailability of curcumin using novel organogel-based nanoemulsions. *J. Agric. Food Chem.*, 60, 21, 5373–5379, 2012.
23. Darder, M., Aranda, P., Ruiz-Hitzky, E., Bionanocomposites: A new concept of ecological, bioinspired and functional hybrid materials. *Adv. Mater.*, 19, 1309–319, 2007.
24. Sozer, N. and Kokini, J.L., Nanotechnology and its applications in the food sector. *Trends Biotechnol.*, 27, 2, 82–89, 2009.
25. Huang, Q., Yu, H., Ru, Q., Bioavailability and delivery of nutraceuticals using nanotechnology. *J. Food Sci.*, 75, 1, 50–57, 2010.
26. Nasri, H., Baradaran, A., Shirzad, H., Rafieian-Kopaei, M., New concepts in nutraceuticals as alternative for pharmaceuticals. *Int. J. Prev. Med.*, 5, 12, 1487–1499, 2014.
27. Souyoul, S.A., Saussy, K.P., Lupo, M.P., Nutraceuticals: A review. *Dermatol. Ther.*, 8, 1, 5–16, 2018.
28. Sadhukhan, P., Saha, S., Dutta, S., Mahalanobish, S., Sil, P.C., Nutraceuticals: An emerging therapeutic approach against the pathogenesis of Alzheimer's disease. *Pharmacol. Res.*, 129, 100–114, 2018.
29. Mannucci, C., Casciaro, M., Sorbara, E.E., Calapai, F., Di Salvo, E., Pioggia, G., Navarra, M., Calapai, G., Gangemi, S., Nutraceuticals against oxidative stress in autoimmune disorders. *Antioxidants (Basel)*, 10, 2, 261, 2021.
30. Janssens, P.L., Hursel, R., Westerterp-Plantenga, M.S., Nutraceuticals for body-weight management: The role of green tea catechins. *Physiol. Behav.*, 162, 83–87, 2016.
31. Sosnowska, B., Penson, P., Banach, M., The role of nutraceuticals in the prevention of cardiovascular disease. *Cardiovasc. Diagn. Ther.*, 7, 21–31, 2017.
32. Pellegrini, M., Senni, C., Bernabei, F., Cicero, A., Vagge, A., Maestri, A., Scorcia, V., Giannaccare, G., The role of nutrition and nutritional supplements in ocular surface diseases. *Nutrients*, 12, 4, 952, 2020.
33. Calvani, M., Pasha, A., Favre, C., Nutraceutical boom in cancer: Inside the Labyrinth of reactive oxygen species. *Int. J. Mol. Sci.*, 21, 6, 1936, 2020.
34. Lama, A., Pirozzi, C., Avagliano, C., Annunziata, C., Mollica, M.P., Calignano, A., Meli, R., Raso, G.M., Nutraceuticals: An integrative approach to starve Parkinson's disease. *Brain Behav. Immun.*, 2, 100037, 2020.
35. Dehdashtian, E., Pourhanifeh, M.H., Hemati, K., Mehrzadi, S., Hosseinzadeh, A., Therapeutic application of nutraceuticals in diabetic nephropathy:

Current evidence and future implications. *Diabetes Metab. Res. Rev.*, 36, 8, 3336, 2020.
36. Chelluboina, B. and Vemuganti, R., Therapeutic potential of nutraceuticals to protect brain after stroke. *Neurochem. Int.*, 142, 104908, 2021.
37. Chauhan, B., Kumar, G., Kalam, N., Ansari, S.H., Current concepts and prospects of herbal nutraceutical: A review. *J. Adv. Pharm. Technol. Res.*, 4, 1, 4–8, 2013.
38. India Nutraceuticals Industry, Official Website of the International Trade Administration, Washington, DC, USA, 2020, https://www.trade.gov/market-intelligence/india-nutraceuticals-industry.
39. Food Safety and Standards Act, Ministry of Law and Justice, Government of India, 2006. https://fssai.gov.in/cms/food-safety-and-standards-act-2006.php#:~:text=It%20is%20an%20Act%20to,and%20wholesome%20food%20for%20human.2006.
40. Nasri, H., Baradaran, A., Shirzad, H., Rafieian-Kopaei, M., New concepts in nutraceuticals as alternative for pharmaceuticals. *Int. J. Prev. Med.*, 5, 12, 1487, 2014.
41. Shinde, N., Bangar, B., Deshmukh, S., Kumbhar, P., Nutraceuticals: A review on current status. *Res. J. Pharm. Technol.*, 7, 1, 110–113, 2014.
42. Madley-Wright, R., New Products. *Nutraceutical World*, 6, 66–71, 2003a.
43. Madley-Wright, R., New Products. *Nutraceutical World*, 6, 118–25, 2003b.
44. Allen, L.V., Nutritional products, in: *Handbook of Nonprescription Drugs*, T.R. Covington, R.R. Berardi, L.L. Young, S.C. Kendall, M.J. Hickey (Eds.), American Pharmaceutical Association, Washington DC, 1997.
45. Ghaffari, S. and Roshanravan, N., The role of nutraceuticals in prevention and treatment of hypertension: An updated review of the literature. *Food Res. Int.*, 128, 108749, 2020.
46. Bhowmik, D., Gopinath, H., Kumar, B.P., Duraivel, S., Kumar, K.S., Nutraceutical-a bright scope and opportunity of Indian healthcare market. *Pharma Innov.*, 1, 11, 29–41, 2013.
47. Patel, J.M., A review of potential health benefits of flavonoids. *Lethbridge Undergrad. Res. J.*, 3, 2, 1–5, 2008.
48. World Health Organization, *Guideline: Calcium supplementation in pregnant women*, pp. 1–30, World Health Organization, Geneva, Switzerland, 2013.
49. Davoodi, H., Esmaeili, S., Mortazavian, A.M., Effects of milk and milk products consumption on cancer: A review. *Compr. Rev. Food Sci. Food Saf.*, 12, 3, 249–264, 2013.
50. Bounous, G., Whey protein concentrate (WPC) and glutathione modulation in cancer treatment. *Altern. Med. Rev.*, 6, 3, 342–342, 2001.
51. Saini, R., Coenzyme Q10: The essential nutrient. *J. Pharm. Bioallied Sci.*, 3, 3, 466–467, 2011.

52. Ruxton, C.H.S., Reed, S.C., Simpson, M.J.A., Millington, K.J., The health benefits of omega-3 polyunsaturated fatty acids: A review of the evidence. *J. Hum. Nutr. Diet.*, 17, 5, 449–459, 2004.
53. Nutakor, C., Essiedu, J.A., Adadi, P., Kanwugu, O.N., Ginger Beer: An overview of health benefits and recent developments. *Fermentation*, 6, 4, 102, 2020.
54. Ma, R.H., Ni, Z.J., Zhu, Y.Y., Thakur, K., Zhang, F., Zhang, Y.Y., Wei, Z.J., A recent update on the multifaceted health benefits associated with ginger and its bioactive components. *Food Funct.*, 12, 2, 519–54, 2021.
55. Hewlings, S.J. and Kalman, D.S., Curcumin: A review of its effects on human health. *Foods*, 6, 10, 92, 2017.
56. Gruenwald, J., Freder, J., Armbruester, N., Cinnamon and health. *Crit. Rev. Food Sci. Nutr.*, 50, 9, 822–834, 2010.
57. Khorasany, A.R. and Hosseinzadeh, H., Therapeutic effects of saffron (*Crocus sativus* L.) in digestive disorders: A review. *Iran J. Basic Med. Sci.*, 19, 5, 455, 2016.
58. Ma, J. and Betts, N.M., Zinc and copper intakes and their major food sources for older adults in the 1994–96 continuing survey of food intakes by individuals (CSFII). *J. Nutr.*, 130, 11, 2838–2843, 2000.
59. Chasapis, C.T., Loutsidou, A.C., Spiliopoulou, C.A., Stefanidou, M.E., Zinc and human health: An update. *Arch. Toxicol.*, 86, 4, 521–534, 2012.
60. He, L., Lang, L., Li, Y., Liu, Q., Yao, Y., Comparison of serum zinc, calcium, and magnesium concentrations in women with pregnancy-induced hypertension and healthy pregnant women: A meta-analysis. *Hypertens. Pregnancy*, 35, 2, 202–209, 2016.
61. Spiro, A. and Buttriss, J., Vitamin D: An overview of vitamin D status and intake in Europe. *Nutr. Bull.*, 39, 4, 322–350, 2014.
62. Meng, X., Li, Y., Li, S., Zhou, Y., Gan, R.Y., Xu, D.P., Li, H.B., Dietary sources and bioactivities of melatonin. *Nutrients*, 9, 4, 367, 2017.
63. Zhang, R., Wang, X., Ni, L., Di, X., Ma, B., Niu, S., Liu, C., Reiter, R.J., COVID-19: Melatonin as a potential adjuvant treatment. *Life Sci.*, 250, 117583, 2020.
64. Codjia, G., Food sources of vitamin A and provitamin A specific to Africa: An FAO perspective. *Food Nutr. Bull.*, 22, 4, 357–360, 2001.
65. Papoutsi, Z., Kassi, E., Chinou, I., Halabalaki, M., Skaltsounis, L.A., Moutsatsou, P., Walnut extract (*Juglans regia* L.) and its component ellagic acid exhibit anti-inflammatory activity in human aorta endothelial cells and osteoblastic activity in the cell line KS483. *Br. J. Nutr.*, 99, 4, 715–722, 2008.
66. Loizou, S., Lekakis, I., Chrousos, G.P., Moutsatsou, P., β-Sitosterol exhibits anti-inflammatory activity in human aortic endothelial cells. *Mol. Nutr. Food Res.*, 54, 4, 551–558, 2010.
67. Ahad, T., Nissar, J., Rather, A.H., Nutraceuticals-health benefits: A review. *Int. J. Eng. Technol. Sci. Res.*, 4, 12, 262–267, 2017.
68. Li, H., Wang, Z., Liu, Y., Review in the studies on tannins activity of cancer prevention and anticancer. *Zhong Yao Cai*, 26, 6, 444–448, 2003.

69. Giordano, P., Scicchitano, P., Locorotondo, M., Mandurino, C., Ricci, G., Carbonara, S., Matteo Ciccone, M., Carotenoids and cardiovascular risk. *Curr. Pharm. Des.*, 18, 34, 5577–5589, 2012.
70. Scicchitano, P., Cameli, M., Maiello, M., Modesti, P.A., Muiesan, M.L., Novo, S., di Cardiologia, S.I., Nutraceuticals and dyslipidaemia: Beyond the common therapeutics. *J. Funct. Foods*, 6, 11–32, 2014.
71. The Royal Society and Royal Academy of Engineering, *Nanoscience and nano technologies: Opportunities and uncertainties*, pp. 1–111, Royal Society and Royal Academy of Engineering, London, 2004.
72. Zanella, M., Ciappellano, S.G., Venturini, M., Tedesco, E., Manodori, L., Benetti, F., Nutraceuticals and nanotechnology. *Agro. Food Ind. Hi Tech.*, 26, 4, 26–31, 2015.
73. Cha, D. and Chinnan, M., Biopolymer-based antimicrobial packaging: A review. *Crit. Rev. Food Sci. Nutr.*, 44, 223–237, 2004.
74. Khan, A., Wen, Y., Huq, T., Ni, Y., Cellulosic nanomaterials in food and nutraceutical applications: A review. *J. Agric. Food Chem.*, 66, 8–19, 2018.
75. Kuzma, J. and VerHage, P., *Nanotechnology in Agriculture and Food Production: Anticipated Applications*, pp. 1–40, Project on Emerging Nanotechnologies, Woodrow Wilson International Center for Scholars, Washington, DC, 2006.
76. Lugani, Y., Sooch, B.S., Singh, P., Kumar, S., Nanobiotechnology applications in food sector and future innovations, in: *Microbial Biotechnology in Food and Health*, pp. 197–225, 2021.
77. Sorrentino, A., Gorrasi., G., Vittoria, V., Potential perspectives of bio-nanocomposites for food packaging applications. *Trends Food Sci. Technol.*, 18, 2, 84–95, 2007.
78. Weiss, J., Takhistov, P., McClements, D.J., Functional materials in food nanotechnology. *J. Food Sci.*, 71, 9, R107–R116, 2006.
79. Chaudhry, Q. and Castle, L., Food applications of nanotechnologies: An overview of opportunities and challenges for developing countries. *Trends Food Sci. Technol.*, 22, 11, 595–603, 2011.
80. Siegrist, M., Cousin, M.E., Kastenholz, H., Wiek, A., Public acceptance of nanotechnology foods and food packaging: The influence of affect and trust. *Appetite*, 49, 2, 459–466, 2007.
81. Diehl, J.F., Will irradiation enhance or reduce food safety? *Food Policy*, 18, 2, 143–151, 1993.
82. Ravishankar, S., Zhang, H., Kempkes, M.L., Pulsed electric fields. *Food Sci. Technol. Int.*, 14, 5, 429–432, 2008.
83. Meena, M., Pilania, S., Pal, A., Mandhania, S., Bhushan, B., Kumar, S., Gohari, G., Saharan, V., Cu-chitosan nano-net improves keeping quality of tomato by modulating physio-biochemical responses. *Sci. Rep.*, 10, 1, 1–11, 2020.
84. Hsu, C.Y., Wang, P.W., Alalaiwe, A., Lin, Z.C., Fang, J.Y., Use of lipid nanocarriers to improve oral delivery of vitamins. *Nutrients*, 11, 1, 68, 2019.

85. Jafari, S.M. and McClements, D.J., Nanotechnology Approaches for increasing nutrient bioavailability. *Adv. Food Nutr. Res.*, 81, 1–30, 2017.
86. Singh, P.K., Jairath, G., Ahlawat, S.S., Nanotechnology: A future tool to improve quality and safety in meat industry. *J. Food Sci. Technol.*, 53, 4, 1739–1749, 2016.
87. Kuswandi, B., Nanotechnology in food packaging, in: *Nanoscience in Food and Agriculture 1. Sustainable Agriculture Reviews*, S. Ranjan, N. Dasgupta, E. Lichtfouse (Eds.), pp. 151–183, Springer, Cham, 2016.
88. Kirwan, M.J. and Strawbridge, J.W., Plastics in food packaging, in: *Food packaging technology*, R. Coles, D. Mc Dowel, M.J. Kirwan (Eds.), pp. 174–240, Blackwell Publishing Ltd, Oxford, 2003.
89. Sharma, C., Dhiman, R., Rokana, N., Panwar, H., Nanotechnology: An untapped resource for food packaging. *Front. Microbiol.*, 8, 1735, 2017.
90. Jafarizadeh-Malmiri, H., Hooda, *et al.*, Nanobiotechnology in food packaging, in: *Nanobiotechnology in Food: Concepts, Applications and Perspectives*, H. Jafarizadeh-Malmiri, Z. Sayyar, N. Anarjan, A. Berenjian (Eds.), pp. 69–79, Springer, Cham, 2019.
91. Malathi, A.N., Santhosh, K.S., Nidoni, U., Recent trends of biodegradable Polymer: Biodegradable films for food packaging and application of nanotechnology in biodegradable food packaging. *Curr. Trends Technol. Sci.*, 3, 2, 73–79, 2014.
92. Lagaron, J.M., Cabedo, L., Cava, D., Feijoo, J.L., Gavara, R., Gimenez, E., Improving packaged food quality and safety. Part 2: Nanocomposites. *Food Addit. Contam.*, 22, 10, 994–998, 2005.
93. Alexandre, M. and Dubois, P., Polymer-layered silicate nanocomposites: Preparation, properties and uses of a new class of materials. *Mater. Sci. Eng. R. Rep.*, 28, 1–63, 2000.
94. Sekhon, B.S., Food nanotechnology - An overview. *Nanotechnol. Sci. Appl.*, 3, 1–15, 2010.
95. Chen, B. and Evans, J.R.G., Thermoplastic starcheclay nanocomposites and their characteristics. *Carbohydr. Polym.*, 61, 4, 455–463, 2005.
96. Fuertes, G., Soto, I., Carrasco, R., Vargas, M., Sabattin, J., Lagos, C., Intelligent packaging systems: Sensors and nanosensors to monitor food quality and safety. *J. Sens.*, 2016, 1–9, 2016.
97. Sundramoorthy, A., Kumar, T.H., Gunasekaran, S., Graphene-based nanosensors and smart food packaging systems for food safety and quality monitoring, in: *Advanced Nanomaterials, Graphene Bioelectronics*, A. Tiwari (Ed.), pp. 267–306, Elsevier, Cambridge, MA, USA, 2018.
98. Momin, J.K., Jayakumar, C., Prajapati, J.B., Potential of nanotechnology in functional foods. *Emir. J. Food Agric.*, 25, 1, 10–19, 2013.
99. Brody, A.L., Case studies on nanotechnologies for food packaging. *Food Technol.*, 7, 102–107, 2007.

100. Al-Naamani, L., Dobretsov, S., Dutta, J., Chitosan-zinc oxide nanoparticle composite coating for active food packaging applications. *Innov. Food Sci. Emerg. Technol.*, 38, 231–237, 2016.
101. Mota de Carvalho, N., Costa, E.M., Silva, S., Pimentel, L., Fernandes, T.H., Pintado, M.E., Fermented foods and beverages in human diet and their influence on gut microbiota and health. *Fermentation*, 4, 4, 90, 2018.
102. Pradhan, N., Singh, S., Ojha, N., Shrivastava, A., Barla, A., Rai, V., Bose, S., Facets of nanotechnology as seen in food processing, packaging, and preservation industry. *Biomed. Res. Int.*, 2015, 365672, 1–17, 2015.
103. Hamad, A.F., Han, J.H., Kim, B.C., Rather, I.A., The intertwine of nanotechnology with the food industry. *Saudi J. Biol. Sci.*, 25, 1, 27–30, 2018.
104. Abbas, K.A., Saleh, A.M., Mohamed, A., MohdAzhan, N., The recent advances in the nanotechnology and its applications in food processing: A review. *J. Food Agric. Environ.*, 7, 3-4, 14–17, 2009.
105. Currier, R.W. and Widness, J.A., A brief history of milk hygiene and its impact on infant mortality from 1875 to 1925 and implications for today: A review. *J. Food Prot.*, 81, 10, 1713–1722, 2018.
106. Huebbe, P. and Rimbach, G.,., Historical reflection of food processing and the role of legumes as part of a healthy balanced diet. *Foods*, 9, 8, 1056, 2020.
107. Ameta, S.K., Rai, A.K., Hiran, D., Ameta, R., Ameta, S.C., Use of nanomaterials in food science, in: *Biogenic Nano-Particles and their Use in Agro-ecosystems*, M. Ghorbanpour, P. Bhargava, A. Varma, D.K. Choudhary (Eds.), pp. 457–488, Springer, Singapore, 2020.
108. Mustafa, F. and Andreescu, S., Nanotechnology-based approaches for food sensing and packaging applications. *RSC Adv.*, 10, 33, 19309–19336, 2020.
109. Das, L., Bhaumik, E., Raychaudhuri, U., Chakraborty, R., Role of nutraceuticals in human health. *J. Food Sci. Technol.*, 49, 2, 173–183, 2012.
110. Sandoval, B., Perspectives on FDA's regulation of nanotechnology: Emerging challenges and potential solutions. *Compr. Rev. Food Sci. Food Saf.*, 8, 4, 375–393, 2009.
111. Severino, P., Andreani, T., Macedo, A.S., Fangueiro, J.F., Santana, M.H.A., Silva, A.M., Souto, E.B., Current state-of-art and new trends on lipid nanoparticles (SLN and NLC) for oral drug delivery. *J. Drug Deliv.*, 2020, 1–10, 2012.
112. Verma, M.L., Dhanya, B.S., Rani, V., Thakur, M., Jeslin, J., Kushwaha, R., Carbohydrate and protein based biopolymeric nanoparticles: Current status and biotechnological applications. *Int. J. Biol. Macromol.*, 154, 390–412, 2020.
113. Wang, Y., Zhang, Q., Zhang, C.L., Li, P., Characterisation and cooperative antimicrobial properties of chitosan/nano-ZnO composite nanofibrous membranes. *Food Chem.*, 132, 1, 419–427, 2012.
114. Neo, Y.P., Ray, S., Jin, J., Gizdavic-Nikolaidis, M., Nieuwoudt, M.K., Liu, D., Quek, S.Y., Encapsulation of food grade antioxidant in natural biopolymer

by electrospinning technique: A physicochemical study based on zein–gallic acid system. *Food Chem.*, 136, 2, 1013–1021, 2013.

115. Baysal, G. and Doğan, F., Investigation and preparation of biodegradable starch-based nanofilms for potential use of curcumin and garlic in food packaging applications. *J. Biomater. Sci. Polym. Ed.*, 31, 9, 1127–1143, 2020.
116. Nile, S.H., Baskar, V., Selvaraj, D., Nile, A., Xiao, J., Kai, G., Nanotechnologies in food science: Applications, recent trends, and future perspectives. *Nano-Micro Lett.*, 12, 45, 2020.
117. Li, M., Liu, H.L., Dang, F., Hintelmann, H., Yin, B., Zhou, D., Alteration of crop yield and quality of three vegetables upon exposure to silver nanoparticles in sludge-amended soil. *ACS Sustain. Chem. Eng.*, 8, 6, 2472–2480, 2020.
118. Peter, A., Tegla, D., Giurgiulescu, L., Cozmuta, A.M., Nicula, C., Cozmuta, L.M., Vagelas, I., Development of Ag/ TiO_2-SiO_2-coated food packaging film and its role in preservation of green lettuce during storage. *Carpathian J. Food Sci. Technol.*, 7, 88–96, 2015.
119. Zahedi, S.M., Karimi, M., Teixeira da Silva, J.A., The use of nanotechnology to increase quality and yield of fruit crops. *J. Sci. Food. Agric.*, 100, 25–31, 2020.
120. Sridhar, A., Ponnuchamy, M., Kumar, P.S., Kapoor, A., Food preservation techniques and nanotechnology for increased shelf life of fruits, vegetables, beverages and spices: A review. *Environ. Chem. Lett.*, 19, 2, 1–21, 2020.
121. Jose, T., George, S.C., Maria, H.J., Wilson, R., Thomas, S., Effect of bentonite clay on the mechanical, thermal, and pervaporation performance of the poly (vinyl alcohol) nanocomposite membranes. *Ind. Eng. Chem. Res.*, 53, 43, 16820–16831, 2014.
122. Yildirim, S., Röcker, B., Pettersen, M.K., Nilsen-Nygaard, J., Ayhan, Z., Rutkaite, R., Radusin, T., Suminska, P., Marcos, B., Coma, V., Active packaging applications for food. *Compr. Rev. Food Sci. Food Saf.*, 17, 1, 165–199, 2018.
123. European Commission, Commission Regulation (EC) No 450/2009 of 29 May 2009 on active and intelligent materials and articles intended to come into contact with food. *Off. J. Eur. Union*, 135, 3–11, 2009.
124. Becerril, R., Nerín, C., Silva, F., Encapsulation systems for antimicrobial food packaging components: An update. *Molecules*, 25, 5, 1134, 2020.
125. Radusin, T.I., Ristić, I.S., Pilić, B.M., Novaković, A.R., Antimicrobial nanomaterials for food packaging applications. *Food Feed Res.*, 43, 2, 119–126, 2016.
126. Turalija, M., Bischof, S., Budimir, A., Gaan, S., Antimicrobial PLA films from environment friendly additives. *Compos. B. Eng.*, 102, 94–9, 2016.
127. Orsuwan, A., Shankar, S., Wang, L.F., Sothornvit, R., Rhim, J.W., Preparation of antimicrobial agar/banana powder blend films reinforced with silver nanoparticles. *Food Hydrocolloids*, 60, 476–485, 2016.
128. Othman, S.H., Abd Salam, N.R., Zainal, N., Kadir Basha, R., Talib, R.A., Antimicrobial activity of TiO_2 nanoparticle-coated film for potential food packaging applications. *Int. J. Photoenergy*, 2014, 945930, 1–6, 2014.

129. Xing, Y., Li, X., Guo, X., Li, W., Chen, J., Liu, Q., Xu, Q., Wang, Q., Yang, H., Shui, Y., Bi, X., Effects of different TiO_2 nanoparticles concentrations on the physical and antibacterial activities of chitosan-based coating film. *Nanomaterials*, 10, 7, 1365, 2020.
130. Young, M., Ozcan, A., Myers, M.E., Johnson, E.G., Graham, J.H., Santra, S., Multimodal generally recognized as safe ZnO/nanocopper composite: A novel antimicrobial material for the management of citrus phytopathogens. *J. Agric. Food Chem.*, 66, 26, 6604–6608, 2017.
131. Mallick, S., Sharma, S., Banerjee, M., Ghosh, S.S., Chattopadhyay, A., Paul, A., Iodine-stabilized Cu nanoparticle chitosan composite for antibacterial applications. *ACS Appl. Mater. Interfaces*, 4, 3, 1313–1323, 2012.
132. Venkatesan, R. and Rajeswari, N., Preparation, mechanical and antimicrobial properties of SiO_2/poly (butylene adipate-co-terephthalate) films for active food packaging. *Silicon*, 11, 5, 2233–2239, 2019.
133. Jafarzadeh, S., Rhim, J.W., Alias, A.K., Ariffin, F., Mahmud, S., Application of antimicrobial active packaging film made of semolina flour, nano zinc oxide and nano-kaolin to maintain the quality of low-moisture mozzarella cheese during low-temperature storage. *J. Sci. Food Agric.*, 99, 6, 2716–2725, 2019.
134. Youssef, A.M., El-Sayed, S.M., El-Sayed, H.S., Salama, H.H., Dufresne, A., Enhancement of Egyptian soft white cheese shelf life using a novel chitosan/carboxymethyl cellulose/zinc oxide bionanocomposite film. *Carbohydr. Polym.*, 151, 9–19, 2016.
135. El-Sayed, S.M., El-Sayed, H.S., Ibrahim, O.A., Youssef, A.M., Rational design of chitosan/guar gum/zinc oxide bionanocomposites based on Roselle calyx extract for Ras cheese coating. *Carbohydr. Polym.*, 239, 116234, 2020.
136. Youssef, A.M., El-Sayed, S.M., El-Sayed, H.S., Salama, H.H., Assem, F.M., Abd El-Salam, M.H., Novel bionanocomposite materials used for packaging skimmed milk acid coagulated cheese (Karish). *Int. J. Biol. Macromol.*, 115, 1002–1011, 2018.
137. Siddiqui, R.A. and Moghadasian, M.H., Nutraceuticals and nutrition supplements: Challenges and opportunities. *Nutrients*, 12, 6, 1593, 2020.
138. Lotha, R., Sundaramoorthy, N.S., Shamprasad, B.R., Nagarajan, S., Sivasubramanian, A., Plant nutraceuticals (Quercetrin and Afzelin) capped silver nanoparticles exert potent antibiofilm effect against food borne pathogen *Salmonella enterica* serovar *Typhi* and curtail planktonic growth in zebrafish infection model. *Microb. Pathog.*, 120, 109–118, 2018.
139. Lotha, R., Shamprasad, B.R., Sundaramoorthy, N.S., Ganapathy, R., Nagarajan, S., Sivasubramanian, A., Zero valent silver nanoparticles capped with capsaicinoids containing *Capsicum annuum* extract, exert potent anti-biofilm effect on food borne pathogen Staphylococcus aureus and curtail planktonic growth on a zebrafish infection model. *Microb. Pathog.*, 124, 291–300, 2018.

140. Anwar, A., Masri, A., Rao, K., Rajendran, K., Khan, N.A., Shah, M.R., Siddiqui, R., Antimicrobial activities of green synthesized gums-stabilized nanoparticles loaded with flavonoids. *Sci. Rep.*, 9, 1, 1–12, 2019.
141. Sánchez-López, E., Gomes, D., Esteruelas, G., Bonilla, L., Lopez-Machado, A.L., Galindo, R., Cano, A., Espina, M., Ettcheto, M., Camins, A., Silva, A.M., Metal-based nanoparticles as antimicrobial agents: An overview. *Nanomaterials*, 10, 2, 292, 2020.
142. Franci, G., Falanga, A., Galdiero, S., Palomba, L., Rai, M., Morelli, G., Galdiero, M., Silver nanoparticles as potential antibacterial agents. *Molecules*, 20, 5, 8856–8874, 2015.
143. Wang, L., Hu, C., Shao, L., The antimicrobial activity of nanoparticles: Present situation and prospects for the future. *Int. J. Nanomedicine*, 12, 1227, 2017.
144. Yun'an Qing, L.C., Li, R., Liu, G., Zhang, Y., Tang, X., Wang, J., Liu, H., Qin, Y., Potential antibacterial mechanism of silver nanoparticles and the optimization of orthopedic implants by advanced modification technologies. *Int. J. Nanomedicine*, 13, 3311, 2018.
145. Mikhailova, E.O., Silver nanoparticles: Mechanism of action and probable bio-application. *J. Funct. Biomater.*, 11, 4, 84, 2020.
146. Tang, S. and Zheng, J., Antibacterial activity of silver nanoparticles: Structural effects. *Adv. Healthcare Mater.*, 7, 13, 1701503, 2018.
147. Chen, M., Yang, Z., Wu, H., Pan, X., Xie, X., Wu, C., Antimicrobial activity and the mechanism of silver nanoparticle thermosensitive gel. *Int. J. Nanomedicine*, 6, 2873, 2011.
148. Kędziora, A., Speruda, M., Krzyżewska, E., Rybka, J., Łukowiak, A., Bugla-Płoskońska, G., Similarities and differences between silver ions and silver in nanoforms as antibacterial agents. *Int. J. Nanomedicine*, 19, 2, 444, 2018.
149. Rudakiya, D.M. and Pawar, K., Bactericidal potential of silver nanoparticles synthesized using cell-free extract of *Comamonas acidovorans: In vitro* and *in silico* approaches. *3 Biotech.*, 7, 2, 1–12, 2017.
150. Yin, I.X., Zhang, J., Zhao, I.S., Mei, M.L., Li, Q., Chu, C.H., The antibacterial mechanism of silver nanoparticles and its application in dentistry. *Int. J. Nanomedicine*, 15, 2555, 2020.
151. Lee, B., Lee, M.J., Yun, S.J., Kim, K., Choi, I.H., Park, S., Silver nanoparticles induce reactive oxygen species-mediated cell cycle delay and synergistic cytotoxicity with 3-bromopyruvate in *Candida albicans*, but not in *Saccharomyces cerevisiae. Int. J. Nanomedicine*, 14, 4801, 2019.
152. He, W., Zhou, Y.T., Wamer, W.G., Boudreau, M.D., Yin, J.J., Mechanisms of the pH dependent generation of hydroxyl radicals and oxygen induced by Ag nanoparticles. *Biomaterials*, 33, 30, 7547–7555, 2012.
153. Li, Y., Qin, T., Ingle, T., Yan, J., He, W., Yin, J.J., Chen, T., Differential genotoxicity mechanisms of silver nanoparticles and silver ions. *Arch. Toxicol.*, 91, 1, 509–519, 2017.

154. Wang, G., Jin, W., Qasim, A.M., Gao, A., Peng, X., Li, W., Feng, H., Chu, P.K., Antibacterial effects of titanium embedded with silver nanoparticles based on electron-transfer-induced reactive oxygen species. *Biomaterials*, 124, 25–34, 2017.
155. Punia, S., Sandhu, K.S., Kaur, M., Siroha, A.K., Nanotechnology: A successful approach to improve nutraceutical bioavailability, in: *Nanobiotechnology in Bioformulations*, R. Prasad, V. Kumar, M. Kumar, D. Choudhary (Eds.), pp. 119–133, Springer, Cham, 2019.
156. Zhao, L., Liu, A., Sun, M., Gu, J., Wang, H., Wang, S., Zhang, J., Guo, C., Duan, R., Zhai, G., Enhancement of oral bioavailability of puerarin by polybutylcyanoacrylate nanoparticles. *J. Nanomater*, 2011, 126562, 1–8, 2011.
157. Yu, H. and Huang, Q., Improving the oral bioavailability of curcumin using novel organogel-based nanoemulsions. *J. Agric. Food. Chem.*, 60, 21, 5373–5379, 2012.
158. Yen, C.C., Chang, C.W., Hsu, M.C., Wu, Y.T., Self-nanoemulsifying drug delivery system for resveratrol: Enhanced oral bioavailability and reduced physical fatigue in rats. *Int. J. Mol. Sci.*, 18, 9, 1853, 2017.
159. Akula, S., Gurram, A.K., Devireddy, S.R., Self-microemulsifying drug delivery systems: An attractive strategy for enhanced therapeutic profile. *Int. Sch. Res. Notices*, 2014, 964051, 2014.
160. Hsu, C.Y., Wang, P.W., Alalaiwe, A., Lin, Z.C., Fang, J.Y., Use of lipid nanocarriers to improve oral delivery of vitamins. *Nutrients*, 11, 1, 68, 2019.
161. Managuli, R.S., Raut, S.Y., Reddy, M.S., Mutalik, S., Targeting the intestinal lymphatic system: A versatile path for enhanced oral bioavailability of drugs. *Expert Opin. Drug Deliv.*, 15, 8, 787–804, 2018.
162. Li, Q., Xia, D., Tao, J., Shen, A., He, Y., Gan, Y., Wang, C., Self-assembled core-shell-type lipid-polymer hybrid nanoparticles: Intracellular trafficking and relevance for oral absorption. *J. Pharm. Sci.*, 106, 10, 3120–3130, 2017.
163. Hu, Q., Bae, M., Fleming, E., Lee, J.Y., Luo, Y., Biocompatible polymeric nanoparticles with exceptional gastrointestinal stability as oral delivery vehicles for lipophilic bioactives. *Food Hydrocolloids*, 89, 386–395, 2019.
164. Abdul Rahim, R., Jayusman, P.A., Muhammad, N., Ahmad, F., Mokhtar, N., Naina Mohamed, I., Mohamed, N., Shuid, A.N., Recent advances in nanoencapsulation systems using PLGA of bioactive phenolics for protection against chronic diseases. *Int. J. Environ. Res. Public Health*, 16, 24, 4962, 2019.
165. Bhattacherjee, A., Dhara, K., Chakraborti, A.S., Argpyrimidine-tagged rutin-encapsulated biocompatible (ethylene glycol dimers) nanoparticles: Synthesis, characterization and evaluation for targeted drug delivery. *Int. J. Pharm.*, 509, 1-2, 507–517, 2016.
166. Nile, S.H., Baskar, V., Selvaraj, D., Nile, A., Xiao, J., Kai, G., Nanotechnologies in food science: Applications, recent trends, and future perspectives. *Nanomicro Lett.*, 12, 1, 1–34, 2020.

3

Nanoencapsulation Systems for Bioactive Compounds

Anindita Deb Pal

Department of Food Science & Nutrition Management, J. D. Birla Institute, Lower Rawdon Street, Kolkata, West Bengal, India

Abstract

Bioactive compounds are capable of bestowing numerous health advantages. However, incorporation of the same into food products and their delivery is a cause of concern mainly due to chemical instability, susceptibility toward oxidative and gastrointestinal degradation as well as insolubility in body fluids. Application of nanotechnology, especially nanoencapsulation with particles of diameters between 1 and 100 nm has been observed to solve the above issue, thereby aiding in increased bioavailability and functionality of the aforesaid compounds. Moreover, their low cost and biodegradability further qualify them as suitable food transportation agents. Different nanoencapsulation systems can be used, including lipid-based carriers, protein nanocapsules, and polysaccharide nanoparticles. Selection of an appropriate nanocarrier is dependent on the type of the biological material, production method, and the kind of food. Lipid nanoparticles can be used to encapsulate several phytochemicals possessing a range of polarities. Moreover, protein and polysaccharide nanovehicles have exhibited regulated and prolonged release of biomolecules. The present article highlights the features of various nanoencapsulation systems along with their advantages and applications as carriers of nutraceuticals. A deeper understanding of the above may aid in successful utilization of nanocapsules for effective distribution of bioactive compounds.

***Keywords*:** Bioactive, bioavailability, carriers, delivery, encapsulation, functional, nanocapsules

**Corresponding author*: deb_anindita@yahoo.com

Tanima Bhattacharya and Shakeel Ahmed (eds.) Nanotechnology in Functional Foods, (61–86)

3.1 Introduction

An enhanced focus of the present population toward food-mediated health benefits has accelerated research in the domain of functional foods and nutraceuticals. These are capable of bestowing significant therapeutic benefits, in addition to nutritional advantages, owing to the presence of several bioactive components, such as polyphenols, flavonoids, sterols, carotenoids, lipids, probiotics, vitamins, and minerals. However, dietary assimilation of functional foods is generally insufficient for distributing the required concentration of these molecules inside the body. Bioactive components face several limitations during their transport in the host system. Being mostly hydrophobic, these compounds are associated with poor solubility in body fluids, sensitivity to enzymes and pH alterations in the gastrointestinal tract (GIT), susceptibility to immune mediated elimination, vulnerability to oxidation and well as rapid elimination from the body. Nonetheless, the hydrophilic bioactive materials show poor penetration across intestinal membranes and swift release rates, thereby limiting their localization, as well as persistence at the target sites. Moreover, these components also experience mechanical degradation, chemical alterations as well changes in the organoleptic properties inside a living host [1]. Therefore, bioactive molecules need to be protected during their transit and delivered to the desired site at a controlled rate for optimizing their beneficial effects. Nanotechnology, especially nanoencapsulation systems, has been identified as a suitable carrier of bioactive compounds [2]. This involves packaging the above constituents inside the capsular walls made up of either protein, carbohydrates, or lipids in varying proportions and combinations. Classification on nanoencapsulation systems based on the kind of raw materials in represented in Figure 3.1. These carrier particles are mostly in the size range of 1 to 100 nm. Selection of an appropriate combination of raw materials for formation of nanovehicles may lead to the development of biocompatible and safe nanocarriers that may in turn offer the required dispersion kinetics, protection against deleterious effects of oxidation, pH and enzymatic action of the GIT, as well as site-specific delivery [3]. Interestingly, the nanometric sizes of these encapsulation systems increase the surface area, hence facilitating not only the solubilization, absorption, and transit of nutraceuticals through the gut epithelium but also their biological actions [4]. Nanoencapsulation systems bestow enhanced efficacy and stability to the bioactive compounds, making it possible for low concentrations of the same to portray the appropriate actions along with minimizing the instances of dose dependent side effects [5].

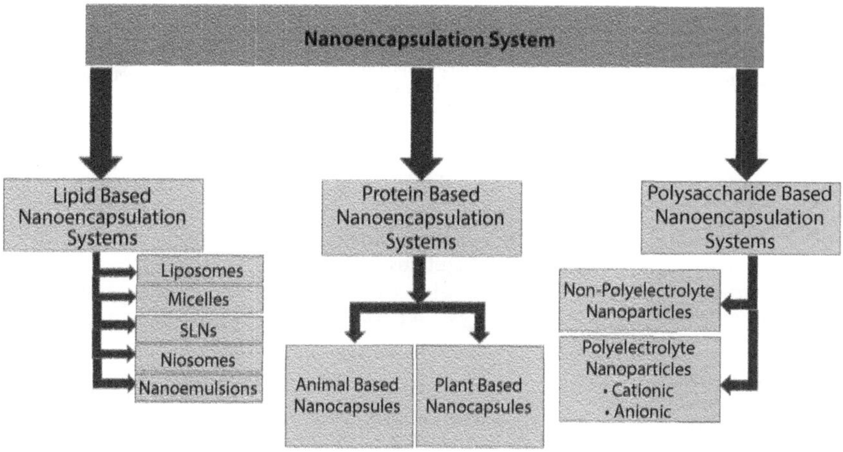

Figure 3.1 Classification of nanoencapsulation systems.

The nanocapsules can be prepared either by the top-down approaches necessitating bringing down materials to the submicron range by several mechanical techniques or the bottoms-up method involving assembly of nanomaterials by self-organization [6].

Bioactive compounds, once ingested and absorbed in the small intestine, are subjected to metabolic alterations through phase I and II transformations including oxidation, reduction, hydrolysis, conjugation as well as ejection via efflux pumps [7]. Moreover, the unabsorbed substances are chemically metabolized by the resident colonic microflora [8]. Therefore, the net availability of bioactive components along with their structural and functional integrity remains dependent on the above factors, which may compromise their accessibility, as well as the efficacy. However, encapsulation of these biochemicals in nanoparticles has been shown to protect the former from chemical alterations and degradations, thereby enhancing their application toward promotion of health [9]. Furthermore, utilization of certain nanocarriers has been shown to facilitate the enhanced uptake of biological substances via the circulatory and lymphatic systems, leading to an overall heightened effect. Moreover, nanovehicles resist the phagocyte-mediated clearance and aid in easy penetration through capillaries to reach the appropriate organ site. Hence, nanocarriers not only help in efficient delivery of biochemicals at the target location but also assist toward preservation of their chemical structures and functional roles. The crucial advantages of nanoencapsulation systems for bioactive constituents are summarized in Figure 3.2. The current chapter discusses the major

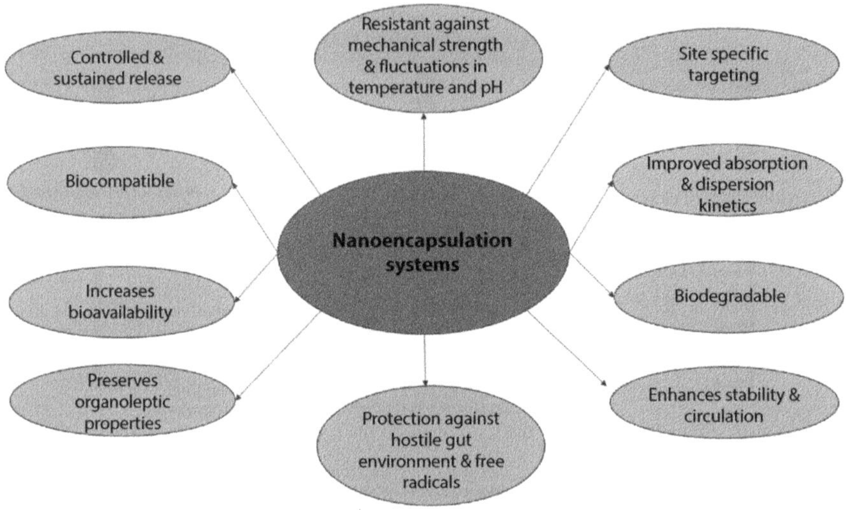

Figure 3.2 Advantages of nanoencapsulation systems for bioactive compounds.

nanocarrier systems along with their associated merits, demerits as well as application in the field of nutraceutical mediated health benefits.

3.2 Nanoencapsulation Systems

3.2.1 Lipid-Dependent Nanoencapsulation Systems

Lipid-dependent nanocapsules constitute one of the most suitable carriers that can be used for delivery of functional foods and nutraceuticals. Nanoliposomes, micelles, solid-lipid nanoparticles, niosomes, and nanoemulsions are some of the principal nanodelivery agents involving lipids. They can be made from a combination of glycerides, sterols, fatty acids, and waxes. The lipid composition and surfactant properties determine the overall particle size and stability of these vehicles [10].

Liposomes

Liposomes consist of lipid bilayers with an interior aqueous space that can be used for the incorporation and delivery of several bioactive compounds, health promoting substances and drugs. These carriers possess the ability to encapsulate biochemicals of varied polarity by encasing them either into the hydrophilic interior or at the interface of the lipid layers [11]. Encapsulation of nutraceuticals via liposomes improves their resistance against oxygen exposure, light, heat, pH, gastrointestinal enzymes, and inhibitors, thus increasing their biological availability [12]. Moreover, utilization of this system has been associated with controlled and targeted release of pharmaceutical agents for prevention and treatment of pathological abnormalities. Being obtained from natural lipids, these are safe and biodegradable carriers that do not elicit any antigenic or pyrogenic response [13]. Liposome-encapsulated curcumin has been shown to exhibit a slow and prolonged release profile of the chemical along with an improved shelf life at room temperature. These carriers have also been documented to bestow increased pH stability, heat resistance and heightened absorption to quercetin and cyclodextrin molecules in the intestinal compartments [14]. Furthermore, liposomes have manifested sustained delivery of fish oils and alpha linolenic acids, displaying properties of better absorption, solubility, and protection against auto-oxidation; along with preventing the release of reactive oxygen and nitrogen species as well as inflammatory mediators from neutrophils and macrophages in human cells [15].

Micelles

Micelles are small spherical nanoencapsulation agents that are formed in aqueous solutions and range between diameters of 10 and 100 nm. These structures have also been recognized as efficient carriers of bioactive constituents due their ability of offering protection to the former against the intrinsic deleterious factors of the host. Additionally, micelles are characterized by good loading capacities and biological stabilities that can be employed for efficient, regulated, and targeted liberation of bioactive compounds [16]. Moreover, micelles can be designed to allow site-specific delivery of biomaterials in response to a variety of stimuli, such as pH and temperature. These transporters have been observed to enhance the solubility, effective concentration, and bio-accessibility of carotenoids derived from tomatoes in a simulated gastrointestinal environment. Interestingly, liposome and micelle-based nanotherapy has now been approved by the Food and Drug Administration (FDA) [17]. Micelles have also been used for transport of several hydrophobic compounds that could otherwise be administered only through the intravenous route by solubilizing them with adjuvants [18]. These nanoparticles enhance the solubilization of different nutraceutical compounds and prolong their dissemination inside a eukaryotic system. Furthermore, they aid in effective internalization, prolonged shelf life and localization of biomolecules at the desired site at a significantly higher concentration, thereby accelerating the overall effectiveness of the pharmacological component.

Solid lipid nanoparticles

Solid lipid nanoparticles (SLN) contain lipid droplets coated with surfactants that are dispersed within an aqueous medium [19]. The lipid phase is either partially or fully solidified, thereby enhancing their ability to protect chemically vulnerable bioactive ingredients [20]. These nanoparticles are characterized by their particle size, distribution kinetics, lipid modifications and the nature of the bioactive substances incorporated in them. SLNs are required to cross the physical barriers and provide a sustained release of biomaterials including essential oils at the target site without imposing toxicological constraints. The essential oils in turn are employed for combating microbial infection with enhanced efficiencies owing to an increased bioavailability, concentration as well as permeability bestowed by the lipid carriers [21, 22]. These nanocarriers have also been observed to enhance the health benefits mediated by ferulic acid, tocopherols, carotenoids, phenolic compounds, buckwheat flavonoids, vitamin E and vitamin D3 [23–25]. Interestingly, SLNs can be employed for delivery through the oral, parenteral, and transdermal routes since they display several desired properties, such as small size, large surface area, and good loading capacities [26]. Moreover, they are also attractive because of their low cost, optimum release rates, biodegradability, and safety. Although associated with conferring stability to the biochemical compound, its efficiency regarding this aspect may sometimes be lower when compared to liposomes and nanoemulsions [27]. Nonetheless, SLNs qualify as suitable lipid-based carriers of bioactive compounds that can be employed for conferring suitable health advantages.

Niosomes

Niosomes constitute another type of lipid-based nanovehicles that are structurally similar to liposomes. These are lamellar structures formed by nonionic surfactants and cholesterol [28]. Their nonionic nature qualifies them as effective delivery agents that possess low toxicity. Niosomes enhance the efficiency of bioactive compounds owing to their ability of assisting prolonged and targeted delivery of the same. Furthermore, these structures are osmotically active and stable in internal fluids, leading to an improved protection of the encapsulated phytochemical compound, ultimately culminating in an increased bioavailability as well as therapeutic index [29].

Nanoemulsions

Nanoemulsions are created by two immiscible liquids that are stabilized by emulgents and display lipophilic surface properties. These carriers are formed by the integration of a dispersed phase, dispersion media as well as an emulsifier [30]. The dispersed phase also called the discontinuous phase is covered by the dispersion media or the continuous phase. The emulsifier forms the interphase conferring stability to the otherwise thermodynamically unstable system, by controlling the collision and interfacial tension between the two phases [31]. Nanoemulsions can be labelled as either oil in water, water in oil or bi-continuous complexes. These oils are derived from various sources including coconut, castor seeds, peanuts, and olives and may comprise of free fatty acids, glycerides, waxes, and oil soluble vitamins [32]. The aqueous phase mainly comprises of water in association with carbohydrates, proteins, and alcohols. The nature of this phase determines the net polarity, ionic charge, and rheological properties of the nanoemulsions [33]. These vehicles not only preserve the sensory attributes but also improve the bioavailability and accessibility of the bioactive components inside the host [34]. Moreover, they facilitate absorption of biochemicals at the intestinal membranes and their passive transport across gut barriers. Furthermore, nanoemulsions reduce the activity of host efflux systems and decrease the degradation of biomolecules, thereby offering increased protection that helps in efficient functionality of the former [35]. Their nanosizes bestow adequate kinetic stability to these structures, the intactness being governed by pH, temperature, ionic strength, and composition. Nanoemulsions have been used for transport of therapeutic compounds, colouring substances, flavour molecules, antioxidants, and antimicrobial constituents since they help in preserving their integrity and promoting stability [36]. Indeed, these have been employed as carrier agents of flavonoids, tannins, stilbenes, astaxanthin as well as vitamins A, D and E, improving their delivery, biological activity, dispersion kinetics as well as bioavailability compared to nonemulsified derivatives [37, 38]. Interestingly, encapsulation and transport of curcumin via nanoemulsions have been approved in phase I clinical trials [39, 40]. Moreover, these vehicles have been successfully utilized in the transport of probiotic microorganisms including members of the Lactic Acid Bacteria (LAB) group to desired portions of the intestinal tract [41]. Additionally, encapsulation of essential oils through nanoemulsions has been documented to increase their functionality, dispersibility, release kinetics, and potency [42]. Furthermore, essential oil-nanoemulsions mediate their anti-pathogenic roles through nonspecific mechanisms including disruption of membrane barriers and inactivation of enzymes, hence being effective even under instances of microbial mutations and antibiotic resistance [43]. The applications of lipid-dependent nanoencapsulation system are tabulated in Table 3.1. Therefore, lipid-based nanoparticles are attractive owing to their low cost, safety, and stability as well as their ability to ensure increased biological availability and target specific release profiles.

3.2.2 Protein-Based Nanoencapsulation Systems

Proteins are excellent food-based delivery systems that are employed for nanoencapsulation of bioactive compounds. The ability of proteins to form various structures makes them suitable for encasing important nutraceuticals and pharmaceuticals [44]. Proteins are easily available in huge quantities at an economical cost and can confer nutritional and therapeutic effects apart from those manifested by the bioactive components incorporated in them [45]. Due to the presence of a variety of functional groups, protein nanocarriers can be used for entrapment of both hydrophilic and hydrophobic substances [46]. Furthermore, being a macronutrient, it can get degraded in the biological system, thereby reducing toxicity concerns, and serving as eco-friendly alternatives [47]. Interestingly, protein nanocarriers have received much attention as oral delivery agents of bioactive substances. Proteins derived from both animals as well as botanical sources can be used for preparation of nanoencapsulation vehicles. The utilization of protein-based nanovehicles for encapsulation of important biological mediators is summarized in Table 3.1.

Animal-based protein nanoencapsulation systems

Animal-based protein nanoencapsulation systems have been associated with several merits including improved absorption and safety [48]. These proteins can be obtained from various sources, such as muscles, eggs, and milk of animals and include gelatin, casein, whey proteins, ovalbumin, and silk fibroins. Gelatin-derived nanovehicles have been studied to facilitate the prolonged and controlled release of biological materials at the desired sites [49]. These nanocarriers can be chemically modified with addition of polyethylene glycol (PEG) and thiol groups to achieve increased absorption in the circulatory and lymphatic systems, improved resistance against digestive enzymes, as well as target specific release of biomaterials including epigallocatechin gallate even under adverse environmental conditions. Interestingly, thiol containing gelatin vehicles has been used for release of flavour and antimicrobial compounds in cheese [50]. Casein is another predominant source of protein nanomaterial associated with high degree of molecular flexibility and thermal stability. It is composed of αS1, αS2, β, and κ-casein that can naturally form colloidal particles for the delivery of bioactive compounds [46]. β-casein displays the maximum nanoparticle forming potential compared to the other fractions. These have been

Table 3.1 Application of nanocarriers for encapsulation of bioactive materials and drugs.

Nanoencapsulation system	Bioactive compound/drug	Reference(s)
Lipid-Based Nanoencapsulation System		
(a) Liposomes	Curcumin	[13]
	Quercetin	[14]
	α-Linolenic acid	[15]
(b) Micelles	Enzymes, DNA	[17]
	Cytotoxic drugs	[16]
(c) Solid lipid nanoparticles	Essential oils	[21, 22]
	Ferulic acid	[23]
	Flavonoids	[23, 25]
	Vitamin D3	[24]
	Chemotherapeutic drugs	[27]
(d) Niosomes	DNA, drugs	[29]
(e) Nanoemulsions	Vitamin E, phytosterols	[37]
	Astaxanthin	[38]
	Curcumin	[39, 40]
Protein-Based Nanoencapsulation System		
Animal-based protein nanoencapsulation system		
(i) Gelatin	Cycloheximide	[49]
	DNA	[50]
(ii) Casein	Chemotherapeutic drugs	[51]
(iii) Whey	β carotene, riboflavin, fatty acids	[54]
(iv) Ovalbumin	Flavonoids	[57]
	Curcumin	[58]

(Continued)

Table 3.1 Application of nanocarriers for encapsulation of bioactive materials and drugs. (*Continued*)

Nanoencapsulation system	Bioactive compound/drug	Reference(s)
(v) Silk proteins	DNA, peptides	[59]
	Vascular endothelial growth factor	[60]
	Curcumin	[61]
Plant-based protein nanoencapsulation system		
(i) Zein	Polyunsaturated fatty acids, curcumin	[63]
	Essential Oils	[65]
	Quercetin	[66]
	Vitamin D3	[67]
	Flavonoids	[68]
(ii) Pea proteins	Ascorbic acid, α- tocopherol	[71]
(iii) Wheat proteins	Triglycerides, curcumin	[76]
(iv) Soy proteins	β carotene	[75]
	Vitamin B12	[76]
Polysaccharide-based nanoencapsulation system		
Nonpolyelectrolyte nanoparticles		
(i) Starch	Insulin, siRNA	[82]
	α-Linoleic acid	[83]
	Flavonoids, anthocyanins	[83, 84]
	Therapeutic drugs	[81]
(ii) Pullulan	Proteins, wound healing compounds	[82]
	Antimicrobial agents	[87]

(*Continued*)

Table 3.1 Application of nanocarriers for encapsulation of bioactive materials and drugs. (*Continued*)

Nanoencapsulation system	Bioactive compound/drug	Reference(s)
(iii) Cyclodextrin	Antimalarial drugs, siRNA	[85]
(iv) Cellulose	Retinol, anticancer drugs	[82]
	Proteins	[89]
(v) Dextran	Chemotherapeutic drugs, wound healing compounds	[82]
	Proteins, growth factors	[91]
(vi) Guar gum	Therapeutic drugs	[82, 92]
Polyelectrolyte nanoparticles		
(i) Cationic		
• Chitosan	Proteins	[94]
	Hydrophobic drugs	[93]
(ii) Anionic		
• Alginate	Essential oils	[98]
	Calcium	[99]
	Chemotherapeutic drugs, wound healing compounds	[95, 96]
	Genes	[97]
• Pectin	Amino acids, drugs, wound healing compounds	[102]
	Polyunsaturated fatty acids, epigallocatechin gallate	[103]
	Probiotic microorganisms, curcumin	[104]
• Hyaluronic acid	Proteins, peptides, nucleotides	[106]
	Anti-cancer drugs	[82]

documented to function in the transport of calcium and phosphorus inside the human body. Being amphiphilic in nature, casein nanovehicles can be employed for packaging as well as transfer of biological molecules of varied polarities [51].

Whey constitutes an additional protein component of milk. It is usually generated as a by-product during formation of cheese. The surface properties along with the ability to form gels, make them suitable candidates for development of nanocapsules [52]. However, these nanoparticles portray low degree of flexibility and stability compared with casein. Nonetheless, they possess the ability to entrap biomolecules into a compact three-dimensional structure. Whey nanoparticles can be synthesized either from whole whey fractions or from specific constituents, such as β-lactoglobulin, α-lactalbumin, and bovine serum albumin (BSA). All nanocarriers manufactured from whey proteins have been documented to display good stability against mechanical stress. Moreover, β-lactoglobulin-based nanovehicles have been found to be resistant against the action of proteolytic enzymes, high temperatures and stomach acidity [53]. Hence, these have been employed for transport of acid labile bioactive molecules. Whey protein nanocarriers have also been utilized for target specific delivery of nutraceutical compounds including β carotene, riboflavin, fatty acids, caffeine as well as probiotic microorganisms [54]. Ovalbumin is an egg-derived protein that has been used for generation of nanoparticles. The functional compounds can be incorporated into the ovalbumin nanocarriers either before or during their assembly [55]. Ovalbumin nanoparticles protect the capsulated contents against the degradative action of pepsin and gastric acids [56]. Furthermore, they can be used for encapsulation of both polar and nonpolar compounds due to their amphiphilic nature. Moreover, these vehicles can be derivatized to achieve the required properties. Additionally, these carriers have also been associated with regulated release kinetics, as well as low toxicity, thereby qualifying as suitable nanocarriers. Ovalbumin-derived nanoparticles have been documented for entrapment and transport of retinol, flavonoids, curcumin, as well as polyunsaturated fatty acids, owing to their ability to engage in hydrophobic linkages with the same [57, 58].

Silk nanoparticles have also been used for encapsulation of important nutraceutical components. Fibroin and sericin constitute the major protein fractions of silk. These proteins contain a plethora of amino acids equipping them with abilities to form both hydrophobic and hydrophilic interactions with a range of bioactive compounds [59]. Moreover, the presence of numerous functional groups allows suitable chemical modification of the above to obtain nanocarriers manifesting the required features.

Addition of PEG to silk fibroin has been shown to not only improve their hydrophobicity but also lower the instances of nonspecific binding. Silk nanoparticles possess high mechanical strength, flexibility as well as resistance to processing treatments. Furthermore, nanocarriers prepared from silk proteins offer protection to the encased bioactive molecules along with promoting their sustained release [60]. Indeed, nanovehicles formed from silk fibroin in combination with chitosan have found application in encapsulation of curcumin, displaying properties of desired solubility, release kinetics and polarity [61]. Silk nanoencapsulation systems are also popular alternatives for incorporation of flavour molecules in food. Recombinant proteins are often employed for preparation of nanoparticles, mainly to overcome the problem of obtaining abundant reserves of the required protein from natural sources. Additionally, recombinant proteins are characterized by greater degrees of purity compared to their conventional counterparts. Interestingly, recombinant silk proteins have been used for achieving protected and directed release of important bioactive ingredients [62]. Therefore, animal proteins can be utilized for nanoencapsulation of essential biological intermediates.

Plant-based protein nanoencapsulation systems

Plant proteins are often employed for construction of nanoencapsulation vehicles, being popular owing to their low cost, sustainability, limited toxicity, nutritional benefits as well as general preference by the vegetarian population [48]. These proteins can be obtained from various sources including corn, soy, and pea. Zein is a proline containing proteins isolated from corns. These proteins are predominantly hydrophobic with a low degree of hydrophilicity. Zein can self-assemble into colloidal particles suitable for nanoencapsulation of nonpolar bioactive components including fat soluble vitamins, polyunsaturated fatty acids, curcumin, lutein, resveratrol, quercetin, and oils [63]. However, they have limited potency in delivery of water-soluble molecules. The stability of these nanotransporters has been reported to increase when coated with compounds, such as sodium caseinate [64]. Zein nanocarriers have been observed to facilitate sustained release of nutraceuticals while protecting them against the deleterious action of gastric acids and enzymes [65–67]. Moreover, owing to their acid resistant properties, these nanovehicles have been used for delivery of antimicrobial and antioxidant compounds in acidic food substances [68]. Furthermore, these nanoparticles have portrayed rapid release rates in the large intestine compared to the small intestine, hence proving as suitable nanocarriers for the colon-targeted release of bioactive constituents.

Pea proteins like globulins further comprising of legumin, vicilin, and convicilin, along with albumins and glutenins constitute good sources for manufacturing nanocarrier agents. Presence of favourable functional groups at the surface of these proteins enables them to form a variety of structures, hence facilitating the formation of nanoparticles for incorporation of bioactive intermediates [69]. Nanocapsules derived from these proteins manifest the properties of water solubility and dispersibility at target sites [70]. Pea proteins are generally used for incorporation of hydrophilic biomolecules. However, they can be utilized for transport of nonpolar intermediates when combined with polysaccharides. Indeed, nanoencapsulation vehicles formed by complexing pea proteins with polysaccharides have been used for delivery of lipids in the eukaryotic system. These nanoparticles display enhanced stability, pH resistance and low toxicity, hence acting as suitable options for transport of sensitive biological intermediates inside the host [71].

Proteins isolated from soy and wheat have also been explored for nanoencapsulation and delivery of nutraceuticals. Indeed, soy protein isolates and wheat gliadin are used for generation of nanovehicles [72]. These proteins require careful isolation and purification to avoid denaturation. Nanoparticles derived from the above exhibit suitable properties including surface adhesion, solubilization, chemical stability, elasticity, mechanical strength, and protection against gastrointestinal degradation [73]. Moreover, the above proteins are readily available at an affordable cost and are easily degraded in the biological system. Soy protein-based nanocarriers have been used for intestinal transport of vitamin B12, β carotene and resveratrol [74, 75]. The aforesaid nanovehicles also protect the bioactive intermediates against oxidation, radiations as well as fluctuations in temperature and acidity. Additionally, wheat gliadin nanoparticles have been employed for successful encasement, protection as well as stable delivery of curcumin [76]. Therefore, protein nanoparticles have displayed widespread application for encapsulation of bioactive components, with their selection being dependent on the structural and functional characteristics of the protein, properties of the bioactive material, nature of the target site, cost as well as availability.

3.2.3 Polysaccharide-Based Nanoencapsulation Systems

Polysaccharide-based nanoparticles have qualified as good delivery agents for pharmaceutical compounds ensuring controlled, targeted, and prolonged release of the same inside the host body [77, 78]. Being amphiphilic in nature, these can assemble in aqueous environments. They possess good affinity toward the mucus membranes present in the gastrointestinal and respiratory organ systems, thereby enhancing the efficient delivery of functional ingredients. Moreover, coating of the above with suitable chemicals, aids in target and stimuli dependent localization of the encapsulated materials [79]. Additionally, properties of polysaccharides, such as stability, safety, and biodegradability maximize their utilization in transport of nutraceutical compounds [80]. Based on the intrinsic charges of the polysaccharides, these nanocarriers can be classified into either nonpolyelectrolyte or polyelectrolytes nanoparticles. The employment of polysaccharide-based nanocarriers for encasement of important pharmacological and functional components has been enlisted in Table 3.1.

Nonpolyelectrolyte nanoparticles

Nonpolyelectrolyte nanoparticles can be prepared from starch, cyclodextrins, cellulose, pullulan, dextran, and guar gum. Starch and its derivates have been employed as nanovehicles owing to their properties of biodegradability, hydrophilicity as well as compatibility in the biological system [81]. Starch has indeed been chemically modified with addition of hydroxyl groups to generate chemically stable nanoparticles that have been utilized for parenteral and colon specific delivery of pharmacological ingredients [82]. Additionally, delivery of bioflavonoids and anthocyanins through amylose nanoparticles has been shown to enhance their bioavailability and protective roles against diabetes and cardiovascular disorders [83, 84]. Moreover, nanoparticles made up of cyclodextrin can be used for colon specific release of bioactive compounds, since they resist digestion and absorption in the upper GIT. In fact, the above carriers have been associated with delivery of numerous biomolecules and anti-malarial drugs [85]. Pullulan, produced from starch by the fungus *Aureobasidium pullulans*, has also been utilized for the development of nanocarriers owing to their biodegradability, structural flexibility, and solubility. Indeed, their usage has been approved in food and beverages since they do not alter the organoleptic properties of the same [86, 87]. Furthermore, cellulose constitutes another polysaccharide that is popularly employed for encapsulation of biological mediators. In fact, chemical modification of cellulose with poly ε-caprolactone and poly L-lactic acid, as well as addition of chitosan to carboxymethyl cellulose has been carried out for generation of stable and soluble nanoparticles [88]. These structures have been used for the delivery of hydrophobic bioactive agents and anticancer drugs. Cellulose nanovehicles have also been used for targeted transport of proteins through the oral route [89]. Moreover, dextran and guar gum-based nanocarriers have been used for promoting specific release of phytochemicals in the colon [90].

These nanoparticles manifest favourable characteristics including water solubility, nonionic nature, low toxicity, high viscosity, and degradation by natural flora; hence facilitating their usage for targeted delivery of growth factors, labile bioactive compounds, and drugs [91, 92].

Polyelectrolyte nanoparticles

Polyelectrolyte nanoparticles include both positively and negatively charged carriers. Cationic nanoparticles are mainly derived from chitosan and have been associated with prolonged release and increased bioavailability of bioactive intermediates, thereby being suitable for release of therapeutic agents in several cases of oral, skin, and ocular disorders [93]. Moreover, they have also been identified to be biocompatible, nontoxic, and mucoadhesive. However, these particles only disperse in dilute acidic environments [94]. Nonetheless, hydrophobic modification of chitosan nanoparticles may solve the solubilization issue.

Anionic nanocarriers are produced majorly from alginate, pectins, and hyaluronic acids. Alginate can be chemically modified to obtain nanovehicles possessing the desired solubility and biological properties [95]. Nanoencapsulation systems derived from alginate display low toxicity and immune resistance along with portraying significant mucus adhesion and biocompatible parameters [96]. Moreover, these vehicles are biodegradable in nature. Furthermore, Van-der-Waals forces amid the bivalent ions and carboxyl residues facilitates the generation of interconnected meshwork that aids in the release of bioactive compounds. Hence, alginate nanoencapsulation carriers are suitable transporters of biochemicals, such as calcium, essential oils, and proteins, helping in their enhanced absorption, prolonged release, and heightened therapeutic actions [97–99]. Additionally, encapsulation of oyster hydrolysates with alginate in combination with liposomes has been studied to not only improve their antihypertensive effect but also aid toward their protection and sustained release [100]. Interestingly, nanoparticles formed by a combination of chitosan and alginate has been found to harbour increased efficiencies compared to those made of chitosan or alginate alone. Pectins on the other hand offer limited protection to bioactive components during their transit through the stomach and small intestine, due to their water solubility. Therefore, calcium pectinate is often used for formation of nanovehicles, being more resistant and stable in comparison to the former [101]. Additionally, pectin derivatized with 4-aminothiophenol, chitosan, hyaluronic acid, and poly (lactide-co-glycolide) have also portrayed good efficiencies as nanocarriers, displaying targeted and specific release of biomolecules including amino acids, drugs, and wound healing compounds [102]. In fact, pectin-based nanotransporters have been used for site-specific and pH dependent release of polyunsaturated fatty acids and epigallocatechin gallate, while protecting them from oxidation [103]. These nanovehicles have also been noticed to increase the stability and efficiency of probiotic *Lactobacillus rhamnosus, Staphylococcus succinus,* and *Enterococcus faecium,* as well as bioactive compounds, such as curcumin and insulin in the gastrointestinal tract [104]. Being nontoxic, nonantigenic, and biodegradable in nature, hyaluronic acid has also been used for formulation of nanoparticles [105]. These nanostructures possess optimum water solubility and stability leading to their utilization for delivery of cancer drugs at target sites. Indeed, previous studies have documented hyaluronic acid nanocarriers to promote increased bioavailability of pharmacological agents, such as proteins, peptides, and nucleotide derivatives [106].

Different polysaccharides can be used as nanovectors for delivery of biochemicals based on the desired site of release. Whereas starch-based nanocarriers can release substances in the mouth due to the digestive action of amylase, other polysaccharides can be used for specific release of intermediates in distinct parts of the gastrointestinal tract [107]. Interestingly, protein-polysaccharide complexes (PPC) have now emerged as significant nanoencapsulation systems, portraying increased retention, bioavailability, and functionality of the entrapped biological ingredients [108]. These are formed by interaction of positively charged proteins with negatively charges groups present on polysaccharides to form stable complexes. Therefore, polysaccharide-based nanoencapsulation systems are suitable alternatives for targeted transport of biological intermediates.

3.3 Limitations of Nanoencapsulation Systems

Although linked with significant advantages, nanoencapsulation systems have also been associated with certain limitations. Lack of complete knowledge regarding the safety and environmental impact of these nanocapsules has raised ethical and regulatory concerns. Besides, little information is available regarding the biological transformation and excretory profile of these nanoparticles. Prolonged utilization of these may lead to problems of bioaccumulation and toxicity because of their increased availability and absorption in the body [109]. Moreover, nanoencapsulation carriers have an enhanced access to the target site. These transporters may also interfere with the default immunological defences of the host for protection of the entrapped material, resulting in alteration of the homeostatic balance [110]. Furthermore, nanoparticles can interfere with the colonic microflora, leading to microbial dysbiosis. Indeed, few nanoparticles have been correlated with physiological manifestations including malfunctioning of the nervous, immune, pulmonary, and endothelial systems [109]. In fact, application of the above carriers has been portrayed to heighten the instances of inflammation as well as damage to the above organ systems, sometimes culminating in cases of Parkinson's and Alzheimer's diseases [111]. Additionally, being small in size, nanovehicles can penetrate into the skin and other mechanical barriers. Previous studies have also documented nanomaterials to be capable of interfering with the central dogma of cellular information, post their nuclear entry. Large-scale preparation of several nanocarriers including liposomes and nanoemulsions are expensive involving massive production costs. Lipid-based carriers suffer from certain disadvantages in terms of their loading capacities and crystallization at low temperatures. Moreover, liposomes have displayed instability in plasma, being themselves sensitive to pH, moisture, and oxidation [112]. Additionally, availability of biocompatible surfactants for generation of nanoemulsions is restricted [113]. Limited *in vivo* studies have been carried out in the above area, hence posing another constraint into determination of the actual efficacy and safety of these carriers. Furthermore, paucity of regulatory guidelines with respect to the production, usage, and disposal of these nanovehicles along with their associated health risks have raised public health concerns. Therefore, a thorough study of the efficacy, safety, and toxicological aspects of nano-encapsulation systems along with development of cost-effective manufacturing techniques may be useful in not only predicting the accurate dose for therapeutic applications but also for enhancing the functionality of the bioactive compounds [10].

3.4 Conclusion

Bioactive components present in functional foods have immense potential for prevention and treatment of certain chronic and acute ailments. However, dietary consumption of these compounds through food sources do not deliver them at concentrations that are required for the desired action. Therefore, isolation of these biochemicals followed by their fortification can not only enhance the nutritional value of food but also bestow significant functional benefits to the host. The bioactive molecules are mostly characterized by suboptimal water solubility as well as susceptibility to the adverse conditions encountered during food processing, packaging as well as their transit through the GIT. Encapsulation of these components in nanoparticles helps in enhancing their dispersibility and resistance toward degradation along with promoting a heightened circulatory time, bioavailability, bio-accessibility, stability, and controlled release. This allows low quantities of the required biochemicals to manifest the therapeutic effects, thereby minimizing chances of concentration dependent toxicity. The nanocarriers also enhance the absorption, transport, and utilization of the bioactive derivatives owing to their small size and high surface areas. These colloidal particles are manufactured from either natural or synthetic polymers and are ideally suited for optimizing delivery of bioactive compounds and drugs in the host. Different food grade and biologically compatible materials including lipids, proteins and polysaccharides can be employed for fabrication of nanoencapsulation systems. Nanocarriers can also be prepared by combination of several raw materials to obtain the desired characteristics. Selection of an appropriate source is dependent on the structural and functional properties of the bioactive material as well as the nature of the raw material that would be best suited for providing adequate protection and potency to the entrapped material. Nanoencapsulation systems have found application in the delivery of important nutraceutical components, such as flavonoids, vitamins, fatty acids, polyphenols, and probiotics. However, nanocarriers are themselves associated with certain safety and toxicological concerns due to their ability to easily penetrate the physical barriers and interfere with host homeostasis. Therefore, a deeper understanding of the nanoencapsulation systems along with adequate toxicological analysis may enhance their application in transport and delivery of biomolecules, leading toward promotion of overall health and wellbeing of the population.

References

1. McClements, D.J., Decker, E.A., Park, Y., Weiss, J., Structural design principles for delivery of bioactive components innutraceuticals and functional foods. *Crit. Rev. Food Sci. Nutr.*, 49, 577, 2009.
2. Singh, H., Nanotechnology applications in functional foods: Opportunities and challenges. *Prev. Nutr. Food Sci.*, 21, 1, 1–8, 2016.
3. Onwulata, C.I., Encapsulation of new active ingredients. *Annu. Rev. Food Sci. Technol.*, 3, 183, 2012.
4. Ezhilarasi, P.N., Karthik, P., Chhanwal, N., Anandharamakrishnan, C., Nanoencapsulation techniques for food bioactive components: A review. *Food Bioprocess Technol.*, 6, 628, 2013.
5. Huang, Q., Yu, H., Ru, Q., Bioavailability and delivery of nutraceuticals using nanotechnology. *J. Food Sci.*, 75, R50, 2010.

6. Momin, J.K., Chitra, J., Prajapati, J.B., Potential of nanotechnology in functional foods. *Emir. J. Food Agric.*, 25, 10, 2013.
7. Johnson, B.M., Charman, W.N., Porter, C.J.H., The impact of P-glycoprotein efflux on enterocyte residence time and enterocyte based metabolism of verapamil. *J. Pharm. Pharmacol.*, 53, 1611, 2001.
8. Xiao, J., Cao, Y., Huang, Q., Edible nanoencapsulation vehicles for oral delivery of phytochemicals: A perspective paper. *J. Agric. Food Chem.*, *65*, 6727, 2017.
9. Sahoo, S.K., Dilnawaz, F., Krishnakumar, S., Nanotechnology in ocular drug delivery. *Drug Discovery Today*, 13, 144, 2008.
10. Martínez-Ballesta, M.C., Gil-Izquierdo, A., García-Viguera, C., Domínguez-Perles, R., Nanoparticles and controlled delivery for bioactive compounds: Outlining challenges for new "Smart-Foods" for health. *Foods*, 7, 72, 2018.
11. Ferriera, C.D. and Nunes, I.L., Oil nanoencapsulation: Development, application, and incorporation into the food market. *Nanoscale Res. Lett.*, 14, 9, 2019.
12. Khan, I., Haque, A., Ahmad, S., Nano-delivery system for bioactive compounds. *MOJ Food Process. Technol.*, *8*, 104, 2020.
13. Mozafari, M.R., Johnson, C., Hatziantoniou, S., Demetzos, C., Nanoliposomes and their applications in food nanotechnology. *J. Liposome Res.*, 18, 309, 2008.
14. Azzi, J., Jraij, A., Auezova, L., Fourmentin, S., Greige-Gerges, H., Novel findings for quercetin encapsulation and preservation with cyclodextrins, liposomes, and drug-in-cyclodextrin-in-liposomes. *Food Hydrocolloids*, 81, 328, 2018.
15. Semenova, M.G., Antipova, A.S., Zelikina, D.V., Martirosova, E.I., Plashchina, I.G., Palmina, N.P., Binyukov, V.I., Bogdanova, N.G., Kasparov, V.V., Shumilina, E.A., Biopolymer nanovehicles for essential polyunsaturated fatty acids: Structure-functionality relationships. *Food Res. Int.*, 88, 70, 2016.
16. Kataoka, K., Harada, A., Nagasaki, Y., Block copolymer micelles for drug delivery: Design, characterization and biological significance. *Adv. Drug Deliv. Rev.*, 47, 113, 2001.
17. He, X. and Hwang, H.M., Nanotechnology in food science: Functionality, applicability, and safety assessment. *J. Food Drug Anal.*, 24, 671, 2016.
18. Matsumura, Y. and Kataoka, K., Preclinical and clinical studies of anticancer agent-incorporating polymer micelles. *Cancer Sci.*, 100, 572, 2009.
19. Weiss, J., Decker, E.A., McClements, D.J., Kristbergsson, K., Helgason, T., Awad, T., Solid lipid nanoparticles as delivery systems for bioactive food components. *Food Biophys.*, 3, 146, 2008.
20. Pardeshi, C., Rajput, P., Belgamwar, V., Tekade, A., Patil, G., Chaudhary, K., Sonje, A., Solid lipid based nanocarriers: An overview. *Acta Pharm.*, 62, 433, 2012.

21. Maryam, I., Huzaifa, U., Hindatu, H., Zubaida, S., Nanoencapsulation of essential oils with enhanced antimicrobial activity: A new way of combating antimicrobial Resistance. *J. Pharmacogn. Phytochem.*, 4, 165, 2015.
22. Oejhlke, K., Behsnilian, D., Mayer-Miebach, E., Weidler, P.G., Greiner, R., Edible solid lipid nanopartilces (SLN) as carrier systems for antioxidants of different lipophilicity. *PLoS One*, 12, e0171662, 2017.
23. Esfanjani, A.F., Assadpour, E., Jafari, S.M., Improving the bioavailability of phenolic compounds by loading them within lipid-based nanocarriers. *Trends Food Sci. Technol.*, 76, 56, 2018.
24. Meghani, N., Patel, P., Kansara, K., Ranjan, S., Dasgupta, N., Ramalingam, C., Kumar, A., Formulation of vitamin D encapsulated cinnamon oil nanoemulsion: Its potential anti-cancerous activity in human alveolar carcinoma cells. *Colloids Surf. B Biointerfaces*, 166, 349, 2018.
25. Zhang, J.M., Wang, D., Wu, Y.H., Li, W., Hu, Y., Zhao, G., Fu, C.M., Fu, S., Zou, L., Lipid-polymer hybrid nanoparticles for oral delivery of tartary buckwheat flavonoids. *J. Agric. Food Chem.*, 66, 4923, 2018.
26. Yang, S., Zhu, J., Lu, Y., Ling, B., Yang, C., Body distribution of camptothecin solid lipid nanoparticles after oral administration. *Pharm. Res.*, 16, 751, 1999.
27. Mukherjee, S., Ray, S., Thakur, R.S., Solid lipid nanoparticles: A modern formulation approach in drug delivery system. *Indian J. Pharm. Sci.*, 71, 349, 2009.
28. Shatalebi, M.A., Mostafavi, S.A., Moghaddas, A., Niosome as a drug carrier for topical delivery of N-acetyl glucosamine. *Res. Pharm. Sci.*, 5, 107, 2010.
29. Uchegbu, I.F. and Vyas, S.P., Non-ionic surfactant based vesicles (niosomes) in drug delivery. *Int. J. Pharm.*, 176, 139, 1998.
30. Setya, S., Talegaonkar, S., Razdan, B.K., Nanoemulsions: Formulation methods and stability aspects. *World J. Pharm. Pharm. Sci.*, 3, 2214, 2014.
31. McClements, D.J. and Rao, J., Food-grade nanoemulsions: formulation, fabrication, properties, performance, biological fate, and potential toxicity. *Crit. Rev. Food Sci. Nutr.*, 51, 285, 2011.
32. Weiss, J., Takhistov, P., McClements, D.J., Functional materials in food nanotechnology. *J. Food Sci.*, 71, 107, 2006.
33. Tadros, T., Izquierdo, P., Esquena, J., Solans, C., Formation and stability of nano-emulsions. *Adv. Colloid Interface Sci.*, 108, 303, 2004.
34. Acosta, E., Bioavailability of nanoparticles in nutrient and nutraceutical delivery. *Curr. Opin. Colloid Interface Sci.*, 14, 3, 2009.
35. McClements, D.J., Decker, E.A., Weiss, J., Emulsion-based delivery systems for lipophilic bioactive components. *J. Food Sci.*, 72, 109, 2007.
36. Bai, A.J. and Rai, V.R., Nanoemulsions and their potential applications in food industry. *Front. Sustain. Food Syst.*, 3, 95, 2019.
37. Cheong, A.M., Tan, C.P., Nyam, K.L., Stability of bioactive compounds and antioxidant activities of kenaf seed oil-in-water nanoemulsions under different storage temperatures. *J. Food Sci.*, 83, 2457, 2018.

38. Alarcon-Alarcon, C., Inostroza-Riquelme, M., Torres-Gallegos, C., Araya, C., Miguel, M., Sanchez-Caamano, C., Protection of astaxanthin from photodegradation by its inclusion in hierarchically assembled nano and microstructures with potential as food. *Food Hydrocolloids*, 83, 36, 2018.
39. Cheng, A.L., Hsu, C.H., Lin, J.K., Hsu, M.M., Ho, Y.F., Shen, T.S., Ko, J.Y., Lin, J.T., Lin, B.R., Ming-Shiang, W., Phase I clinical trial of curcumin, a chemopreventive agent, in patients with high-risk or pre-malignant lesions. *Anticancer Res.*, 21, 2895, 2001.
40. Wang, X., Jiang, Y., Wang, Y.W., Huang, M.T., Ho, C.T., Huang, Q., Enhancing anti-inflammation activity of curcumin through O/W nanoemulsions. *Food Chem.*, 108, 419, 2008.
41. Hou, R.C.W., Lin, M.Y., Wang, M.M.C., Tzen, J.T.C., Increase of viability of entrapped cells of *Lactobacillus delbrueckii ssp. bulgaricus in artificial sesame oils emulsions. J. Dairy Sci.*, 86, 424, 2003.
42. Edris, A.E., Pharmaceutical and therapeutic potentials of essential oils and their individual volatile constituents: A review. *Phytother. Res.*, 21, 308, 2007.
43. Jemaa, M.B., Neves, M.A., Falleh, H., Isoda, H., Ksouri, R., Nakajima, M., Nanoencapsulation of *Thymus capitatus* essential oil: Formulation process, physical stability characterization and antibacterial efficiency monitoring. *Ind. Crops Prod.*, 113, 414, 2018.
44. Chen, L., Remondetto, G.E., Subirade, M., Food protein-based materials as nutraceutical delivery systems. *Trends Food Sci. Technol.*, 17, 272, 2006.
45. Samaranayaka, A.G.P. and Li-Chan, E.C.Y., Food-derived peptidic antioxidants: A review of their production, assessment, and potential applications. *J. Funct. Foods*, 3, 229, 2011.
46. Elzoghby, A.O., Samy, W.M., Elgindy, N.A., Protein-based nanocarriers as promising drug and gene delivery systems. *J. Control. Release*, 161, 38, 2012.
47. Bourbon, A.I., Pereira, R.N., Pastrana, L.M., Vicente, A.A., Cerqueira, M.A., Protein-Based Nanostructures for Food Applications. *Gels*, 5, 9, 2019.
48. Fathi, M., Donsi, F., McClements, D.J., Protein-based delivery systems for the nanoencapsulation of food ingredients. *Compr. Rev. Food Sci. Food Saf.*, 17, 920, 2018.
49. Saxena, A., Sachin, K., Bohidar, H.B., Verma, A.K., Effect of molecular weight heterogeneity on drug encapsulation efficiency of gelatin nano-particles. *Colloids Surf. B*, 45, 42, 2005.
50. Kommareddy, S. and Amiji, M., Poly(ethylene glycol)–modified thiolate gelatin nanoparticles for glutathione-responsive intracellular DNA delivery. *Nanomedicine: NBM*, 3, 32, 2007.
51. Bacha, M., Mandelbaum, A., Portray, I., Perlstein, H., Even-Chen, S., Barenholz, Y., Danino, D., Development and characterization of a novel drug nanocarrier for oral delivery, based on self-assembled β-casein micelles. *J. Control. Release*, 160, 164, 2012.
52. Gunasekaran, S., Ko, S., Xiao, L., Use of whey proteins for encapsulation and controlled delivery applications. *J. Food Eng.*, 83, 31, 2007.

53. Takagi, K., Teshima, R., Okunuki, H., Sawada, J.I., Comparative study of *in vitro* digestibility of food proteins and effect of preheating on the digestion. *Biol. Pharm. Bull.*, 26, 969, 2003.
54. Madalena, D.A., Ramos, O.L., Pereira, R.N., Bourbon, A.I., Pinheiro, A.C., Malcata, F.X., Teixeira, J.A., Vicente, A.A., *In vitro* digestion and stability assessment of β lactoglobulin/riboflavin nanostructures. *Food Hydrocolloids*, 58, 89, 2016.
55. Abeyrathne, E.D., Lee, H.Y., Ahn, D.U., Egg white proteins and their potential use in food processing or as nutraceutical and pharmaceutical agents–A review. *Poult. Sci.*, 92, 3292, 2013.
56. Huntington, J.A. and Stein, P.E., Structure and properties of ovalbumin. *J. Chromatogr. B*, 756, 189, 2001.
57. Li, Z. and Gu, L., Fabrication of self-assembled (-)-epigallocatechin gallate (EGCG) ovalbumin-dextran conjugate nanoparticles and their transport across monolayers of human intestinal epithelial Caco-2 cells. *J. Agric. Food Chem.*, 62, 1301, 2014.
58. Feng, J., Wu, S.S., Wang, H., Liu, S.B., Improved bioavailability of curcumin in ovalbumin-dextran nanogels prepared by Maillard reaction. *J. Funct. Foods*, 27, 55, 2016.
59. Numata, K. and Kaplan, D.L., Silk-based delivery systems of bioactive molecules. *Adv. Drug Deliv. Rev.*, 62, 1497, 2010.
60. Kundu, J., Chung, Y.I., Kim, Y.H., Tae, G., Kundu, S.C., Silk fibroin nanoparticles for cellular uptake and control release. *Int. J. Pharm.*, 388, 242, 2010.
61. Gupta, V., Aseh, A., Rios, C.N., Aggarwal, B.B., Mathur, A.B., Fabrication and characterization of silk fibroin-derived curcumin nanoparticles for cancer therapy. *Int. J. Nanomed.*, 4, 115, 2009.
62. Tokareva, O., Michalczechen-Lacerda, V.A., Rech, E.L., Kaplan, D.L., Recombinant DNA production of spider silk proteins. *Microb. Biotechnol.*, 6, 651, 2013.
63. Huang, X.X., Huang, X.L., Gong, Y.S., Xiao, H., McClements, D.J., Hu, K., Enhancement of curcumin water dispersibility and antioxidant activity using core-shell protein-polysaccharide nanoparticles. *Food Res. Int.*, 87, 1, 2016.
64. Davidov-Pardo, G., Joye, I.J., Espinal-Ruiz, M., McClements, D.J., Effect of Maillard conjugates on the physical stability of zeinnanoparticles prepared by liquid antisolvent coprecipitation. *J. Agric. Food Chem.*, 63, 8510, 2015.
65. Wu, Y., Luo, Y., Wang, Q., Antioxidant and antimicrobialproperties of essential oils encapsulated in zein nanoparticles prepared byliquid-liquid dispersion method. *LWT - Food Sci. Technol.*, 48, 283, 2012.
66. Aytac, Z., Ipek, S., Durgun, E., Uyar, T., Antioxidant electrospun zein nanofibrous web encapsulating quercetin/cyclodextrin inclusioncomplex. *J. Mater. Sci.*, 53, 1527, 2017.
67. Lin, Y., Wang, Y.H., Yang, X.Q., Guo, J., Wang, J.M., Corn protein hydrolysate as a novel nano-vehicle: Enhanced physicochemical stability and *in vitro* bioaccessibility of vitamin D3. *LWT - Food Sci. Technol.*, 72, 510, 2016.

68. Donsì, F., Voudouris, P., Veen, S.J., Velikov, K.P., Zein-based colloidal particles for encapsulation and delivery of epigallocatechin gallate. *Food Hydrocolloids*, 63, 508, 2017.
69. Nienhaus, K., Wang, H., Nienhaus, G.U., Nanoparticles for biomedical applications: Exploring and exploiting molecular interactions at the nano-bio interface. *Mater. Today Adv.*, 5, 100036, 2020.
70. Donsì, F., Senatore, B., Huang, Q., Ferrari, G., Development of novel pea protein-based nanoemulsions for delivery of nutraceuticals. *J. Agric. Food Chem.*, 58, 10653, 2010.
71. Nesterenko, A., Alric, I., Silvestre, F.O., Durrieu, V., Vegetable proteins in microencapsulation: A review of recent interventions and their effectiveness. *Ind. Crops Prod.*, 42, 469, 2013.
72. Martins, J.T., Bourbon, A.I., Pinheiro, A.C., Fasolin, L.H., Vicente, A.A., Protein-based structures for food applications: From macro to nanoscale. *Front. Sustain. Food Syst.*, 2, 77, 2018.
73. Zhang, X., Do, M.D., Dean, K., Hoobin, P., Burgar, I.M., Wheat gluten-based natural polymer nanoparticle composites. *Biomacromolecules*, 8, 345, 2007.
74. Zhang, J., Field, C.J., Vine, D., Chen, L., Intestinal uptake and transport of vitamin B12-loaded soy protein nanoparticles. *Pharm. Res.*, 32, 1288, 2014.
75. Deng, X.X., Zhang, N., Tang, C.H., Soy protein isolate as a nanocarrier for enhanced water dispersibility, stability and bioaccessibility of *β*-carotene. *J. Sci. Food Agric.*, 97, 2230, 2017.
76. Yang, S., Dai, L., Sun, C., Gao, Y., Characterization of curcumin loaded gliadin-lecithin composite nanoparticles fabricated by antisolvent precipitation in different blending sequences. *Food Hydrocolloids*, 85, 185, 2018.
77. Lemarchand, C., Gref, R., Couvreur, P., Polysaccharide-decorated nanoparticles. *Eur. J. Pharm. Biopharm.*, 58, 327, 2004.
78. Polyakov, N.E. and Kispert, L.D., Water soluble biocompatible vesicles based on polysaccharides and oligosaccharides inclusion complexes for carotenoid delivery. *Carbohydr. Polym.*, 128, 207, 2015.
79. Liu, Z., Jiao, Y., Wang, Y., Zhou, C., Zhang, Z., Polysaccharides-based nanoparticles as drug delivery systems. *Adv. Drug Deliv. Rev.*, 60, 1650, 2008.
80. Gopinath, V., Saravanan, S., Al-Maleki, A.R., Ramesh, M., Vadivelu, J., A review of natural polysaccharides for drug delivery applications: Special focus on cellulose, starch and glycogen. *Biomed. Pharmacother.*, 107, 96, 2018.
81. Narayanan, D., Nair, S., Menon, D., A systematic evaluation of hydroxyethyl starch as a potential nanocarrier for parenteral drug delivery. *Int. J. Biol. Macromol.*, 74, 575, 2015.
82. Sivapragasam, N., Thavarajah, P., Ohm, J., Ohm, J., Margaret, K., Thavarajah, D., Novel starch based nano scale enteric coatings from soybean meal for colon-specific delivery. *Carbohydr. Polym.*, 111, 273, 2014.
83. Feng, T., Wang, K., Liu, F.F., Ye, R., Zhu, X., Zhung, H.N., Xue, Z.M., Structural characterization and bioavailability of ternary nanoparticles consisting of

amylose, α-linoleic acid and β-lactoglobulin complexed with naringin. *Int. J. Biol. Macromol.*, 99, 365, 2017.
84. Kumari, A., Yadav, S.K., Yadav, S.C., Biodegradable polymeric nanoparticles based drug delivery systems. *Colloids Surf. B Biointerfaces*, 75, 1, 2010.
85. Davis, M.E., The first targeted delivery of sirna in humans via a self-assembling, cyclodextrin polymer-based nanoparticle: From concept to clinic. *Mol. Pharmaceutics*, 6, 659, 2009.
86. Leathers, T.D., Biotechnological production and applications of pullulan. *Appl. Microbiol. Biotechnol.*, 62, 468, 2003.
87. Synowiec, A., Gniewosz, M., Kraniewska, K., Przybyl, J.L., Baczek, K., Weglarz, Z., Antimicrobial and antioxidant properties of pullulan film containing sweet basil extract and an evaluation of coating effectiveness in the prolongation of the shelf life of apples stored in refrigeration conditions. *Innov. Food Sci. Emerg.*, 23, 171, 2014.
88. Teramoto, Y. and Nishio, Y., Cellulose diacetate-graft-poly(lactic acid)s: Synthesis of wide-ranging compositions and their thermal and mechanical properties. *Polymer*, 44, 2701, 2003.
89. Song, Y. and Chen, L., Effect of net surface charge on physical properties of the cellulose nanoparticles and their efficacy for oral protein delivery. *Carbohydr. Polym.*, 121, 10, 2015.
90. Shukla, R.K. and Tiwari, A., Carbohydrate polymers: Applications and recent advances in delivering drugs to the colon. *Carbohyr. Polym.*, 88, 399, 2012.
91. Parraga, J.E., Zorzi, G.K., Diebold, Y., Seijo, B.A., Sanchez, A., Nanoparticles based on naturally occurring biopolymers as versatile delivery platforms for delicate bioactive molecules: An application for ocular gene silencing. *Int. J. Pharm.*, 477, 12, 2014.
92. Soumya, R.S., Ghosh, S., Abraham, E.T., Preparation and characterization of guar gum nanoparticles. *Int. J. Biol. Macromol.*, 46, 267, 2010.
93. Duceppe, N. and Tabrizian, M., Advances in using chitosan-based nanoparticles for *in vitro* and *in vivo* drug and gene delivery. *Expert Opin. Drug Deliv.*, 7, 1191, 2010.
94. Amidi, M., Mastrobattista, E., Jiskoot, W., Hennink, W.E., Chitosan-based delivery systems for protein therapeutics and antigens. *Adv. Drug Deliv. Rev.*, 62, 59, 2010.
95. Salatin, S. and Jelvehgarim, M., Natural polysaccharide based nanoparticles for drug/gene delivery. *Pharm. Sci., 23*, 84, 2017.
96. Sun, J. and Tan, H., Alginate-based biomaterials for regenerative medicine applications. *Mater. (Basel)*, 6, 1285, 2003.
97. You, J.O. and Peng, C.A., Calcium-alginate nanoparticles formed by reverse microemulsion as gene carriers. *Macromol. Symp.*, 219, 147, 2005.
98. Bilia, A.R., Guccione, C., Isacchi, B., Righeschi, C., Firenzuoli, F., Bergonzi, M.C., Essential oils loaded in nanosystems: A developing strategy for a successful therapeutic approach. *Evid.-Based Complement. Alternat. Med.*, 2014, 651593, 2014.

99. Guo, H.H., Hong, Z.A., Yi, R.Z., Core-shell collagen peptide chelated calcium/calcium alginate nanoparticles from fish scales for calcium supplementation. *J. Food Sci.*, 80, N1595, 2015.
100. Xie, C.L., Lee, S.S., Choung, S.Y., Kang, S.S., Choi, Y.J., Preparation and optimisation of liposome-in-alginate beads containing oyster hydrolysate for sustained release. *Int. J. Food Sci. Technol.*, 51, 2209, 2016.
101. Fernandez-Hervas, M.J. and Fell, J.T., Pectin/chitosan mixtures as coatings for colon-specific drug delivery: An *in vitro* evaluation. *Int. J. Pharm.*, 169, 115, 1998.
102. Perera, G., Barthelmes, J., Bernkop-Schnürch, A., Novel pectin-4-aminothiophenole conjugate microparticles for colon-specific drug delivery. *J. Control. Release*, 145, 240, 2010.
103. Peres, I., Rocha, S., Gomes, J., Morais, S., Pereira, M.C., Coelho, M., Preservation of catechin antioxidant properties loaded in carbohydrate nanoparticles. *Carbohydr. Polym.*, 86, 147, 2011.
104. Sun, Q.C., Zhang, Z.P., Zhang, R.J., Gao, R.C., McClements, D.J., Development of functional or medical foods for oral administration of insulin for diabetes treatment: Gastroprotective edible microgels. *J. Agric. Food Chem.*, 66, 4820, 2018.
105. Arpicco, S., Milla, P., Stella, B., Dosio, F., Hyaluronic acid conjugates as vectors for the active targeting of drugs, genes and nanocomposites in cancer treatment. *Molecules*, 19, 3193, 2014.
106. Oh, E.J., Park, K., Kim, K.S., Kim, J., Yang, J., Kong, J., Lee, M.Y., Hoffman, A.S., Hahn, S.K., Target specific and long-acting delivery of protein, peptide, and nucleotide therapeutics using hyaluronic acid derivatives. *J. Control. Release*, 141, 2, 2010.
107. Zou, L.Q., Zheng, B.J., Zhang, R.J., Zhang, Z.P., Liu, W., Liu, C.M., Xiao, H., McClements, D.J., Food-grade nanoparticles for encapsulation, protection and delivery of curcumin: Comparison of lipid, protein, and phospholipid nanoparticles under simulated gastrointestinal conditions. *RSC Adv.*, 6, 3126, 2016.
108. Wei, Z. and Huang, Q., Assembly of protein–polysaccharide complexes for delivery of bioactive ingredients: A perspective paper. *J. Agric. Food Chem.*, 67, 1344, 2019.
109. Hoet, P.H.M., Hohlfeld, I.B., Salata, O.V., Nanoparticles – known and unknown health risks. *J. Nanobiotechnol.*, 2, 12, 2004.
110. Maurer-Jones, M.A., Bantz, K.C., Love, S.A., Marquis, B.J., Haynes, C.L., Toxicity of therapeutic nanoparticles. *Nanomedicine (Lond)*, 4, 219, 2009.
111. Chaudhry, Q., Scotter, M., Blackburn, J., Ross, B., Boxall, A., Castle, L., Aitken, R., Watkins, R., Applications and implications of nanotechnologies for the food sector. *Food Addit. Contam.*, 25, 241, 2008.
112. Akbarzadeh, A., Rezaei-Sadabady, R., Davaran, S., Joo, S.W., Zarghami, N., Hanifehpour, Y., Samiei, M., Kouhi, M., Nejati-Koshki, K., Liposome: Classification, preparation, and applications. *Nanoscale Res. Lett.*, 8, 102, 2013.
113. Chau, C.F., Wu, S.H., Yen, G.C., The development of regulations for food nanotechnology. *Trends Food Sci. Technol.*, 18, 269, 2007.

4

Nanotechnology in Honey: Future and Perspectives Honey as Nanoparticles

Merve Keskin[1], Gülsen Kaya[2] and Saban Keskin[1]*

[1]Vocational School of Health Services, Bilecik Şeyh Edebali University, Bilecik, Turkey
[2]Scientific and Technological Research Center, Inonu University, Malatya, Turkey

Abstract

With the development of nanotechnology, many industries related to this technology have developed rapidly. Nanotechnology includes the synthesis of particles smaller than 100 nm and its applicability in various fields. Various physical and chemical methods are used for the synthesis of nanoparticles (NPs). These methods are expensive and take a long time. Therefore, the conception of "green nanotechnology" was introduced. Green synthesis of nanoparticles involves the formation of nanoparticles from many natural sources. There are many biological resources available for use in nanoparticle synthesis (eggs, honey, various herbal materials, etc.). Recently, honey-mediated green synthesis has become a topic of interest in research. Honey is known as one of life's oldest food sources thanks to its various medicinal properties. Honey-mediated green synthesis is a new method used to synthesize metallic nanoparticles. The fact that honey performs as a stabilizing and a reducing agent makes this synthesis more convenient. This synthesis is inexpensive and does not contain toxic substances harmful to health in health applications. Also, it is a reproducible, biocompatible, fast, safe and simple method, and honey-mediated green synthesis can provide numerous valuable end products in various fields.

Keywords: Honey, nanotechnology, apitherapy, eco-friendly

**Corresponding author*: merveozdemirkeskin@gmail.com

Tanima Bhattacharya and Shakeel Ahmed (eds.) Nanotechnology in Functional Foods, (87–102)

4.1 Introduction

Nanotechnology is a technology that covers production, the characterization and usage of nanoparticles of 1 to 100 nm size. Many technological developments, a wide variety of organic and inorganic to achieve the desired properties design and characterize nanoparticles using materials make it possible [1]. In this context, it is thought that silver nanoparticles are hope in cancer treatment. In recent years, silver nanoparticles (AgNP) can be used in a wide range of areas, especially in the health field, due to some physical and chemical properties. Silver nanoparticles can be synthesized by physical, chemical, and biological methods. In recent years, green synthesis (biosynthesis) is more preferred compared with other methods. Because it is environmentally friendly, cost-effective, and being easily synthesized in large quantities [2]. It is reported that green biosynthesis silver nanoparticles has an effect on decreased of growth cells in colon cancer [3].

Controlled drug release systems date back to the 1800s, and the first simple implants were prepared as pellets placed under the skin by compression of the active substance. The first nanoparticle drug delivery system was developed by Speiser and Birrenbach in 1969 [4, 5]. Speiser and Birrenbach tested carrier systems for vaccine and antineoplastic active ingredients by polymerizing monomers in the aqueous phase in their laboratories at that time, forming micellar sacs. In the 1980s, it started to focus on cancer therapy with studies showing nano drug release systems, low toxicity, and high chemotherapy effect [6]. Intelligent drug delivery systems that provide the release of the drug in the right periods, at the right concentrations, in the right place, have a wide variety from implanted electronic devices to polymer-based materials, with the development of nanotechnology, to nanoparticles and nanocapsules (Figure 4.1).

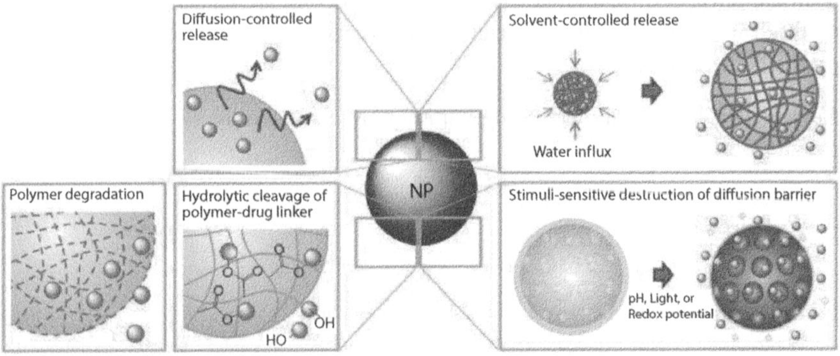

Figure 4.1 Controlled drug release properties [7].

Honey is an aromatic and viscous substance and is a very important nutrient for humans (Figure 4.2). Having a very high nutritional value in terms of energy value, honey is used in human nutrition. The role of honey in protecting human health is closely related to its purity and composition. The composition of honey varies depending on many factors, such as the geographical characteristics of the region, where it is produced, plant flora, harvest time, and production method [8]. Honey is a natural bee product that provides healing. It is an intense mixture of sugars (glucose and fructose). It contains different ingredients like various minerals, proteins, vitamins, enzymes, and phenolic compounds [9, 10]. The bioactive components of honey have antioxidant, antimicrobial, anti-inflammatory, antidepressant, hepato protective, and immunomodulatory activity.

Figure 4.2 Strained and honeycomb honey.

4.1.1 History of Honey

Honey is an important bee product known since antique times. It is known that people in ancient societies used honey as the first food source and treated diseases with honey. For example, corn honey has been used to treat arthritis. Honey has an antibacterial effect against various bacteria [11]. In the past, the widespread use of antibiotics in the medical world caused the antibiotic effect of honey to be ignored. Later, with the rapid increase in resistance to antibiotics, people sought traditional treatment. However, honey continued to be used as a treatment agent as traditional and complementary medicine.

4.1.2 Types of Honey

There are many types of honey that vary according to the geographical characteristics and production method of the region where it is produced.

Honey definitions, such as mono floral and hetero floral honeys, flower honeys and honeydew honeys, pure honeys and fraudulent honeys, are frequently used terminologies. Since honey is a complex mixture, adulteration is quite easy, and it is very difficult to determine it. Today, the purity of honey according to food codex (TSE, CEU, IHC, etc.) depends on water content, conductivity, acidity, pH, color value, sugar ratios (% C4/C13), proline amount, diastase activity, hydroxyl methyl furfural value, and insoluble substances [12]. They serve to determine the naturalness and purity of honey. It does not provide knowledge about the bioactivity potential of the honey being tested and thus does not reflect on the high quality of the honey. Biological activities of honey are anti-microbial (bacterial, fungal, or viral), antioxidant, anti-inflammatory, and antitumor properties. The basis of these characteristics is the race of the honey bee, its geographical region and the plant flora as nectar sources [8, 12]. According to the source, honeys are divided into two coarse groups as "flower honey" obtained from plant nectars and "honeydew honey" obtained from the secretions of plant sucking insects (Hemiptera) living on plants. In addition to this rough grouping, the classification of honeys according to their origin is also of great importance [13]. Flower honey (blossom) is honey that bees create by transforming the sugary substances collected from the nectar in the plant flowers in their stomachs, and the honeys that are better known among the public are this type of honey. Honeys such as linden, chestnut, clover, agave, lavender, thyme, clover, citrus, acacia, and rhododendron are flower honeys. Honeydew, on the other hand, consists of two forms. The first is honey formed by the consumption of sugars that are formed by extracting the sugary mixtures obtained by some insects by sucking the sap of plants (phyleome) by honey bees. Pine honey (*Pinus* L.) is a secretion honey formed in this way and is formed by the bees evaluate the liquid consisting of leaves of leaf plants or scissor (*Marchalina hellenica*) in the leaves of pine trees. The source of honeydew honey, including pine honey, is the honeydew (nectar). Slime is the rectal secretion of insects that feed on phloem sap with intense sugar content [14]. After the insects get the necessary nutrients for themselves, they expel the remaining sugary substance (the insect's extract) as feces. Bees take this sugary substance and bring it to the hive and turn it into honey. Honey produced in this way is called secretion honey [14].

4.1.3 Physical and Chemical Characterization of Honey

Honey, as a natural blend, contains 95-98% carbohydrates in dry matter. The main ingredient is fructose and glucose (60-70%) and also contains about 20 to 25 different mono-, di-, tri-, and oligosaccharides. [12]. Honey also contains (1–3%) different vitamins (A, B, C) and many substances beneficial to human health. These compounds provide the biological activity of honey [15]. The quality criteria for honey are summarized in Table 4.1.

Dark-colored honeys have higher biological active compounds than light colored honey. Dark honeys, like Manuka honey in New Zealand, chestnut honey in Eastern Europe, an Arabic type of acacia honey, are very valuable for alternative medicine applications. Thus, commercial values are higher than light colored honey. It has been reported that oak honey produced in Turkey has a high phenolic content [16]. *Kernes guercus*, *Lachnus iliciphilus*, and *Thelaxes dryophila* secretions could be the source of oak honey. Sweating oak leaves under stress conditions, such as rapid temperature changes during day and night, could be an alternative method for production of honey. The perspiration of the leaves causes the secretion of various sugars, which are a significant food source for bees in honey production. Thus, honey extract and nectar honey are different

Table 4.1 Quality criteria of honey for 100 g of honey [12].

	Blossom honey	**Honeydew honey**
Water (max)	20 g	20 g
Fructose (max)	45 g	40 g
Glucose (max)	40 g	32 g
Fructose + Glucose (min)	60 g	45 g
Sucrose (max)	4.8 g	4.7 g
Insoluble in water (max, g)	0.1 g	0.1 g
Free acidty (max)	50 meq/kg	50 meq/kg
Minerals	0.5	2
Amino acids	0.4	0.7
pH (max)	4.5	6.5
Diastases (min)	8	8
Hydroxy methyl furfural (HMF, max)	40 ppm	40 ppm
Proline (min)	300 ppm	300 ppm
Naphthalane (max)	10 ppb	10 ppb

from each other in terms of chemical composition, physical properties and melissopalynological analysis. It is a well-known fact that different types of honey show different physicochemical properties [17].

4.1.4 Biologically Active Properties of Honey

Honey has an importance in Apitherapy, the treatment method made with bee products, due to its content. Other bee products except honey, used for apitherapeutic purposes are propolis, pollen, royal jelly, and bee venom. Bee products are used by people from past to the present. Before the discovery of antibiotics, honey was the most commonly used natural remedy for the treatment of wounds. In recent years, the high number of side effects of synthetic drugs, allergic conditions and resistance to antibiotics has led human beings to seek natural treatment methods. New alternative therapies have been developed, such as phytotherapy with herbal preparations and apitherapy with bee products. In countries, such as China, Romania, and Egypt, clinics and apitherapy centers that treat diseases only with bee products are becoming more and more common. In animal researches and clinical studies conducted in different laboratories around the world, it is observed that honey has shown very promising developments in terms of its therapeutic potential. "Medical Grade Honey" is currently available for use by patients in developed countries, such as the USA, UK, New Zealand, Australia and some European countries. These honeys are graded mainly on the basis of their antibacterial activities and as they are certified only by certain laboratories, high quality, and pure honey presentation is assured to consumers. Highly rich content of polyphenolic compounds, vitamins, enzymes and minerals found in the structure of honey, also called secondary metabolite agents, are responsible for its biologically active properties such as antioxidant, antimicrobial, antiviral, anti-inflammatory, antitumoral, and so on [8-12]. The true value and medicinal properties of honey come from the presence of these secondary metabolites that make up it. One of the most important features of honey is that it has antimicrobial properties. In addition, honey is used in the treatment of burns, wounds, gastrointestinal disorders, skin ulcers, cataracts, eye ailments [18, 19]. In recent clinical studies, effective results are emerging with the use of honey in the treatment of ulcers on infected wounds [20, 21].

Containing high carbohydrate content, honey is a high-energy natural product. The energy of 100 grams of honey is approximately 304 calories. Syrup of honey is consumed by athletes [11], the elderly and tuberculosis patients due to its high energy content. The most important feature of honey is that some of its energy value is that it has various biological activities.

4.2 Green Synthesis of Nanoparticles

Recent advances in nanotechnology focusing on simple, environmentally friend and inexpensive synthesis methods. Synthesis of NPs generally follows two approaches. These are "Top Down" approach and "Bottom Up" approaches. In the "Top Down" approach, NPs are obtained by the formation of smaller size particles from large particles. Meanwhile, various physical and chemical methods are used [22]. In "Bottom-Up" synthesis, NPs are produced from small particles such as atoms and molecules. Green synthesis of nanoparticles is an environmentally friendly and safety form of synthesis consuming natural sources. This green approaching has sight a new period in safety nanotechnology. Microorganisms, various biological resources (honey, egg, milk, etc.), and plant extracts are widely used in the synthesis of NPs by biological methods [23].

The fact that honey is both a stability and a reducing agent allows nanoparticles to be used in green synthesis. Honey-mediated synthesis has many advantages over methods using other biological sources. The comparison of green synthesis of physical and chemical methods in the synthesis of nanoparticles and microorganisms mediated by green synthesis and more efficient green synthesis mediated by honey is shown in Figure 4.3.

Nanoparticle formation is a relatively faster process than microorganism method, and the bacteria or fungi must be cultured carefully. It needs time delay for the nanoparticles to be transformed by microorganisms. It is also a difficult process to separate nanoparticles from microorganisms.

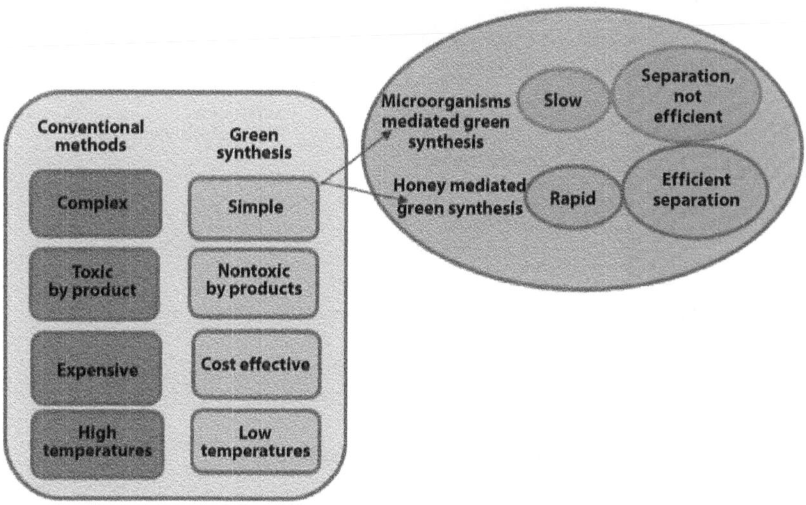

Figure 4.3 Comparison of traditional and green synthesis methods [24].

The reducing and stabilizing agents used in the biosynthesis of various metallic nanoparticles and the ambient conditions (temperature) are given in Table 4.2. While a general procedure is used for palladium, gold, and silver nanoparticles, specific conditions have been established for carbon and platinum nanoparticles (Figure 4.4).

Table 4.2 Different nanoparticles synthesized by honey-mediated method.

Used metal type	Reducing agent	Stabilization agent	Temperature
Gold	a) Vit. C b) H_2O_2 and gluconic acid	Proteins	Room temperature [25, 26]
Silver	Honey	Proteins	Room temperature [26–33]
Palladium	Honey	Honey	Room temperature [30]
Platinum	Honey	Proteins in honey	100°C [31]

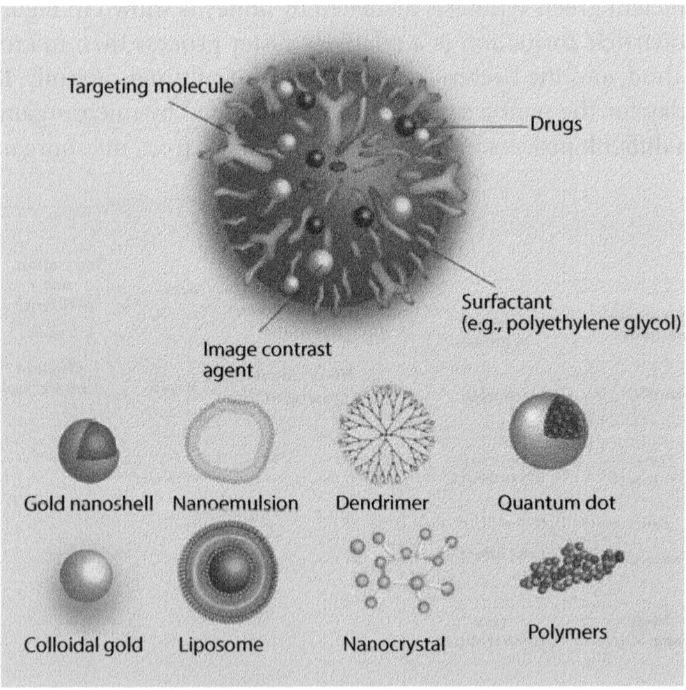

Figure 4.4 Different carriers for drug release [25].

4.3 Green Synthesis of Honey-Mediated Nanoparticles

Honey-mediated nanoparticles are generally obtained using gold, silver, palladium, and platinum (Table 4.2).

Gold nanoparticles were synthesized by reducing the honey solutions prepared by diluting in different volumes with HAuCl4 compound [32]. Vitamin C (the reducing agent) and fructose, the main component of honey, are thought to act together as the primary reducing agent. In nanoparticle synthesis, fructose is thought to be a possible reducing agent for the reaction, while proteins in honey are thought to be responsible for stability [25]. Green synthesis of nanoparticles using honey as both a reducing and stabilizing agent at room temperature has been reported [32].

Silver nanoparticles were synthesized in which honey was used as a reducing and stabilizing agent under room conditions [33]. Honey has thus replaced the previously used harmful reducing agents. In another study, it was reported that a method mediated by sunlight and honey was used in the synthesis of Ag nanoparticles [34]. It has been determined that silver nanoparticles synthesized through honey are effective corrosion inhibitors and remain stable at room temperature for a long time. While honey proteins are known to stabilize nanoparticles, fructose acts as a reducing agent [28].

In a synthesis method using honey, Palladium nanoparticles in the range of 5.0 to 40.0 nm were synthesized. The synthesized nano-catalyst showed high regeneration property. These nanoparticles act as a catalyst for crosslinking in Suzuki reactions. Honey-mediated palladium nanoparticles have potency for applications in a variety of areas, as nanotechnology, catalytic reactions, and biosensors [30].

For photo acoustic systems, honey has been used as a precursor to synthesize carbon nanoparticles [35]. These polysorbate and polyethylene glycol C nanoparticles are smaller than nanoparticles such as silica-coated gold particles [36], nano-tubes (SWNT) and Cu [34] used for node of lymph. In one study, a fluorescent carbon point was synthesized with a quantum yield (QY) of 19.8%. Carbon nanodots showed higher stability, among other advantages, such as harmlessness, high quantum efficiency. However, the synthesized carbon spots act as a detector for the determination of Fe^{3+}. Researchers reported that there are various potential practices of carbon spots in biosensing and bioimaging [37].

A current, low-cost, honey-mediated method has been reported to obtain Pt nanoparticles and nanowires [31]. In their study, they synthesized

2.2 nm sized honey mediated Pt nanoparticles at 100 °C [31]. It has been observed that platinum nanoparticles functionalized with honey are a cubic crystal. FTIR spectrums showed that proteins bind to Pt nanoparticles via the carboxylate group. These particles were able to remain stable in water for a long time and exhibit catalytic properties.

4.4 Bioactivity of Honey Nanoparticles

Metallic and carbon nanoparticles obtained with honey-mediated green methods have many applications today. It has been determined that gold and silver nanoparticles synthesized by these methods show strong antibacterial activity [26]. Ag nanoparticles exhibit antimicrobial activity and anticorrosive activity [20]. Palladium nanoparticles known to exhibit catalytic activity can be used in biosensors [30]. Carbon nanoparticles are used in many fields as bio-detection, fluorescent staining, real-time photoacoustic imaging, and so on [37].

The effect of sidr honey on immune response and anticancer activity was tested in studies conducted in 2020 [38]. Sidr honey from three sources was diluted and tested to synthesize AgNPs and for their biological activity. The data obtained showed that AgNPs obtained by using this type of honey can regulate the growth of spleen cells and are effective on bacteria. While Sidr honey has anticancer activity against human hepatocellular carcinoma, it is not effective against tumor cells.

In one study, AuNps were synthesized to inhibit bacterial strains [39]. It has reported they had antibacterial effect against Staphylococcus aureus. Nanoparticles synthesized by honey-mediated method have significant antimicrobial activity against various microbial species. The lowest minimum inhibitory concentration (MIC) was reported against Staphylococcus aureus [26].

Silver nanoparticles could be used in various applications in different areas such as medicine, biosensors, catalysis, electronics, energy storage, antimicrobial protection, and biomedical treatments [37, 40, 41].

In a study, it was shown that the addition of honey mediated silver nanoparticles protected corn grains caused significant reductions in aflatoxin B1 toxin [42]. In addition, antibacterial and anticancer effects have been detected for biogenic AgNPs [43, 44]. As a result of the attraction between positively charged Ag ions and the negative charge of the bacterial membrane and the reaction of Ag ions with vital enzymes through their high affinity, DNA replication is disrupted and the antibacterial activity of AgNPs occurs [45].

It was realized the green synthesis of honey mediated silver nanoparticles obtained from two different flower sources had antimicrobial and anticancer activities [46]. Biogenic AgNPs exhibited cytotoxic effect against HepG2 cells and antibacterial effect against some Gram (-) and Gram (+) bacteria. In a study, gold honey mediated nanoparticles showed inhibitory effect on L929 mouse fibroblast cells [47]. Neupane *et al.* (2019) syntheses honey mediated silver nanoparticles. They found that nanoparticles had both antimicrobial and antioxidant activities [48].

4.5 Future and Perspectives

Thanks to the physical and chemical technologies used for a long time in the production of nanoparticles, high-resolution, desired small particles can be produced in a short time, but their toxic content is high, particle stability is not good and the technologies used are expensive, so new technologies should be investigated. In this sense, with the start of researches, the study of materials and living things that already exist in nature, containing perfectly designed nanoscale dimensions, has inspired scientists, and the production of inorganic materials using living structures has begun to be researched. As a result of the researches, the term Green Nanotechnology has emerged, which is environmentally friendly, with low toxic substance content, and is based on the production of nanoparticles from living cells. This term is expressed as working methods that investigate easily applicable methods that reduce the problem of waste products, do not harm human health, within the scope of nanotechnology science.

Honey, whose antimicrobial, anticancer and wound healing properties are known, may be limited due to the amount of sugar it contains. With the production of honey-based nanoparticles, it is possible to eliminate the negative situations caused by sugars and to increase the use of honey in the field of health. From this aspect, it is important to obtain honey-based nanoparticles and it is thought that their usage area will increase in the future.

4.6 Conclusion

Nanoparticles synthesized by the green method using honey have many advantages over other methods. As no harmful substances are formed, the reaction takes place at relatively low temperatures. These methods have emerged as alternatives to other synthesis methods (chemical and physical) used. Honey-mediated methods, in which honey perform as a reducing and stabilizing factor, are environmentally friendly methods and provide advantages such as low cost, time saving and simple synthesis method. Nanoparticles functionalized with honey have a variety of catalytic activity. These nanoparticles have many biological activities and properties, such as biodetection and bioimaging, besides being resistant to abrasion. It is thought that nanoparticles synthesized with green methods through honey will provide valuable end products that can be used in many applications seeking efficient, inexpensive, environmentally friendly alternatives.

References

1. Albrecht, M.A., Evan, C.W., Raston, C.R., Green chemistry and the health implications of nanoparticles. *Green Chem.*, 8, 417–32, 2006.
2. Thakkar, K.N., Mhatre, S.S., Parikh, R.Y., Biological synthesis of metallic nanoparticles. *Nanomedicine*, 6, 257–262, 2010.
3. Kuppusamy, P., Ichwan, S.J., Al-Zikri, P.N. *et al.*, *In vitro* anticancer activity of Au, Ag nanoparticles synthesized using Commelina nudiflora L. Aqueous extract against HCT-116 colon cancer cells. *Biol. Trace Elem. Res.*, 173, 297–305, 2016.
4. Kreuter, J., Nanoparticles – A historical perspective. *Int. J. Pharm.*, 331, 1, 1–10, 2007.
5. Park, S. and Ruoff, R.S., Chemical methods for the production of graphenes. *Nat. Nanotechnol.*, 4, 217–224, 2009.
6. Grislain, L., Couvreur, P., Lenaerts, V., Roland, M., Deprez-Decampeneere, D., Speiser, P., Pharmokinetics and distribution of a biodegradable drug-carrier. *Int. J. Pharm.*, 15, 3, 335–345, 1993.
7. Lee, J.H. and Yeo, Y., Controlled drug release from pharmaceutical nanocarriers. *Chem. Eng. Sci., 125*, 75–84, 2015.
8. Keskin, Ş., Mayda, N., Keskin, M., Özkök, A., Investigation of Bilecik honeys in terms of melissopalynology and chemical analyses. *Gıda/Food*, 45, 2, 275–289, 2020.
9. Özmen, N. and Alkın, E., Antimicrobial properties of honey and its effects on human health. *Uludağ Arıcılık Dergisi*, 4, 155–160, 2006.
10. Spilioti, E., Jaakkola, M., Tolonen, T., Lipponen, M., Virtanen, V., Chinou, I., Kassi, E., Karabournioti, S., Moutsatsou, P., Phenolic acid composition, anti-atherogenic and anticancer potential of honeys derived from various regions in Greece. *PLoS One*, 9, 4, e94860, 2014.
11. Bogdanov, S., Jurendic, T., Robert Sieber, R., Gallmann, P., Honey for nutrition and health: A review. *J. Am. Coll. Nutr.*, 27, 6, 677–689, 2017.
12. Keskin, M., Keskin, Ş., Kolaylı, S., Health-promoting benefits of honey, in: *Preparation of Phytopharmaceuticals for the Management of Disorders*, pp. 303–306, Academic Press, Elsevier, 2021.

13. Turkish Honey Codex, Türk Gıda Kodeksi Bal Tebliği, (Codex number: 2012/58). https://www.resmigazete.gov.tr/eskiler/2012/07/20120727-12.htm
14. Zander, E. and Koch, A., *Der Honig*, 201 p, Eugen Ulmer Verlag, Stuttgort, 1994.
15. Oddo, P. and Piro, R., Main European unifloral honeys: Descriptive sheets. *Apidologie*, 35, 38–81, 2004.
16. Can, Z., Yıldız, O., Şahin, H., Turumtay, E., Silici, S., Kolaylı, S., An investigation of turkish honeys, their physico-chemical properties, antioxidant capacities and phenolic profiles. *Food Chem.*, 180, 133–141, 2015.
17. Pita-Calvo, C. and Vazquez, M., Differences between honeydew and blossom honeys: A review. *Trends Food Sci. Technol.*, 59, 79–87, 2017.
18. Castaldo, S. and Capasso, F., Propolis, an old remedy used in modern medicine. *Fitoterapia*, 73, 1–6, 2002.
19. Orhan, F., Sekerel, B.E., Kocabas, C.N., Sackesen, C., Adalioglu, G., Tuncer, A., Complementary and alternative medicine in children with asthma. *Ann. Allergy Asthma Immunol.*, 90, 611–615, 2003.
20. Molan, P.C., The antibacterial activity of honey: The nature of antibacterial activity. *Bee World*, 73, 5–28, 1992.
21. Jull, A., Walker, N., Parag, V., Molan, P., Rodgers, A., Randomized clinical trial of honey-impregnated dressings for venous leg ulcers. *Br. J. Surg.*, 95, 175–182, 2008.
22. Singh, N. and Bath, P.K., Quality evaluation of different types of Indian honey. *Food Chem.*, 58, 1–2, 129–133, 1997.
23. Imtiyaz Hussain, N.B., Singh, A.S., Himani Singh Singh, S.C., Green synthesis of nanoparticles and its potential. *Appl. Biotechnol. Lett.*, 38, 545–560, 2016.
24. Balasooriya, E.R., Jayasinghe, C.D., Jayawardena, U.A., Ruwanthika, R.W.D., Mendis de Silva, R., Udagama, P.V., Honey mediated green synthesis of nanoparticles: new era of safe nanotechnology. *J. Nanomaterials*, 2017, 5919836, 2017.
25. Jafarızad, A., *The use of gold nanoparticles and gold nanoparticles/reduced graphene oxide composite structures as a drug delivery system in cancer therapy* (Doctoral dissertation), Department of Chemistry, Ataturk University, Erzurum, Turkey, 2016.
26. Philip, D., Honey mediated green synthesis of gold nanoparticles. *Spectrochim. Acta Part A: Mol. Biomol. Spectrosc.*, 73, 4, 650–653, 2009.
27. Sreelakshmi, C., Datta, K.K.R., Yadav, J.S., Subba Reddy, B.V., Honey derivatized Au and Ag nanoparticles and evaluation of its antimicrobial activity. *J. Nanosci. Nanotechnol.*, 11, 8, 6995–7000, 2011.
28. Haiza, H., Azizan, A., Mohidin, A.H., Halin, D.S.C., Green synthesis of silver nanoparticles using local honey. *Nano Hybrids*, 4, 87–98, 2013.
29. Obot, I.B., Umoren, S.A., Johnson, A.S., Sunlight-mediated synthesis of silver nanoparticles using honey and its promising anticorrosion potentials for mild steel in acidic environments. *J. Mater. Environ. Sci.*, 4, 6, 1013–1018, 2013.

30. Reddy, S.M., Datta, K.K., Sreelakshmi, C., Eswaramoorthy, M., Reddy, B.V.S., Honey mediated green synthesis of Pd nanoparticles for suzuki coupling and hydrogenation of conjugated olefins. *Nanosci. Nanotechnol. Lett.*, 4, 4, 420-425, 2012.
31. Venu, R., Ramulu, T.S., Anandakumar, S., Rani, V.S., Kim, C.G., Bio-directed synthesis of platinum nanoparticles using aqueous honey solutions and their catalytic applications. *Colloids Surf. A: Physicochem. Eng. Asp.*, 384, 1–3, 733-738, 2011.
32. Olaitan, P.B., Adeleke, O.E., Ola, I.O., Honey: a reservoir for microorganisms and an inhibitory agent for microbes. *Afr. Health Sci.*, 7, 3, 159–165, 2007.
33. Bar, H., Bhui, D.K., Sahoo, G.P., Sarkar, P., Pyne, S., Misra, A., Green synthesis of silver nanoparticles using seed extract of Jatropha curcas. *Colloids Surf. A: Physicochem. Eng. Asp.*, 348, 1–3, 212–216, 2009.
34. González-Miret, M.L., Terrab D Hernanz, A., Fernández- Recamales, M.A., Heredia, F.J., Multivariate correlation between color and mineral composition of honeys and by their botanical origin. *J. Agric. Food Chem.*, 53, 7, 2574–2580, 2005.
35. Wu, L., Cai, X., Nelson, K., A green synthesis of carbon nanoparticles from honey and their use in real-time photoacoustic imaging. *Nano Res.*, 6, 5, 312–325, 2013.
36. Luke, G.P., Bashyam, A., Homan, K.A., Makhija, S., Chen, Y.S., Emelianov, S.Y., Silica-coated gold nanoplates as stable photoacoustic contrast agents for sentinel lymph node imaging. *Nanotechnology*, 24, 45, 2013.
37. Yang, X., Zhuo, Y., Zhu, S., Luo, Y., Feng, Y., Dou, Y., Novel and green synthesis of high-fluorescent carbon dots originated from honey for sensing and imaging. *Biosens. Bioelectron.*, 60, 292–298, 2014.
38. Ghramh, H.A., Ibrahim, E.H., Kilany, M., Study of anticancer, antimicrobial, immunomodulatory, and silver nanoparticles production by Sidr honey from three different sources. *Food Sci. Nutr.*, 8, 445–455, 2020.
39. Mittal, A.K., Chisti, Y., Banerjee, U.C., Synthesis of metallic nanoparticles using plant extracts. *Biotechnol. Adv.*, 31, 2, 346–356, 2013.
40. Kannan, N. and Subbalaxmi, S., Green synthesis of silver nanoparticles using Bacillus subtillus IA751 and its antimicrobial activity. *J. Nanosci. Nanotechnol.*, 1, 2, 87–94, 2011.
41. Vidhu, V.K., Aromal, S.A., Philip, D., Green synthesis of silver nanoparticles using Macrotyloma uniflorum. *Spectrochim. Acta A*, 83, 392–397, 2011.
42. Ali, A.M., Hussein, H.Z., Majeed, S.H., Evaluation efficiency of silver nanoparticles enriched by honey in the detoxification of aflatoxin B1. *EC Microbiol.*, 13, 3, 85–91, 2017.
43. Mohammed, A.E., Alqahtani, A., Almutairi, A., Alshamari, B., Aabed, K., Antibacterial and cytotoxic potential of biosynthesized silver nanoparticles by some plant extracts. *Nanomaterials*, 8, 382, 1–15, 2018.

44. Alqahtani, M.A.M., Mohammed, A.E., Daoud, S.I., Alkhalifah, D.H., Lichens (Parmotrema clavuliferum) extracts: Bio-mediator in silver nanoparticles formation and its antibacterial potential. *J. Bionanosci.*, 11, 410–415, 2017.
45. Shao, W., Liu, X., Min, H., Dong, G., Feng, Q., Zuo, S., Preparation, characterization, and antibacterial activity of silver nanoparticle-decorated graphene oxide nanocomposite. *ACS Appl. Mater. Interfaces*, 7, 6966–6973, 2015.
46. Al-Brahim, J.S. and Mohammed, A.E., Antioxidant, cytotoxic and antibacterial potentials of biosynthesized silver nanoparticles using bee's honey from two different floral sources in Saudi Arabia. *Saudi J. Biol. Sci.*, 27, 363–373, 2020.
47. Boldeiu, A., Simion, M., Mihalache, I., Radoi, A., Banu, M., Varasteanu, P., Kusko, M., Comparative analysis of honey and citrate stabilized gold nanoparticles: *In vitro* interaction with proteins and toxicity studies. *J. Photochem. Photobiol. B: Biol.*, *197*, 111519, 2019.
48. Neupane, B.P., Chaudhary, D., Paudel, S., Timsina, S., Chapagain, B., Jamarkattel, N., Tiwari, B.R., Himalayan honey loaded iron oxide nanoparticles: Synthesis, characterization and study of antioxidant and antimicrobial activities. *Int. J. Nanomed.*, *14*, 3533–3541, 2019.

5

Nanoparticles in Nondairy Yogurt Formulation Techniques for Amelioration of Lifestyle Disorder Diseases

Tanima Bhattacharya[1*], Rokeya Akter[2,3] and Md. Habibur Rahman[3,4]

[1]Hubei Collaborative Innovation Center for Advanced Organic Chemical Materials, Ministry of Education Key Laboratory for the Synthesis and Application of Organic Functional Molecules, College of Chemistry and Chemical Engineering, Hubei University, Wuhan, P. R. China
[2]Department of Pharmacy, Jagannath University, Sadarghat, Dhaka, Bangladesh
[3]Department of Global Medical Science, Yonsei University Wonju College of Medicine, Yonsei University, Wonju, Gangwon-do, South Korea
[4]Department of Pharmacy, Southeast University, Banani, Dhaka, Bangladesh

Abstract

The ever-escalating health care costs and public consciousness in recent years have given more attention to probiotic foods, like milk and dairy products. On the other hand, lactose intolerance and other medical and ethical reason for not using diary-based products have brought nondairy products to the limelight. These products include flavonoid-rich fruit beverages. They can be used as a vehicle for probiotics or mixed with dairy products, as well as nondairy products. Yogurts are plant seed protein-based, which may further be fortified with essential fatty acids and other phenolic compounds to improve their nutritive value. The inherent stability issues of these products can be overcome by implementing nanotechnology. The aim of this chapter is to introduce latest nondairy yogurts to evaluate newly formulated nanoparticles-induced nondairy yogurt as functional food in curing lifestyle disorder diseases.

Keywords: Nondairy yogurt, seed protein, nanoparticles, functional foods, lifestyle disorder diseases

**Corresponding author*: btanima1987@gmail.com

Tanima Bhattacharya and Shakeel Ahmed (eds.) Nanotechnology in Functional Foods, (103–136)

Abbreviations

BYG	Bio-yogurt
Cell	ethyl cellulose, Vit.
Cli	clinical, Et.
COVID	Corona Virus Disease
CVD	cardiovascular diseases
D	Vitamin D
DANY	Dairy Analog Nano yogurt
DDS	drug delivery system
FCM	fermented coconut milk
FHM	fermented hemp milk
GABA	Gamma Amino Butyric Acid
GBR	Germinated Brown Rice
GERD	gastro esophageal reflux disease
GI	Gastro-Intestinal
HDL	high density lipoprotein
HIV	Human Immunodeficiency Virus
LAB	*Lactobacillus* species
LDL	low density lipoprotein
MUFA	Monounsaturated fatty Acid
NE	Nanoemulsion
NP	nanoparticles
OPC	Oat Protein Concentrate
Pre	preclinical
PUFA	Poly unsaturated fatty Acids
RFCY	reduced fat coconut yogurt
SPNE	soya protein nanoparticles
VLDL	Very Low Density Lipoprotein
WFCY	whole fat coconut yogurt

5.1 Introduction

Milk and dairy-related products are among the five food groups that need to be consumed to meet the nutritional requirement of a healthy life [1]. This group contains oleic acid, conjugated linoleic acid, omega-3 fatty acids, short-chain and medium-chain fatty acids, bioactive compounds, vitamins, and minerals that promote good health in humans [2, 3]. However, whole milk and some dairy proteins may cause some healthy side effects [4]. In some infants, especially after weaning, milk indigestion may lead to bloating, flatulence, vomiting, and diarrhea. This condition in babies and young children is considered as lactose intolerance which is due to insufficient amount of lactase enzyme in the body [5, 6]. The primary function of this enzyme in the gut is to breakdown lactose in the milk to simple sugars glucose and galactose. Another health side effect of whole milk is associated with excessive intake of fat that may be linked to a host of cardiovascular diseases, including hypertension. Infantile or childhood obesity may result due to the uncontrolled consumption of buffalo milk. Dairy product intakes are generally inversely proportional to the incidence of gastric cancer, colorectal cancer, breast cancer but remain inconsistent with prostate cancer, ovarian cancer, and lung cancer [7]. Consumptions of traditional fermented dairy products, including curd or yogurt, cheese, kefir, shrikhand, are associated with enhanced gut health and reduced risk of intestinal diseases [8, 9]. Although the common source of dairy products is milk from cow [10, 11], goat, [12] and sheep [13], recent studies also focused on milk from various other sources, such as camel [14], mare [15], and ewe bovine milk [16]. Recent developments also included techniques for the preparation for milk and milk products from different animal sources [17, 18]. Nowadays, there is also a trend of introducing dairy substitutes [19] that offer the health benefit of cow milk, but more akin to weight management [20], and less side effect, such as allergy [21]. When milk is not consumed for whatever reason (lack of resource, medical or ethical reason, or others), nondairy beverages may be adopted as a replacement. These include beverages like fruits, vegetables, legumes, oilseeds, and various nuts. Alternative dairy sources are rich in phytosterols and polyphenolic compounds of enormous health benefits [21]. In this chapter, we are highlighting some of the most common dairy substitutes, which are categorized as functional foods, such as soya [22], coconut [23], sesame [24], pea [25], almond [26], and cashew [27]. The prime focuses are several types of functional fermented products, especially yogurts [28] fortified with essential nutrients for enhanced immune functions. Nanotechnology has already expanded to almost every industry sector including the food, agriculture, and pharmaceuticals [29]. Since nanotechnology has also put is footprints in dairy technology, we herein highlight the stability, functionality, bioavailability, and efficacy of some dairy and nondairy products through nanoparticles preparation of bioorganic [29] and metal origin [30]. Figure 5.1 represents beneficial effects of functional foods on human health.

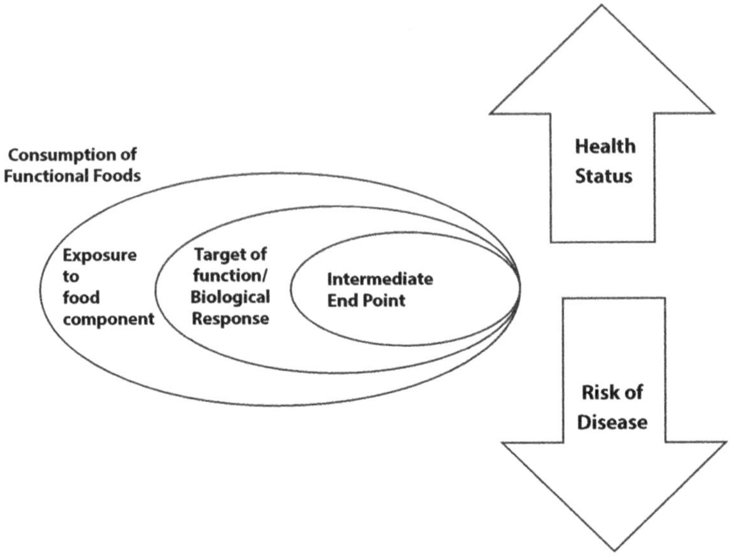

Figure 5.1 Schematic diagram of effect of functional foods.

5.2 Nondairy Products as Functional Foods and Their Role in Immune Function

Public awareness and consumer acceptance have been pivotal to the recent flourishing market of functional food (FF). FF is considered to be such components of human food that caters to better health and nutritional status. Dairy or nondairy products containing probiotics [31], prebiotics and symbiotic [32] may be considered as FFs. Probiotics, such as bio yogurts, are living microorganisms, which are the starter culture [33]. This microorganism alters the microflora of the host especially the gut microbiome (by implantation or colonization), which later on exerts beneficial effects on the health of the host. The probiotic word was first derived from Greek Language "for life" and was first used by Lilly and Stillnerin 1965. Most common probiotics used in bio yogurt are *Lactobacillus* sp (LAB) [34], *Bifidobacterium* sp [35], though some other microorganisms [36] are under consideration. Prebiotics are a nondigestible food ingredient that beneficially affects the host by selectively stimulating the growth and/or activity of one or a limited number of bacteria in the colon, and thus improves host health. Glen Gibson and Marcel Roberfroid first coined introduced the term in 1995. This group acts as feed to the gut microbiome and the short-chain fatty acids microbial byproducts are released and are

Table 5.1 Nutrient composition of cow milk and other plant-based milk (Source: USDA).

Nutrient	Cow milk	Almond milk	Cashew milk	Coconut milk	Oat milk	Rice milk	Soya milk
Energy (kcal)	122	41–59	60	70–424	130	130	100
Protein(g)	8.05	1.0–1.1	1.01	0–3.03	4.01	1.01	8.0
Fat(g)	4.83	2.50–4.01	2.5	4.5–42.44	2.5	1.99	4.01
Saturated Fat(gm)	3.067	0	0	4.5–30.3	0	0	0.51
Carbohydrate(gm)	11.71	4.01–6.0	9	7.99–9.09	24	27	7.9
Fiber(gm)	0	0.9–1.0	0	1.0	1.9	0	1.1
Sugar(gm)	12.35	3–5	7.01	3.03–7.01	19	13.99	6.0
Calcium(mg)	293	450–451	450	101	350	300	299
Iron(mg)	0.05	0.71–0.72	0.36	6.5	1.8	1.08	1.07
Magnesium(mg)	27	N/A	17	41	N/A	N/A	39

(Continued)

Table 5.1 Nutrient composition of cow milk and other plant-based milk (Source: USDA). (*Continued*)

Nutrient	Cow milk	Almond milk	Cashew milk	Coconut milk	Oat milk	Rice milk	Soya milk
Potassium(mg)	342	180	29	41–491	120	N/A	299
Sodium(mg)	115	149–161	170	46	115	60	119
Riboflavin(mg)	0.451	0.034–0.424	0.509	N/A	0.509	N/A	0.510
Vitamin B12 (mcg)	1.29	3.01	3	3	N/A	N/A	2.99
Vitamin A (IU)	464	499–500	499	499	499	499	501
Vitamin D (IU)	120	89–100	101	101	101	101	119
Information: USDA							

circulated by the blood to induce positive health effects such as immunomodulation locally (in the intestine) and all over in the body. The two major prebiotics are fructo-oligosaccharides and galacto-oligosaccharides [33]. These are responsible for lowering low-density lipoprotein (LDL), increasing collagen, water, and calcium retention, reducing dementia, lowering atherosclerosis, and decreasing atopic dermatitis. Symbiotics are a combination of prebiotics and probiotics [36]. The present chapter throws some light on the utilization of these probiotics, prebiotics, and symbiotic as ingredients for plant based yogurts. The emphasis once again would be on modified FFs through nanotechnology approach, which are applied to improve the immune system. Nutrient composition of cow milk and other plant-based milk is brief in Table 5.1.

5.2.1 Soybean

Milk is a rich source of B vitamins, carotenes, and minerals like zinc and potassium. It can be considered as wholesome food. Some saturated fats [37] and allergens [38] in milk may however lead to adverse effects in humans. Avoidance of food components of vital importance to the body cannot be the only option, and the better option remains to be diary substitutes. In this connection, the significance of the golden miracle bean legume, soyabean, to extract its milk and use it as milk substitute [39] is evident. It is to be thought that China is the birth place of soybean milk utilization and introduced in the early 20th century for its medicinal purposes [40]. Soybean and related legumes contain isoflavones (phytoestrogen) in their nonglycosylated forms, which are known as aglycons [41]. Isoflavones reduce blood cholesterol [42], ease menopause problems, reduce risks of osteoporosis [43] in geriatrics, and provide protection against certain breast and prostate cancers. Soybeans contain 20% oil and 40% good quality protein. They are known as complete protein sources as they contain all the indispensable amino acids in the proportion that are beneficial for growth, development, and repairing tissues. There are several contradictory studies on soy isoflavones in reducing LDL and cardiovascular disease (CVD) risks. One research report drew an interrelationship of hypercholesterolemia, oxidation of LDL, and isoflavone intake in reducing atherosclerosis [44]. It was revealed that the higher level of induction of auto antibodies against oxidized LDL in hyperlipidemic albino rats could be prevented by consumption of soya yogurt with or without isoflavones supplement. It is also observed that probiotics with soya milk fermented with *Bifidobacterium*, LAB are more effective in health status improvement in humans, as well as reduction in atherosclerosis and hypertension. Of the soybean isoflavones, diadzin and genistein are important in reducing the oxidation of LDL [45]. Since this yogurt lacks vitamin C (ascorbic acid), increasing its content through fortification for better bioavailability is also adopted [46]. In order to determine the relationship between soya yogurts and enhanced immunologic effect, Sengupta *et al.* [47] conducted a series of experiments. The studies reflect that soya yogurt fortified with eicosapentaenoic acid (EPA) and docosahexaenoic acid DHA (from fish oil) and gamma linolenic acid (GLA) from evening primrose oil reduces hypercholesterolemia and improve brain health probably through modulation of the cholesterol/lipid-related genes Sterol regulatory element-binding proteins (SREBP), peroxisome proliferator-activated receptors (PPAR), cytochrome P450, family 1 (CYPa1), and β-hydroxy β-methylglutaryl-CoA (HMGCoA). No definite plaque or sudden death was also observed in the Bulb mice. Current research efforts try to modify the functionality of probiotic soya yogurt in the human digestive system. In one study by Saito *et al.*, *L. fermentum* and *L. plantarum* were isolated and used in soya yogurt fermentation and their viability was observed during storage time. For both strains, the result showed a decrease in the cell count by the log 1 cycle at days 1 and 9. On 18th day of storage for *L. fermentum*, it was log32 and for *L. plantarum* it was log3 cycles at pH 2.5. When saliva gastric exposure was carried out at pH 3, a reduction of 0.5 to 1 log is observed in *L. plantarum* on day 1, day 9, and day 18. At pH 3 on the 27th day, higher resistance is observed in *L. fermentum* than *L. plantarum*. The simulated gastrointestinal digestion under these conditions revealed maintenance of probiotic population profile in storage and cell viability in the order of log1 cycle. This study reflects that small intestinal resistance is more important than gastric survival [48]. Similarly, a comparative study on digestibility (buccal and duodenal) report of soya yogurt and soya milk was conducted [49]. It is observed that at pH 6.5, the protein solubility of both soy yogurt and soy milk remained similar. It differed when the pH is maintained at 5.1 to 5.5 (similar to the gastric environment). The soya aggregates used in the study were large in size (50–100 μm) and gelation resulted in low protein bioaccessibility. This data revealed that the pH of the final product is an important index for soya protein quality after ingestion [50].

Improvement of the functional properties of soyabean is now becoming a necessity in the research community. New innovative fermentation techniques are reported on bioconversion of gamma amino butyric acid (GABA) and isoflavones by *L. brevis* in soyabean yogurt powder. The enzymes glutamate decarboxylase and β-glucosidase activities increased; however, the isoflavone glycoside and malonylglycoside contents decreased after fermentation for 72 h followed by highest Daidezein (179.93 μg/g),

Table 5.2 Dairy analogs and their potential health benefits.

DA	Probiotics	Prebiotic matrix	Pre/Cli	Health benefits	Reference
Soya	*Bifidobacterium*	N/A	Pre	Gut health	[45]
Soya	*L. fermentum, L. Plantarum*	N/A	Pre	Increasing gut health	[48]
Soya	*LAB + Spirulinaplatensis*	N/A	Cli	Lowers cholesterol	[52]
Oats	*LAB*	N/A	Pre	Bioavailability	[59]
Coconut	*ABT culture, LAB*	N/A	Pre	Improves central nervous system	[60]
Soya,	*LAB, Bifidobacteria,*	N/A	Pre	Improves gut microbiome	[61]
Coconut	*L. bulgaricus, L. delbrukeii*	Coconut powder	Pre	Antibacterial against pathogens	[62]
Almond	*LAB*	Innulin	Pre	Improves digestion	[63]
Sesame	*LAB*	N/A	Pre	Antioxidant, anti-inflammatory	[63]
Rice	*Bifidobacterium*	N/A	Pre	Provides innate immunity	[64]
Rice	*LAB, thermophiles starter culture*	GBR flour	Pre	Lowers VLDL, LDL	[65]

glycitein (44.10 μg/g), and genistein (126.24 lg/g) contents [51]. Sengupta *et al.* [52] also reported the nutrient bioavailability and stability of soya yogurt powder fortified with *Spirulinaplatensis*. Here, the study highlighted the effect of different drying techniques (like hot air oven, microwave, and freeze-drying) on the nutritional profile of the fortified yogurt. It was reported that freeze-dried yogurt exhibits better functional and antioxidant properties, which is a similar report by Kitawaki *et al.* [53]. Kitawaki *et al.* conducted in vivo analysis and concluded that no accumulation of lipid in the hepatic system was observed. In the case of Spirulina-fortified yogurt, no assured result of reduced lipid accumulation was observed. Extensive studies are thus required before coming to a definite conclusion on the association between Spirulina-fortified yogurt consumption and reduced adipose tissue depositions. The enhanced functionality of several soya yogurts is discussed in the following sections (Table 5.2).

5.2.2 Sesame

As already mentioned in the preceding sections, plant-based milk and related products are highly consumed nowadays. Although soy milk is the most common, it contains antinutrient factors that may cause flatulence in some people [54]. The queen of "oilseeds that is sesame, generally then takes a lead role in making dairy substitutes. This oilseed contains sesamin, sesamolin, and traces of sesamol group of phytosterols, which are known for their antioxidant, antihypertensive, immune regulatory, and anticancer effects [55]. It is also rich in calcium and phosphorous, while the fatty acids composition includes oleic, linoleic and stearic acids. Sesame has a role in neuro-disorder diseases modulation. Although it is an important dairy substitute (DS) for making fermented yogurts, very limited reports are available so far. Sesame yogurts processing also experienced problems in texture which makes them flowy, thereby creating wide acceptance problem in the market. The research conducted by Abdullah Afaneh (2013), involved incorporating and enriching them with gum systems. Sesame milk fortified with 0.2% cheese powder, 0.6%carboxymethyl cellulose, and 0.2% guar gum, resulted in a firm yogurt texture with the highest organoleptic score. Intestinalimmune function depends on the functionality and resistance of the probiotics. Sesame containing yogurt was also reported to exhibit the highest probiotic count, enhanced proteolysis activity and free radical scavenging activity [56]. However, sesame milk with its rich bioactive compounds composition remains to be rarely used in the consumer market because it experiences destabilization [57]. Unacceptable flavor also attributes to the less popularity of sesame milk and milk products. A report of fortification with dragon fruit to irradicate the odor of the sesame is cited. It is observed that the lowest pH is obtained with 10% sesame and 2% dragon fruit. The total solids, sucrose contents increased in the fortified product when compared with the control [58]. To improve the utilization of this plant milk, researchers are currently working on developing new formulations and more studies on its functionalization are needed [66].

5.2.3 Oats

Milk contains high-fat content and cholesterol. The total fat contents of cow milk is 4% to 5%, whereas buffalo milk contains 7-8%. Daily consumption of this milk may increase total blood cholesterol and LDL. Hence, people may suffer from CVD from consuming a huge amount of full-fat curd, yogurt, or cream-based products. Oats contain β-glucan [67] a nutraceutical that regulates in maintaining gut health by delaying the gastric emptying time [68] and increase transit time resulting in reducing cholesterol [69]. It has a huge amount of soluble fiber that binds with bad cholesterol and lessens the chance of plaque depositions on the arterial wall. Recently, diet management has been a huge focus due to COVID-19 crisis where people confined at home with little/seldom exercises are under increased risk of comorbidities. The various types of FFs, which are of benefit to maintain human health during the crisis period, should be given paramount importance. Oats have gained importance not only as breakfast cereals [70] but also as confectionaries and desserts. In this section, their roles as prebiotic, probiotic matrix, as well as symbiotic food substance, are discussed. There are very limited reports, so far, on "yogurt" (oat+yogurt), though it is now a booming sector of research. Since there is burgeoning trend of consumption of oat milk [71], some eye-catching reports are available. Despite benefits, oats fibers have some drawbacks in preparing yogurt. During processing steps of milk extraction [72], due to high starch content, gelatin [73] occurs thereby unstable emulsion results, which is not desirable for yogurt formation. Even anti-nutrients are also needed to be removed properly before processing. A technique on prestabilization of the oat matrix for probiotic growth was presented to overcome the stability issue of oat yogurt emulsion [74]. Here, oat protein concentrate (OPC), a byproduct obtained after β-glycan processing, was used to prepare the yogurt. The impact of fermentation and lactic acid bacteria growth on protein hydrolysis was observed. A thermally induced starch gel formation strengthens the matrix. The fermentation leads to the production of lactic acid and thereby firming the proteogel properties. A slight increase in tyrosine and lysine was observed when compared with the nonfermented product. This study revealed starch-protein interactive

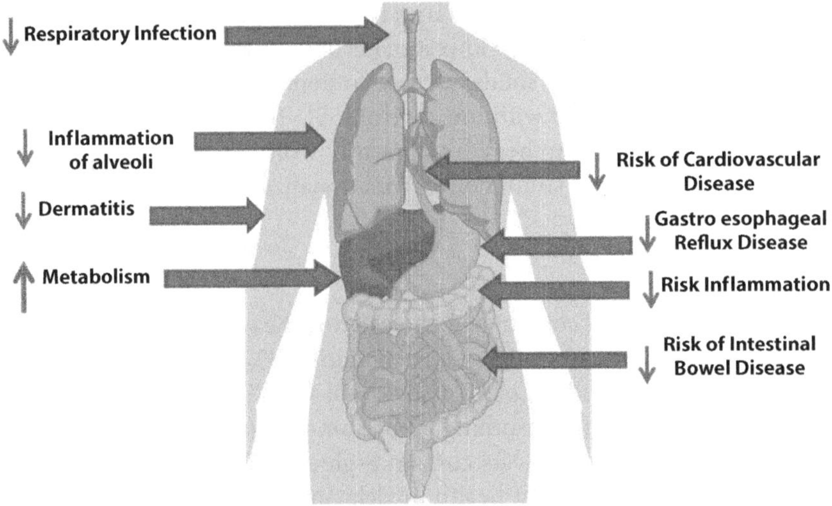

Figure 5.2 Schematic diagram of the effect of probiotics in human immune function.

structure during the fermentation of OPC. Improvement on the methodology, as well as in vitro, in vivo study for the functionality was suggested to be the way forward [75]. Figure 5.2 gives an outline of effects of probiotics in human health.

5.2.4 Coconut

Yogurt, the cultured fermented milk, is useful for people who are lactose intolerant [76]. *Lactobacillus* species (LABsp) and *Streptococcus thermophiles* are the probiotics being mostly used. Some studies suggest that non-dairy matrixes like combinations of coconut milk, mungbean milk, pea curd [77] are used for the lactic acid fermentation. These combinations still restricted consumption by the strict vegans. To overcome this, food technologist sorted out by designing new techniques, where only plant milk are used and the product is cost-effective and feasible [78]. Coconut milk or coconut sap is mixed with water homogenized, pasteurized, starter culture added and finally transformed into yogurt [79]. A sensory evaluation score of 3.68/4 resulted in the overall acceptance and even the acidity exhibited viability of the cells in consumable amount. The optimization was done by response surface methodology [80]. Coconut milk yogurt fortified with citrus fruits, pulp and flavor-showed higher nutritive value (energy, 142.49 ± 2.63; carbohydrate, 9.55 ± 0.12,) than the plain yogurt made with cow milk (carbohydrate, 4.46 ± 0.05; protein, −7.26 ± 0.11

and fat, −9.82 ± 0.02) [81]. Recent evaluation of the viability of combined probiotic ABT (*L. acidophilus, B. bifidum*, and *S. thermophiles*) culture is observed in yogurt prepared from 50:50 ratio of cow and coconut milk. It is observed that the addition of coconut milk in yogurt preparation increases these probiotics, and the resistance to the gastric environment is expected to be stable. The yogurt also yielded a high amount of total solids, volatile fatty acids, medium-chain fatty acids (MCFA) especially lauric acids [82]. The product is expected to improve brain function due to MCFA [83]. Thereby, authors suggest in vivo model study for the same. Some reports even suggest modifying symbiotic bio-yogurt with coconut flour (rich indigestible fiber which facilitates the growth of probiotic cultures). The symbiotic yogurt exhibited its efficacy as antibacterial agent against foodborne pathogen Klebsiella and in decreasing order against *Staphyloccocus* sp., *Bacillus cereus*, and *E. coli.* No antifungal activity was observed in the coconut yogurt during storage conditions when compared WITH the control yogurt [60]. Another report of symbiotic yogurt where raisins incorporated in cow milk yogurt (CMY) is reported [84]. Here, the inter-relationship of the probiotics (oligosaccharide from raisins) enhanced the probiotic viability without an altered organoleptic score. This may improve intestinal gut health and have a probable chance of reducing lesion risk. A trend in the functionalization of yogurt is not new. There are reports of many coupled yogurt where, fortifications were done to enrich the nutrition value by adding carotene, zeaxanthin, or other phenolic compounds [85]. Similarly, to enhance phytophenols content, coconut milk and soya milk were combined to form yogurt of densely rich MCFA followed by short-chain fatty acids, polyunsaturated, and monounsaturated fatty acids. Caprylic acid, lauric, myristic and palmitic acids content were highest among the fatty acids. The isoflavone rich yogurt did not compromise the beneficial effect due to its high content of lauric acids. The product contained a magnificent ratio of essential and nonessential amino acids and almost all above 100% in the protein score, except lysine (lack of the amino acid in coconut milk). The count of the probiotics (*Streptococcus thermophiles* and *Lactobacillus delbrueckii* subsp. Bulgaricus) is less but no fungal or mold growth was observed during the 15 days of storage. The results indicate that the quality indicators of shelf life do not exceed the definite parameters [61]. Further *ex vivo* or *in vivo* modeling is extremely important for this study. The plant milk research to prove as probiotic matrix thereby fulfilling therapeutic criterion is at the highest peak nowadays. Recent trend and development in this field promotes reducing fats in yogurt. We already cited the highest amount of medium-chain fatty acids in coconut yogurt. A group of researchers developed a reduced fat coconut milk yogurt and observed its stability and acceptance [86]. The fatty acid profile of the reduced fat coconut yogurt (RFCY) is significantly different from that of the whole fat coconut yogurt (WFCY) from coconut milk. The water holding capacity, yield stress, and consistency coefficient were high in RFCY than WFCY. RFYC showed better viability of *L. bulgaricus* and *S. thermophiles* compared WITH that of WFCY. The main problem faced by fermented final product from plant milk, i.e., yogurt is the acceptance and the resistance of the probiotic microorganisms. A study reported the viability study of LABsp in coconut milk made the streamline for developing therapeutics. The comparison of the viability of isolated *L. casei* subsp. rhamnosus observed in between fermented hemp milk 10.92log (cfu/ml) and fermented coconut milk 13.26log (cfu/ml) during the first-day storage, respectively. At the 15-day storage, fermented coconut milk (eFCM) was 11.26log (cfu/ml), and fermented hemp milk (FHM) was 8.28log (cfu/ml) and consequently, 9.41log (cfu/ml) and 7.41log (cfu/ml) for FCM and FHM respectively. Earlier reports stated that if the viable count is above 8log (cfu/1 ml), then the fermented product proved to be the probiotic matrix and can effectively fulfill the criteria of being used in therapeutic [87]. Therefore, these characteristics cater as an immune-boosting substance for further research.

5.2.5 Nuts

Hyperallergenicity, flatulence, lactose intolerance [88] are generally considered as drawback for soya milk and milk products. Thus, there is arising importance for the usage of nuts as an alternative milk analog [89]. Nuts are good sources of vitamins and minerals and recent reports of energy balance are also cited as reason for consumers preferring nutmeats [88]. But, after extraction of the milk, mostly protein and minerals are lost; thereby fortification is required [90]. Almonds are low-calorie nuts with high sugar content, whereas cashew is rich in saturated fats and cholesterol. Consumers preferring creamier texture opt for cashew milk over almond milk. Almond milk contains fewer amounts of minerals like manganese, magnesium, and copper. They are devoid of B vitamins and zinc. Cashew contains more than 50% of calcium. Both almonds and cashews are densely rich in Vitamin E (tocopherol, tocotrienol) [91]. Brazil nuts, hazelnuts [92], and tigernuts, are also used. The combined milk yogurts [91, 93] are more focused to fill the nutrition gaps. Very limited reports are available however, on these variations. Mostly, these nuts are used as protein concentrates in other probiotic matrix of plant origin. The polyunsaturated fatty acid (PUFA) and monounsaturated fatty acids (MUFA) from these nuts may exhibit effects in reducing the level of blood cholesterol and lower the risk of CVD [94]. Though almond milk has PUFA and MUFA, which results in anti-inflammatory and antiproliferative effects, the carbohydrate content needs to be modified. Thereby, fortification of almond yogurt with inulin was carried out and the probiotic count was observed. *Lactobacillus* probiotic showed a viable count of > 10 cfu/ml and survived in vitro gastric environment even after 3 weeks of storage [95, 96]. In vitro studies suggest lower inflammatory response and enhanced uptake of iron in the intestinal epithelial cells when almond yogurt containing probiotics *L. rhamnosus*, *B. bifidum*, and *B. longum* are used. It also enhanced the energetic metabolism of enterocytes that are beneficial to allergenicity and intolerance [88]. Low fat cashew yogurt is challenging nowadays. Researchers developed cashew yogurt of 2.63 g lipid/100 g yogurt using *B. clausii*. The yogurt also showed a consistent viability of probiotics till the 60th day of storage above 8 cfu/ml, and after 150 days, the viability decreased to approximately 4 cfu/ml [97]. This functional yogurt did not compromise its creaming index even being low-fat. This can be suggested for obesity management as well. Recent trends also continue to focus on nutrient fortification. It is observed that nuts (almonds, cashew, pistachio, walnuts, and hazels) all together are responsible for increasing amino acids, alpha-tocopherol, folic acid, and selenium, which are very important in reducing abdominal obesity, CVD, hypertension and high-fasting glycemia [98]. Furthermore, stability researches are conducting to make novel Pm (probiotic matrix) used as a functional food. Other nuts like pine nuts, tiger nuts are also being used nowadays, which make wide variations of yogurt for vegans [99].

5.2.6 Rice

Among all nondairy products, rice milk from brown rice or rice starch [100] is the most hypoallergenic [101, 102]. It has fewer amounts of fat and fiber and is of very thin consistency. To stabilize it, mostly oil and surfactant are used to set the consistency and to get a proper microemulsion without phase separation. Rice bran is sometimes used as a fortifying agent for other plant-based yogurt (PBY). Rice is a very cheap and common staple food across the world. Brown rice is rich in nutrients like essential amino acids, fatty acids, vitamins, minerals, and other nonnutrient phytochemicals concentrated in the germ layer of the endosperm. Rice bran layer contains phenolic compounds like oryzanol, anthocyanin, and other phenolic compounds which reduce LDL cholesterol and exhibit anti-oxidant, antiproliferative, anti-inflammatory, antidiabetic and for other conditions including hear diseases [103–105]. The fiber of rice bran are beneficial in curing chronic disease like diverticulosis [106], CVD [107], and colon cancer [108]. It was reported [104, 109] that the incorporation of rice anthocyanin (Figure 5.3) rice protein, rice fiber, or rice as a whole facilitates functionality like the stability of emulsion and growth of probiotics [105] and immunity. The nutritional property increases due to fermentation by amylolyticlactic acid bacteria thereby increasing the digestibility of starch in children. It was also observed that incorporation of Jasmine rice in a symbiotic yogurt made from gelatin-cow milk has low retrograding properties [64]. Even after storage of 21 days, the count of probiotic Bifidobacterium reported being >8 cfu/ml, in an in vitro gastric environment. This is an important criterion for the treatment of intestinal disease to enhance immunity. Probiotics have already been known to be associated with innate immunity. Changes in the gut microbiome significantly affect mucosal immune response. Study to assure the positive effect on the intestinal mucosa of mice model by rice bran was reported [110]. Increase in immunoglobulin A at locally and systematically observed in the rice bran-induced feed. *Lactobacillus* sp colonization also increased and antigen-presenting cells were also enhanced. It may be thus concluded from the report that bran-induced microbiological changes may enhance the IgA response in boosting the immune system. Rice bran oil containing gamma-aminobutyric, gamma oryzanol, and tocopherol, to cotrienol has proved to be hypocholesterolemic. A study by Sengupta *et al.* (2016) reported soya bio yogurt fortified with rice bran oil (RBO) and soybean meal oligosaccharides (SMO) good source of sesamoline when administered in Swiss albino mice model, showed a favorable effect in reducing blood cholesterol, triglycerides, and LDL-c cholesterol.

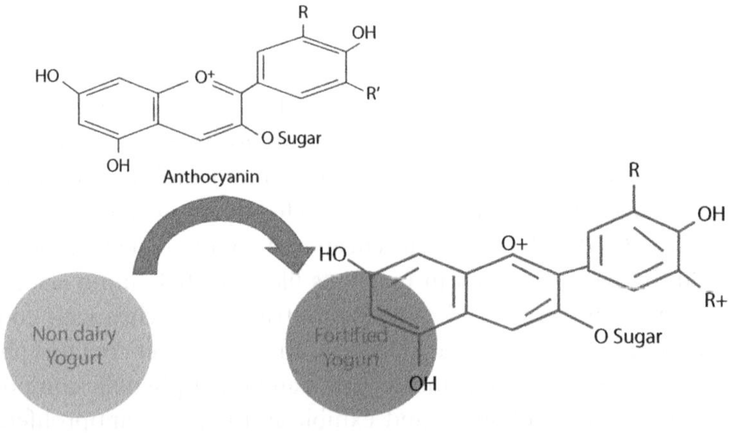

Figure 5.3 Schematic diagram of nondairy yogurt incorporated with anthocyanin from black rice.

The induced diet caused a significant decrease in lowering the atherogenic index. Consumption of bio-fortified soy yogurt significantly improved the HDL-c in serum, superoxide dismutase, and catalase level in liver homogenate and significantly decreased the serum alanine aminotransferase, aspartate aminotransferase, alkaline phosphatase, acid phosphatase, and gamma-glutamyltransferase activities. This bio-yogurt (BYG) maybe specifically useful for CVD patients. Germinated brown rice (GBR) has now been the centre of attraction of research. Researchers are claiming that GBR has enriched GABA and G-oryzanol due to germination of sprouts. In addition to the mouth feel, digestion in the intestine is improved also due to free sugars and oligosaccharides formed from the hydrolysis of starch. During germination, anti-nutrients phytate is also decompound to phosphoric and inositols to provide energy. The initial microbial density of the end product showed above 7 cfu/ml. When the fermentation and storage continued for several days, the viability of probiotics S. thermophiles and L .bulgaricus observed was above 9 cfu/ml (which we have already reported as the prime criterion for using as pharmaceuticals for enhancing intestinal immune factor. The significant increase of gamma-Amino butyric acid (GABA) and oleic acid resulted when BYG formed with GBR may be used to lower the VLDL and LDL in CVD patients [111]. Similar reports to confirm that Lab sp fermentation on GBR increases their bioactive compounds was cited [112]. In this study, only GBR derived flour was used for yogurt preparation using a thermophiles starter culture, without the addition of any form of milk. A significant rise of GABA and oleic acid and a decrease in saturated fatty acids were observed. Continuous exploration of various types of rice is gaining interest as alternative dairy beverage [113].

5.3 Nanoparticles Bio-Yogurt Boosting Immunity

Nanoscience is a science of production, design, and applications of structures, devices, and delivery of systems with a controlled particle size (10-9m) [114]. This technology, which flourished in the last few decades, has Burge-going applications in various field of research and manufactured products. The science deals with manipulation of the particle size at the atomic and molecular level. The particle's large surface area is of prime importance in the enhancement of various functionalities of the matrices. It is observed that milk being excellent nutritive food material, its functionality of delivering bioactive compounds can be monitored better and in a precise manor by using of this technology. The basis of the formation of nanoparticles is top-down and bottom-up approach [115]. Generally, the top-down approach is the physical method consisting of grinding, milling, machining whereas self-assembly and self-organization derived from biology is the basis of food nanobiotechnology [116]. Nanotechnology is appreciated in food packaging as nanosensors(as active and smart packaging) [117, 118], edible coatings [119], and to improve the functionality of various foods [120]. Many researches are done in manifesting minerals particle size less than 100 nm, which can easily cross the intestinal layer for boosting immunity in athletes. Nutraceuticals compounds like bioactive proteins, flavonoids, polyunsaturated fatty acids are more efficacious and target the site of disease effectively when encapsulated in nanoparticles form [121]. This activity not only provides nutrition but also health benefits. The bioactivity of the substance in the brain depends on effective crossing of the blood-brain barrier (BBB). The reduction of bioactive particle size from macro to nano vastly affects their property [117]. The particle size is interdependent with the bioavailability, delivery, and solubility properties. The stability of nutraceuticals during processing, distribution, and storage is improved by this technology. Nanoparticles encapsulated in vesicles or walled particles improves the functional bioactive agents retention time, internalization, transit time through the intestine, controlled release in a specific manner at the target site of action [118]. This technique retards degradations of the delivery systems like associated colloids, nanofibers, nanocapsules, and nanotubes [122]. We herein discuss several research outputs of nanoparticles (NP)-based bio-yogurt in boosting immunity in Table 5.3.

Nanoemulsions (NE) and nanoencapsulation are the most effective techniques for delivering bioactive compounds in a matrix. Nanoemulsions is an emulsion of an intimate liquid or liquid crystal into one another with discrete disperse liquid particles of 20- to 200 nm in dimension. Nano emulsions are much stable than micro emulsions and also protects the bioactive components [127]. There are two types of NE i.e. water in oil and oil in water. These NEs are mostly used in health care and therapeutics but its major drawback is the cost of production. The most promising nanostructures are liposomes, solid lipid nanoparticles, casein micelles, micro emulsions and polymeric NP. The bioactive is much stable when they are encapsulated in these NPs. Liposomes are NP of 50 nm diameter consisting of a bilayer of lipids with an aqueous core. Solid-lipid NP is a lipid that solidifies at room temperature to form crystals with a diameter of 500 to 1000 nm. Protein nanoparticles are also a new trend of forming nanogels with consistency and self-assembly for drug delivery system (DDS). Herein, we summarize recent advances in preparation techniques, as well as DDS of the nanocarrier, which will affect the immune system.

Table 5.3 Dairy analog nano yogurt (DANY) with health benefits.

DANY	Probiotics	Dimension (nm)	Morphology	Pre/Cli	Health benefits	References
Soya	LAB	72–586	Spherical	Cli	Hypocholesterolimic	[123]
Soya, Curcumin	LAB	220–287	Spherical	Pre	Antibacterial, Antifungal	[124]
Soya, EthylCel.	LAB	70	Spherical	Pre	Enhanced Immunity	[125]
Coconut	Bifidobacterium	81.96–83.53	Spheroid	Pre	Antagonistic Action Food borne Pathogens	[62]
GoldNP, Maghemite NP	LAB	13–500	spheres,prisms	Pre	GI function enhanced	[126]

Functionalized yogurt already existed in the last few decades. Fortification of yogurt with various kinds of bioactive macromolecules is not also a new concept. Reports are there to enrich the lacking nutrient in yogurt matrix, like the addition of ascorbic acid [128], maleic acid from apple and fruit waste fiber from pomace to improve the set type and gel matrix of the yogurt [129]. Researchers continue to modify their designs and new yogurt production and one study reported that plant-based oil bodies are used for milk analogs [130]. Nanotechnology improved the bioavailability of bioactive compounds. Recent trend involves in studies of gastrointestinal in vivo and in vitro reports, where positive results of gut health were observed. The NP-induced plant-based milk and their products increase the stability of the bioactive components in blood nutrients. For example, soya protein nanocarrier successfully encapsulated curcumin which is a plant-based anti-bacterial and anti-fungal flavonoid. The NP loaded with the bioactive component was 220 to 287 nm in diameter. The system was processed by desolvation technique which was followed by evaporation and the encapsulation efficacy and release efficacy was 97.2% and 2.7%, respectively [124]. Similar report observed by Park *et al.* 2019, effective nutrient delivery by nanoemulsion of curcumin soya bean oil in a milk model food matrix [131]. Researchers then started trying to work on the release efficacy of the soya protein and other protein nanoparticles for therapeutic delivery. Optimizations of various antioxidants NP were also observed. We already discussed that gamma oryzanol, which is a very essential phytochemicals in rice bran, is beneficial in increasing high-density lipoprotein (HDL) (good cholesterol) but its poor water solubility hinders its efficacy when it is used as nutraceuticals or fortifiers in food products. Ethyl cellulose and encapsulating, material (good as prebiotic) improves the bioavailability of oryzanol. The particle size of 70 nm remains unchanged during storage. These nanocarrier loaded particles may be used in the near future to fortify products with water-insoluble antioxidants [125]. Soya nanoparticles emulsion of 72 to 586 nm particle size showed good antioxidant properties when bio yogurt was prepared. Viscosity, penetration value, and water holding capacity were decreased when there was an increase in reduction size SPNE (soya protein nanoparticles) [123]. This study lacks enough data on the information on how the NP is affecting the intestinal health; or immune system, storage time and nanoemulsion stability, accumulation internalization of the NP, and ultimate metabolic fate need a vivid inspection. Therefore, current studies advocate more explorations not only on SPNE needs to be done but also other variable proteins like pea protein [132] need to be emphasized. While modifying the yogurt with bioavailability, researchers sometimes had to compromise some unique features like odor or color. Even it is reported that functionalized yogurt are sometimes reported unacceptance due to color changes. Fortification with minerals evidently creates this problem. So, a group of researchers tried to develop zinc, insulin, calcium nanoparticles bio yogurt with no odor or altered rheological properties. The in vivo experimental design is underway, but the digestibility results showed the better solubility of the yogurt when compared to the macro-mineral fortification. Researchers are also developing newer methods for developing green nanoparticles from fruits and vegetables. Recently, ball milling of coconut nanoparticles flour (81.96–83.53 nm) was utilized as a probiotic matrix of Bifidobacterium. Coconut rich in peptide and isoflavones are selected as the matrix. Ball milling technique is a highly efficient and safe technique and nutrients retention is the highest after the processing. *B. bifidobacterium* has an antagonistic effect on enterogenic pathogens thereby increasing the innate immunity of human body. The study reported that the viability count changes to 2 log cycles after first-day storage to the 60th day storage. Only 0.5 log cycle changes were observed in *Lactobacillus delbrueckii* sp. *Bulgaricus* and *Streptococcus thermophiles* after 30 days of storage [133]. Study on two general nanomedicine substances gold NP and magnetite NP fortified yogurt to cure anemia, acts as nano theranostic for cancer. In the study by Hwang *et al.* biogenesis of Gold NP and MGNP by bacteria (*Streptococcus acidophilus* and *Lactobacillus* sp) was first carried out. Then, these NP were added to increase functionality. Another setup was done by generating a magnetic field for even distribution of NP where *L. fermentum* was the probiotic. The *L. fermentam* and MGNP already have reports of oral drug delivery and that was why the researchers selected this couple. Uneven distribution was observed when the magnetic field was removed. It is observed that at pH 4, the probiotics remained alive without altering the texture setting impact. In vivo experiments are highly recommended for detecting their theranostic efficacy [126]. Figure 5.4 shows the mechanism of action for gold NP and magnetite NP fortified yogurt.

Vitamin D is an oil-soluble vitamin formed from its precursors in the skin layer upon exposure to sunlight. There is a chance of melanoma development due to avoidance of sun exposure (vitamin D deficiency), which is also a great deal of concern at the moment as many people are confined to their home due to COVID-19 pandemic. So, this kind of emergency may harm the vitamin D metabolism which is important for calcium metabolism and absorption, insulin activity, and the innate immune system of the human body. Vitamin D enriches sources are scarce in number. So, fortification of foods can be the most promising way to intake vitamin D. Liposomes proved to cater the

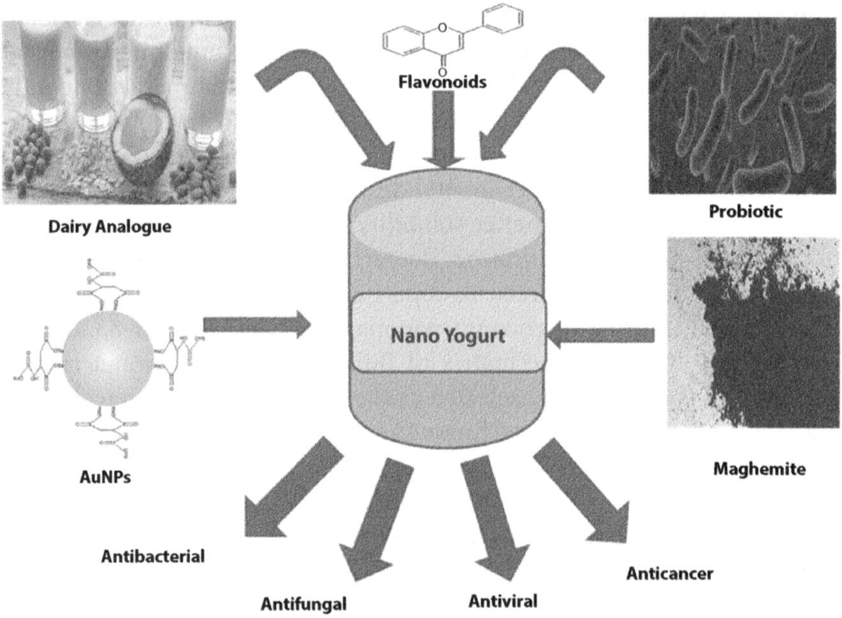

Figure 5.4 Schematic diagram of nanoparticles induced yogurt and their benefits.

delivery in a successful manner in hydrogel matrix-like yogurt [134]. Natural anticancer colorant delivery also improves the color and texture as well as viability of the probiotics [135]. Thus, this modified functional food may be given to patients suffering from osteoarthritis, osteoporosis, and osteocancer patients for enhanced metabolism preventing bone fractures or tissue rupture.

5.4 Future Prospect and Conclusion

In the present situation of the COVID-19 pandemic [136] people have to focus on building their immunity until effective vaccine to combat the coronavirus is available. Previous studies developed an interrelationship of probiotics with respiratory tract infections [137]. Researchers not only should focus on developing vaccines but also can modify foods with those probiotics. In this section, discussion on future application of nanomedicine for fatal diseases like, cancer [138], COVID-19 [139], and HIV [140] are presented to throw some light on overcoming the co-morbidities like CVD and Diabetes. In the preceding sections, we have discussed the role of functional ingredients of modified yogurt on public health. Nanotechnology induced plant-based yogurt are now at the quantum leap of research. Vivid researches inputs are needed and more studies are required to isolate new probiotics from different sources and to assess their health promoting effects. New plant-based milk analogs should be explored for wide application. Inducing nanotechnology and implementing in the betterment of functional foods must also be viewed with eagle eye due to wide variations in outcomes. Specifically, data from in vivo modeling is not reported to large extent. There are certain complications in between in vivo studies that depend on species, internal body system, and gender which are not the case in vitro studies. Thus, there is a great deal of demand for animal modeling to understand the metabolic interactions with probiotic-prebiotics–symbiotic matrices. Epidemiological studies should also be carried out. Mortality rates of infants and mothers in less-developed countries should be controlled and proper utilization of the probiotic foods of low-cost may be used as a weapon to alleviate such problems. Bio yogurt induced with nano flavonoids can further be used to increase the immune system. Recent reports showed that flavonoid nanocapsules (rutin entrapped in casein micelles) intake, as a supplement with probiotic yogurt, is not so much health beneficial as per the nanocapsules embedded bio-yogurt matrix. The nanoparticles-induced BYG has good flavonoid protein interactions which result in withstanding strongly against the harsh acidic gastric environment and increasing the immunity of the GI tract [141]. More variations like pea, hemp flaxseed milk are under research. Combining the use of these phenolic groups with probiotics may be beneficial if considered for vaccination strategy of coronavirus. More awareness about the importance of dairy milk substitutes should be outlined.

References

1. Prentice, A.M., Dairy products in global public health. *Am. J. Clin. Nutr.*, 99, 5, 1212S–1216S, 2014.
2. Zheng, N., Gao, Y., Zhu, W., Meng, D., Allan Walker, W., Short chain fatty acids produced by colonizing intestinal commensal bacterial interaction with expressed breast milk are anti-inflammatory in human immature enterocytes. *PLoS One*, 15, 2, e0229283, 2020.
3. Gad, A.S. and Sayed, A.F., Increasing health benefits of milk fat by Partial Replacement with vegetable oils fortified with carotenoids. *J. Am. Sci.*, 6, 8, 352–360, 2010.
4. Matsuo, H., Yokooji, T., Taogoshi, T., Common food allergens and their IgE-binding epitopes. *Allergol. Int.*, 64, 4, 332–343, 2015.
5. Corgneau, M., Scher, J., Ritie-Pertusa, L., Le, D.T.L., Petit, J., Nikolova, Y., Banon, S., Gaiani, C., Recent advances on lactose intolerance: Tolerance

thresholds and currently available answers. *Crit. Rev. Food Sci. Nutr.*, 57, 15, 3344–3356, 2017.
6. Prado, F.C., Parada, J.L., Pandey, A., Soccol, C.R., Trends in non-dairy probiotic beverages. *Food Res. Int.*, 41, 2, 111–123, 2008.
7. Thorning, T.K., Raben, A., Tholstrup, T., Soedamah-Muthu, S.S., Givens, I., Astrup, A., Milk and dairy products: Good or bad for human health? An assessment of the totality of scientific evidence. *Food Nutr. Res.*, 60, 1, 32527, 2016.
8. Mane, R., Padghan, P., Patil, S., Patil, R., Studies on physico-chemical properties of shrikhand from buffalo milk blended with sweet corn milk. *Asian J. Dairy Food Res.*, 36, 4, 295–297, 2017.
9. Liu, J.-R., Wang, S.-Y., Chen, M.-J., Yueh, P.-Y., Lin, C.-W., The anti-allergenic properties of milk kefir and soymilk kefir and their beneficial effects on the intestinal microflora. *J. Sci. Food Agric.*, 86, 15, 2527–2533, 2006.
10. Pereira, P.C., Milk nutritional composition and its role in human health. *Nutrition*, 30, 6, 619–627, 2014.
11. Fei, X.U., Kang, R.E.N., Yang, Y-z, Guo, J-p, Liu, Y-m, Lu, Y-q, Li, X-b, Immunoassay of chemical contaminants in milk: A review. *J. Integr. Agric.*, 14, 11, 2282–2295, 2015.
12. Olumee-Shabon, Z. and Boehmer, J.L., Detection of casein phosphopeptides in goat milk. *J. Proteome Res.*, 12, 6, 3034–3041, 2013.
13. Balthazar, C.F., Pimentel, T.C., Ferrão, L.L., Almada, C.N., Santillo, A., Albenzio, M., Mollakhalili, N. *et al.*, Sheep milk: Physicochemical characteristics and relevance for functional food development. *Compr. Rev. Food Sci. Food Saf.*, 16, 2, 247–262, 2017.
14. Mohan, G., Gupta, V., Raj, A., Kaur, R., Consumer acceptance of camel milk in emerging economy. *J. Int. Food Agribusiness Mark.*, 32, 1, 54–68, 2020.
15. Park, Y.W. and Haenlein, G.F.W., Therapeutic, hypo-allergenic and bioactive potentials of goat milk, and manifestations of food allergy, in: *Handbook of Milk of Non-Bovine Mammals*, pp. 151–179, 2017.
16. Pietrzak-Fiećko, R. and Kamelska-Sadowska, A.M., The comparison of nutritional value of human milk with other mammals' milk. *Nutrients*, 12, 5, 1404, 2020.
17. Hoonsopon, D. and Puriwat, W., The effect of reference groups on purchase intention: Evidence in distinct types of shoppers and product involvement. *Australas. Mark. J.*, 24, 2, 157–164, 2016.
18. Coutinho, N.M., Silveira, M.R., Rocha, R.S., Moraes, J., Ferreira, M.V.S., Pimentel, T.C., Freitas, M.Q. *et al.*, Cold plasma processing of milk and dairy products. *Trends Food Sci. Technol.*, 74, 56–68, 2018.
19. Chalupa-Krebzdak, S., Long, C.J., Bohrer, B.M., Nutrient density and nutritional value of milk and plant-based milk alternatives. *Int. Dairy J.*, 87, 84–92, 2018.

20. Mozaffarian, D., Dairy foods, obesity, and metabolic health: The role of the food matrix compared with single nutrients. *Adv. Nutr.*, 10, 5, 917S–923S, 2019.
21. Vanga, S.K. and Raghavan, V., How well do plant based alternatives fare nutritionally compared to cow's milk? *J. Food Sci. Technol.*, 55, 1, 10–20, 2018.
22. Takagi, A., Kano, M., Kaga, C., Possibility of breast cancer prevention: Use of soy isoflavones and fermented soy beverage produced using probiotics. *Int. J. Mol. Sci.*, 16, 5, 10907–10920, 2015.
23. Patil, U. and Benjakul, S., Coconut milk and coconut oil: Their manufacture associated with protein functionality. *J. Food Sci.*, 83, 8, 2019–2027, 2018.
24. Ahmadian-Kouchaksaraei, Z., Varidi, M., Varidi, M.J., Pourazarang, H., Influence of processing conditions on the physicochemical and sensory properties of sesame milk: A novel nutritional beverage. *LWT-Food Sci. Technol.*, 57, 1, 299–305, 2014.
25. Yang, M., Li, N., Tong, L., Fan, B., Wang, L., Wang, F., Liu, L., Comparison of physicochemical properties and volatile flavor compounds of pea protein and mung bean protein-based yogurt. *LWT*, 152, 112390, 2021.
26. Kundu, P., Dhankhar, J., Sharma, A., Development of non dairy milk alternative using soymilk and almond milk. *Curr. Res. Nutr. Food Sci. J.*, 6, 1, 203–210, 2018.
27. Schuster, M.J., Wang, X., Hawkins, T., Painter, J.E., Comparison of the nutrient content of cow's milk and nondairy milk alternatives: What's the difference? *Nutr. Today*, 53, 4, 153–159, 2018.
28. Yeom, H.W., Evrendilek, G.A., Jin, Z.T., Zhang, Q.H., Processing of yogurt-based products with pulsed electric fields: Microbial, sensory and physical evaluations. *J. Food Process. Preserv.*, 28, 3, 161–178, 2004.
29. Santillán-Urquiza, E., Ruiz-Espinosa, H., Angulo-Molina, A., Ruiz, J.F.V., Méndez-Rojas, M.A., Applications of nanomaterials in functional fortified dairy products: Benefits and implications for human health. In *Nutrient Delivery* (pp. 293–328). Academic Press, 2017.
30. Gammariello, D., Conte, A., Buonocore, G.G., Del Nobile, M.A., Bio-based nanocomposite coating to preserve quality of Fior di latte cheese. *J. Dairy Sci.*, 94, 11, 5298–5304, 2011.
31. Turkmen, N., Akal, C., Özer, B., Probiotic dairy-based beverages: A review. *J. Funct. Foods*, 53, 62–75, 2019.
32. Shimizu, K., Yamada, T., Ogura, H., Mohri, T., Kiguchi, T., Fujimi, S., Asahara, T. *et al.*, Synbiotics modulate gut microbiota and reduce enteritis and ventilator-associated pneumonia in patients with sepsis: A randomized controlled trial. *Crit. Care*, 22, 1, 1–9, 2018.
33. Davani-Davari, D., Negahdaripour, M., Karimzadeh, I., Seifan, M., Mohkam, M., Masoumi, S.J., Berenjian, A., Ghasemi, Y., Prebiotics: Definition, types, sources, mechanisms, and clinical applications. *Foods*, 8, 3, 92, 2019.

34. Myintzaw, P., Pennone, V., McAuliffe, O., Begley, M., Callanan, M., Correlation of organic acid tolerance and genotypic characteristics of Listeria monocytogenes food and clinical isolates. *Food Microbiol.*, 104, 104004, 2022.
35. Salazar, N., Dewulf, E.M., Neyrinck, A.M., Bindels, L.B., Cani, P.D., Mahillon, J., de Vos, W.M. *et al.*, Inulin-type fructans modulate intestinal Bifidobacterium species populations and decrease fecal short-chain fatty acids in obese women. *Clin. Nutr.*, 34, 3, 501–507, 2015.
36. Ashaolu, T.J., A review on selection of fermentative microorganisms for functional foods and beverages: The production and future perspectives. *Int. J. Food Sci. Technol.*, 54, 8, 2511–2519, 2019.
37. Key, T.J., Appleby, P.N., Bradbury, K.E., Sweeting, M., Wood, A., Johansson, I., Kühn, T. *et al.*, Consumption of meat, fish, dairy products, and eggs and risk of ischemic heart disease: A prospective study of 7198 incident cases among 409 885 participants in the Pan-European EPIC Cohort. *Circulation*, 139, 25, 2835–2845, 2019.
38. Hochwallner, H., Schulmeister, U., Swoboda, I., Spitzauer, S., Valenta, R., Cow's milk allergy: From allergens to new forms of diagnosis, therapy and prevention. *Methods*, 66, 1, 22–33, 2014.
39. Pua, A., Tang, V.C.Y., Goh, R.M.V., Sun, J., Lassabliere, B., Liu, S.Q., Ingredients, processing, and fermentation: Addressing the organoleptic boundaries of plant-based dairy analogues. *Foods*, 11, 6, 875, 2022.
40. Kant, R. and Broadway, A., The benefits of consuming soya milk-A review. *Trends Biosci.*, 8, 5, 1159–1162, 2015.
41. Ben-Harb, S., Panouille, M., Huc-Mathis, D., Moulin, G., Saint-Eve, A., Irlinger, F., Bonnarme, P., Michon, C., Souchon, I., The rheological and microstructural properties of pea, milk, mixed pea/milk gels and gelled emulsions designed by thermal, acid, and enzyme treatments. *Food Hydrocolloids*, 77, 75–84, 2018.
42. Ramdath, D.D., Padhi, E.M.T., Sarfaraz, S., Renwick, S., Duncan, A.M., Beyond the cholesterol-lowering effect of soy protein: A review of the effects of dietary soy and its constituents on risk factors for cardiovascular disease. *Nutrients*, 9, 4, 324, 2017.
43. Landete, J.M., Arqués, J., Medina, M., Gaya, P., de Las Rivas, B., Muñoz, R., Bioactivation of phytoestrogens: Intestinal bacteria and health. *Crit. Rev. Food Sci. Nutr.*, 56, 11, 1826–1843, 2016.
44. Cavallini, D.C.U., Abdalla, D.S.P., Vendramini, R.C., Bedani, R., Bomdespacho, L.Q., Pauly-Silveira, N.D., de Valdez, G.F., Rossi, E.A., Effects of isoflavone-supplemented soy yogurt on lipid parameters and atherosclerosis development in hypercholesterolemic rabbits: A randomized double-blind study. *Lipids Health Dis.*, 8, 1, 1–10, 2009.
45. Shilpa, V., Subrota, H., Deepika, Y., Biofunctionality of probiotic soy yoghurt. *Food Nutr. Sci.*, 2011.
46. Earle, M.D., Innovation in the food industry. *Trends Food Sci. Technol.*, 8, 5, 166–175, 1997.

47. Sengupta, S., Koley, H., Dutta, S., Bhowal, J., Antioxidant and hypocholesterolemic properties of functional soy yogurts fortified with ω-3 and ω-6 polyunsaturated fatty acids in Balb/c mice. *Eur. J. Lipid Sci. Technol.*, 121, 5, 1800397, 2019.
48. Fernández, M.F., Boris, S., Barbes, C., Probiotic properties of human lactobacilli strains to be used in the gastrointestinal tract. *J. Appl. Microbiol.*, 94, 3, 449–455, 2003.
49. Rui, W., Luo, A.-L., Shuo, Z., Wen, H., Bing, D., Yihan, S., Kefei, W. *et al.*, Analysis of Stellar Spectra from LAMOST DR5 with Generative Spectrum Networks. *Publ. Astron. Soc. Pac.*, 131, 996, 024505, 2019.
50. Rui, X., Zhang, Q., Huang, J., Li, W., Chen, X., Jiang, M., Dong, M., Does lactic fermentation influence soy yogurt protein digestibility: A comparative study between soymilk and soy yogurt at different pH. *J. Sci. Food Agric.*, 99, 2, 861–867, 2019.
51. Takimoto, A., Nair, P.K., Alavalapati, J.R.R., Socioeconomic potential of carbon sequestration through agroforestry in the West African Sahel. *Mitigation Adapt. Strategies Global Change*, 13, 7, 745–761, 2008.
52. Sengupta, S. and Bhowal, J., Effects of different methods of drying on antioxidant and microscopic characteristics of spirullina platensis enriched soy yogurt. *J. Microbiol. Biotechnol. Food Sci.*, 1001–1006, 2021.
53. Kitawaki, R., Nishimura, Y., Takagi, N., Iwasaki, M., Tsuzuki, K., Fukuda, M., Effects of lactobacillus fermented soymilk and soy yogurt on hepatic lipid accumulation in rats fed a cholesterol-free diet. *Biosci. Biotechnol. Biochem.*, 73, 7, 1484–1488, 2009.
54. Thummar, A.C. and Ramani, V.M., Phenotypic and genotypic characterization of probiotic bacteria isolated from Gir cow milk for their fortification in soya food. *J. Pure Appl. Microbiol.*, 10, 3, 2249–2259, 2016.
55. Prasad, N., Sanjay, K.R., Prasad, D.S., Vijay, N., Kothari, R., Nutrition & food. *J. Nutr.*, 2, 2, 1000127, 2012.
56. Arab, R., Freidja, M.L., Oomah, B.D., Benali, S., Madani, K., Boulekbache-Makhlouf, L., Quality parameters, probiotic viability and sensory properties of probiotic stirred sesame yogurt. Ann. Univ. Dunarea de Jos of Galati. Fascicle VI. *Food Technol.*, 44, 1, 9–25, 2020.
57. Naziri, E., Koupantsis, T., Mantzouridou, FTh, Paraskevopoulou, A., Tsimidou, M.Z., Kiosseoglou, V., Influence of thermal treatment on the stability of vegetable "milk" obtained by ultrafiltration of aqueous oil body extracts from various sources. *Eur. J. Lipid Sci. Technol.*, 119, 7, 1600362, 2017.
58. Suhartatik, N., Widanti, Y.A.W., Anwar, S., Yoghurt susu wijen dengan pewarna alami ekstrak buah naga merah. *Media Ilm. Teknol. Pangan*, 5, 1, 43–48, 2018.
59. Brückner-Gühmann, M., Vasil'eva, E., Culetu, A., Duta, D., Sozer, N., Drusch, S., Oat protein concentrate as alternative ingredient for non-dairy yoghurt-type product. *J. Sci. Food Agric.*, 99, 13, 5852–5857, 2019.

60. EL-Moslamy, H., Shokry, H., Rezk, H., Abdel-Fattah, R., Bioprocess strategies and characterization of anti-multidrug resistant human pathogens copper/copper oxide nanoparticles from citrus peel waste extracts. *J. Nanomater. Mol. Nanotechnol.*, 13, 21–23, 2017.
61. Yilmaz-Ersan, L. and Topcuoglu, E., Evaluation of instrumental and sensory measurements using multivariate analysis in probiotic yogurt enriched with almond milk. *J. Food Sci. Technol.*, 59, 1, 133–143, 2022.
62. Salama, H.H., Abdelhamid, S.M., Dairouty, R.M.K.E., Coconut bio-yoghurt phytochemical-chemical and antimicrobial-microbial activities. *Pak. J. Biol. Sci.: PJBS*, 22, 11, 527–536, 2019.
63. Bernat, N., Cháfer, M., Chiralt, A., González-Martínez, C., Probiotic fermented almond "milk" as an alternative to cow-milk yoghurt. *Int. J. Food Stud.*, 4, 2, 2015.
64. Kumari, A.G.I.P., Ranadheera, C.S., Prasanna, P.H.P., Senevirathne, N.D., Vidanarachchi, J.K., Development of a rice incorporated synbiotic yogurt with low retrogradation properties. *Int. Food Res. J.*, 22, 5, 2032, 2015.
65. Mehendale, R., Joshi, M., Patravale, V.B., Nanomedicines for treatment of viral diseases. *Crit. Rev. Ther. Drug Carr. Syst.*, 30, 1, 2013.
66. Yadav, D.K., Veerakyathappa, H.C., Patki, P.E., Sharma, G.K., Optimisation of processing parameters for the preparation of vegetable curd from decorticated sesame seeds (Sesamum indicum). *Def. Life Sci. J.*, 5, 1, 38–44, 2020.
67. Faure, A.M., Koppenol, W.H., Nyström, L., Iron (II) binding by cereal beta-glucan. *Carbohydr. Polym.*, 115, 739–743, 2015.
68. Wolever, T.M.S., Tosh, S.M., Spruill, S.E., Jenkins, A.L., Ezatagha, A., Duss, R., Johnson, J., Chu, Y.F., Steinert, R.E., Increasing oat β-glucan viscosity in a breakfast meal slows gastric emptying and reduces glycemic and insulinemic responses but has no effect on appetite, food intake, or plasma ghrelin and PYY responses in healthy humans: A randomized, placebo-controlled, crossover trial. *Am. J. Clin. Nutr.*, 111, 2, 319–328, 2020.
69. Andrade, E.F., Lima, A.R.V., Nunes, I.E., Orlando, D.R., Gondim, P.N., Zangeronimo, M.G., Alves, F.H.F., Pereira, L.J., Exercise and beta-glucan consumption (Saccharomyces cerevisiae) improve the metabolic profile and reduce the atherogenic index in type 2 diabetic rats (HFD/STZ). *Nutrients*, 8, 12, 792, 2016.
70. Lyskjær, L., Overvad, K., Tjønneland, A., Dahm, C.C., Substitutions of oatmeal and breakfast food alternatives and the rate of stroke. *Stroke*, 51, 1, 75–81, 2020.
71. Behera, S.S., Ray, R.C., Zdolec, N., Lactobacillus plantarum with functional properties: An approach to increase safety and shelf-life of fermented foods. *BioMed. Res. Int.*, 2018.
72. Deswal, A., Deora, N.S., Mishra, H.N., Optimization of enzymatic production process of oat milk using response surface methodology. *Food Bioprocess Technol.*, 7, 2, 610–618, 2014.

73. Barbosa, B.T., Rodrigues, J.F., Bastos, S.C., Sensory optimization of nutritionally enriched strawberry yogurt. *Br. Food J.*, 2017.
74. Sethi, S., Tyagi, S.K., Anurag, R.K., Plant-based milk alternatives an emerging segment of functional beverages: A review. *J. Food Sci. Technol.*, 53, 9, 3408–3423, 2016.
75. Tian, L., Scholte, J., Scheurink, A.J.W., van den Berg, M., Bruggeman, G., Bruininx, E., de Vos, P., Schols, H.A., Gruppen, H., Effect of oat and soybean rich in distinct non-starch polysaccharides on fermentation, appetite regulation and fat accumulation in rat. *Int. J. Biol. Macromol.*, 140, 515–521, 2019.
76. Szilagyi, A. and Ishayek, N., Lactose intolerance, dairy avoidance, and treatment options. *Nutrients*, 10, 12, 1994, 2018.
77. Wu, H., Rui, X., Li, W., Chen, X., Jiang, M., Dong, M., Mung bean (Vigna radiata) as probiotic food through fermentation with Lactobacillus plantarum B1-6. *LWT-Food Sci. Technol.*, 63, 1, 445–451, 2015.
78. Tangyu, M., Muller, J., Bolten, C.J., Wittmann, C., Fermentation of plant-based milk alternatives for improved flavour and nutritional value. *Appl. Microbiol. Biotechnol.*, 103, 23, 9263–9275, 2019.
79. Asghar, M.T., Yusof, Y.A., Noriznan, M., Yaacob, M.E., Ghazali, H.M., Chang, L.S., Manaf, Y.N., Effect of processing method on vitamin profile, antioxidant properties and total phenolic content of coconut (Cocos nucifera L.) sugar syrup. *Int. J. Food Sci. Technol.*, 55, 7, 2762–2770, 2020.
80. Yaakob, H., Ahmed, N.R., Daud, S.K., Malek, R.A., Rahman, R.A., Optimization of ingredient and processing levels for the production of coconut yogurt using response surface methodology. *Food Sci. Biotechnol.*, 21, 4, 933–940, 2012.
81. Priya, S.R., Preparation and quality assessment of yoghurt prepared from dairy milk and coconut (cocos nucifera, L) milk. *CORD*, 32, 1, 10–10, 2016.
82. Ismai, M.M., Chemical and microbial characterizations of bio-yoghurt made using ABT culture, cow milk and coconut milk. *EC Microbiol.*, 5, 109–124, 2017.
83. Paul, A.A., Kumar, S., Kumar, V., Sharma, R., Milk analog: Plant based alternatives to conventional milk, production, potential and health concerns. *Crit. Rev. Food Sci. Nutr.*, 60, 18, 3005–3023, 2020.
84. Amirah, A.S., Nor Syazwani, S., Radhiah, S., Anis Shobirin, M.H., Nor-Khaizura, M.A.R., Wan Zunairah, W.I., Shazini, N.R., Influence of raisins puree on the physicochemical properties, resistant starch, probiotic viability and sensory attributes of coconut milk yogurt. *Food Res.*, 4, 1, 77–84, 2020.
85. de Campo, C., Assis, R.Q., da Silva, M.M., Costa, T.M.H., Paese, K., Guterres, S.S., de Oliveira Rios, A., Flôres, S.H., Incorporation of zeaxanthin nanoparticles in yogurt: Influence on physicochemical properties, carotenoid stability and sensory analysis. *Food Chem.*, 301, 125230, 2019.

86. Chetachukwu, A.S., Thongraung, C., Yupanqui, C.T., Development of reduced-fat coconut yoghurt: Physicochemical, rheological, microstructural and sensory properties. *Int. J. Dairy Technol.*, 72, 4, 524–535, 2019.
87. Szparaga, A., Tabor, S., Kocira, S., Czerwińska, E., Kuboń, M., Płóciennik, B., Findura, P., Survivability of probiotic bacteria in model systems of non-fermented and fermented coconut and hemp milks. *Sustainability*, 11, 21, 6093, 2019.
88. Bernat, N., Cháfer, M., Chiralt, A., Laparra, J.M., González-Martínez, C., Almond milk fermented with different potentially probiotic bacteria improves iron uptake by intestinal epithelial (Caco-2) cells. *Int. J. Food Stud.*, 4, 1, 2015.
89. Bridges, M., Moo-ove over, cow's milk: The rise of plant-based dairy alternatives. *Pract. Gastroenterol.*, 21, 20–27, 2018.
90. Williams, S.R.O., Chaves, A.V., Deighton, M.H., Jacobs, J.L., Hannah, M.C., Ribaux, B.E., Morris, G.L., Wales, W.J., Moate, P.J., Influence of feeding supplements of almond hulls and ensiled citrus pulp on the milk production, milk composition, and methane emissions of dairy cows. *J. Dairy Sci.*, 101, 3, 2072–2083, 2018.
91. Rai, S.R., Pachisia, J., Singh, S., A study on the acceptability of plant-based milk and curd among the lactose intolerant people residing in Kolkata. *Int. J. Health Sci. Res.*, 8, 38–43, 2018.
92. Bernat, N., Chafer, M., Rodríguez-García, J., Chiralt, A., González-Martínez, C., Effect of high pressure homogenisation and heat treatment on physical properties and stability of almond and hazelnut milks. *LWT-Food Sci. Technol.*, 62, 1, 488–496, 2015.
93. Muzykiewicz, A., Zielonka-Brzezicka, J., Klimowicz, A., The antioxidant potential of flesh, albedo and flavedo extracts from different varieties of grapefruits. *Acta Sci. Pol. Technol. Aliment.*, 18, 4, 453–462, 2019.
94. Bowen, K.J., Kris-Etherton, P.M., West, S.G., Fleming, J.A., Connelly, P.W., Lamarche, B., Couture, P. *et al.*, Diets enriched with conventional or high-oleic acid canola oils lower atherogenic lipids and lipoproteins compared to a diet with a western fatty acid profile in adults with central adiposity. *J. Nutr.*, 149, 3, 471–478, 2019.
95. Bernat, N., Cháfer, M., Chiralt, A., González-Martínez, C., Development of a non-dairy probiotic fermented product based on almond milk and inulin. *Food Sci. Technol. Int.*, 21, 6, 440–453, 2015.
96. Mazzaglia, A., Legarová, V., Giaquinta, R., Lanza, C.M., Restuccia, C., The influence of almond flour, inulin and whey protein on the sensory and microbiological quality of goat milk yogurt. *LWT*, 124, 109138, 2020.
97. Shi, H., Kraft, J., Guo, M., Physicochemical and microstructural properties and probiotic survivability of symbiotic almond yogurt alternative using polymerized whey protein as a gelation agent. *J. Food Sci.*, 85, 10, 3450–3458, 2020.

98. Bustamante, J., van Stempvoort, S., García-Gallarreta, M., Houghton, J.A., Briers, H.K., Budarin, V.L., Matharu, A.S., Clark, J.H., Microwave assisted hydro-distillation of essential oils from wet citrus peel waste. *J. Cleaner Prod.*, 137, 598–605, 2016.
99. Pandey, S. and Poonia, A., Plant-based milk substitutes: A novel non-dairy source, in: *Innovations in Food Technology*, pp. 63–71, Springer, Singapore, 2020.
100. Kiss Firmino Dourado Costa, K., Resende Oliveira, É., Chaves Ribeiro, A.E., Ribeiro Oliveira, A., Lázara Ribeiro Rodrigues de Barros Vinhal, G., Ferreira de Oliveira, T., ... Soares Junior, M.S. Texture profile of fermented rice extracts with probiotic strains and different contents of waxy maize starch, and sensory acceptance of flavoured selected extract. *Int. J. Food Sci. Technol.*, 55, 2, 490–499, 2020.
101. Dupont, C., Bocquet, A., Tomé, D., Bernard, M., Campeotto, F., Dumond, P., Essex, A. *et al.*, Hydrolyzed rice protein-based formulas, a vegetal alternative in cow's milk allergy. *Nutrients*, 12, 9, 2654, 2020.
102. Zhou, L., Wang, J., Li, D., Li, Y., An electrochemical aptasensor based on gold nanoparticles dotted graphene modified glassy carbon electrode for label-free detection of bisphenol A in milk samples. *Food Chem.*, 162, 34–40, 2014.
103. Henderson, A.J., Ollila, C.A., Kumar, A., Borresen, E.C., Raina, K., Agarwal, R., Ryan, E.P., Chemopreventive properties of dietary rice bran: Current status and future prospects. *Adv. Nutr.*, 3, 5, 643–653, 2012.
104. Nontasan, S., Moongngarm, A., Deeseenthum, S., Application of functional colorant prepared from black rice bran in yogurt. *APCBEE Proc.*, 2, 62–67, 2012.
105. Vickers, N.J., Animal communication: When i'm calling you, will you answer too? *Curr. Biol.*, 27, 14, R713–R715, 2017.
106. Nandi, I. and Ghosh, M., Studies on functional and antioxidant property of dietary fibre extracted from defatted sesame husk, rice bran and flaxseed. *Bioact. Carbohydr. Dietary Fibre*, 5, 2, 129–136, 2015.
107. Nie, Y., Luo, F., Wang, L., Yang, T., Shi, L., Li, X., Shen, J., Xu, W., Guo, T., Lin, Q., Anti-hyperlipidemic effect of rice bran polysaccharide and its potential mechanism in high-fat diet mice. *Food Funct.*, 8, 11, 4028–4041, 2017.
108. Zarei, I., Oppel, R.C., Borresen, E.C., Brown, R.J., Ryan, E.P., Modulation of plasma and urine metabolome in colorectal cancer survivors consuming rice bran. *Integr. Food Nutr. Metab.*, 6, 3, 2019.
109. Hasani, S., Khodadadi, I., Heshmati, A., Viability of Lactobacillus acidophilus in rice bran-enriched stirred yoghurt and the physicochemical and sensory characteristics of product during refrigerated storage. *Int. J. Food Sci. Technol.*, 51, 11, 2485–2492f, 2016.
110. Yang, X., Twitchell, E., Li, G., Wen, K., Weiss, M., Kocher, J., Lei, S., Ramesh, A., Ryan, E.P., Yuan, L., High protective efficacy of rice bran against human rotavirus diarrhea via enhancing probiotic growth, gut barrier function and innate immunity. *Sci. Rep.*, 5, 1, 1–12, 2015.

111. Kim, M.-H., Ahn, S.-I., Lim, C.-M., Jhoo, J.-W., Kim, G.-Y., Effects of germinated brown rice addition on the flavor and functionality of yogurt. *Korean J. Food Sci. Anim. Resour.*, 36, 4, 508, 2016.
112. Cáceres, P.J., Peñas, E., Martínez-Villaluenga, C., García-Mora, P., Frías, J., Development of a multifunctional yogurt-like product from germinated brown rice. *LWT*, 99, 306–312, 2019.
113. Koonyosying, P., Flemming, B., Kusirisin, W., Lerttrakarnnon, P., Utama-ang, N., Fucharoen, S., Srichairatanakool, S., Production, iron analysis and consumer perception of functional Thai Sinlek iron rice (Oryza sativa) drink. *Int. J. Food Sci. Technol.*, 56, 4, 1972–1986, 2021.
114. Martínez-Ballesta, M.C., Gil-Izquierdo, Á, García-Viguera, C., Domínguez-Perles, R., Nanoparticles and controlled delivery for bioactive compounds: Outlining challenges for new "smart-foods" for health. *Foods*, 7, 5, 72, 2018.
115. Lugani, Y., Sooch, B.S., Singh, P., Kumar, S., Nanobiotechnology applications in food sector and future innovations. In *Microbial Biotechnology in Food and Health* (pp. 197–225). Academic Press, 2021.
116. Jafarizadeh-Malmiri, H., Sayyar, Z., Anarjan, N., Berenjian, A., *Nanobiotechnology in food: Concepts, applications and perspectives*, Springer, Berlin, Germany, 2019.
117. Li, X., Tsibouklis, J., Weng, T., Zhang, B., Yin, G., Feng, G., Cui, Y. *et al.*, Nano carriers for drug transport across the blood–brain barrier. *J. Drug Targeting*, 25, 1, 17–28, 2017.
118. Beik, J., Abed, Z., Ghoreishi, F.S., Hosseini-Nami, S., Mehrzadi, S., Shakeri-Zadeh, A., Kamran Kamrava, S., Nanotechnology in hyperthermia cancer therapy: From fundamental principles to advanced applications. *J. Controlled Release*, 235, 205–221, 2016.
119. Prakash, A., Baskaran, R., Vadivel, V., Citral nanoemulsion incorporated edible coating to extend the shelf life of fresh cut pineapples. *LWT*, 118, 108851, 2020.
120. Bano, A., Gupta, A., Sharma, S., Sharma, R., Recent developments in nanocarrier-based nutraceuticals for therapeutic purposes, in: *Biogenic Nano-Particles and Their Use in Agro-ecosystems*, pp. 371–391, Springer, Singapore, 2020.
121. Shishir, M.R.I., Xie, L., Sun, C., Zheng, X., Chen, W., Advances in micro and nano-encapsulation of bioactive compounds using biopolymer and lipid-based transporters. *Trends Food Sci. Technol.*, 78, 34–60, 2018.
122. Eatemadi, A., Daraee, H., Karimkhanloo, H., Kouhi, M., Zarghami, N., Akbarzadeh, A., Abasi, M., Hanifehpour, Y., Joo, S.W., Carbon nanotubes: Properties, synthesis, purification, and medical applications. *Nanoscale Res. Lett.*, 9, 1, 1–13, 2014.
123. Sengupta, S., Bhattacharyya, D.K., Goswami, R., Bhowal, J., Emulsions stabilized by soy protein nanoparticles as potential functional non-dairy yogurts. *J. Sci. Food Agric.*, 99, 13, 5808–5818, 2019.

124. Teng, Z., Luo, Y., Wang, Q., Nanoparticles synthesized from soy protein: Preparation, characterization, and application for nutraceutical encapsulation. *J. Agric. Food Chem.*, 60, 10, 2712–2720, 2012.
125. Ghaderi, S., Ghanbarzadeh, S., Mohammadhassani, Z., Hamishehkar, H., Formulation of gammaoryzanol-loaded nanoparticles for potential application in fortifying food products. *Adv. Pharm. Bull.*, 4, Suppl 2, 549, 2014.
126. Garcés, V., González, A., Sabio, L., Sánchez-Arévalo, C.M., Gálvez, N., Dominguez-Vera, J.M., Magnetic and golden yogurts. Food as a potential nanomedicine carrier. *Materials*, 13, 2, 481, 2020.
127. Singh, H., Nanotechnology applications in functional foods; opportunities and challenges. *Prev. Nutr. Food Sci.*, 21, 1, 1, 2016.
128. Taneva, I. and Panayotov, P., Analysis of vitamin C enriched yoghurt by direct extraction of rosehip fruit in cow's milk during storage. *Ukr. J. Food Sci.*, 7, 1, 61–69, 2019.
129. Wang, X., Kristo, E., LaPointe, G., The effect of apple pomace on the texture, rheology and microstructure of set type yogurt. *Food Hydrocolloids*, 91, 83–91, 2019.
130. McClements, D.J., Newman, E., McClements, I.F., Plant-based milks: A review of the science underpinning their design, fabrication, and performance. *Compr. Rev. Food Sci. Food Saf.*, 18, 6, 2047–2067, 2019.
131. Park, S.J., Hong, S.J., Garcia, C.V., Lee, S.B., Shin, G.H., Kim, J.T., Stability evaluation of turmeric extract nanoemulsion powder after application in milk as a food model. *J. Food Eng.*, 259, 12–20, 2019.
132. Doan, C.D. and Ghosh, S., Formation and stability of pea proteins nanoparticles using ethanol-induced desolvation. *Nanomaterials*, 9, 7, 949, 2019.
133. Garcia, S.L.A., da Silva, G.M., Medeiros, J.M.S., de Queiroga, A.P.R., de Queiroz, B.B., de Farias, D.R.B., Correia, J.O., Florentino, E.R., Buriti, F.C.A., Correction: Influence of co-cultures of Streptococcus thermophilus and probiotic lactobacilli on quality and antioxidant capacity parameters of lactose-free fermented dairy beverages containing Syzygium cumini (L.) Skeels pulp. *RSC Adv.*, 10, 29, 16905–16905, 2020.
134. Jafari, S.M., Vakili, S., Dehnad, D., Production of a functional yogurt powder fortified with nanoliposomal vitamin D through spray drying. *Food Bioprocess Technol.*, 12, 7, 1220–1231, 2019.
135. Almeida, H.H.S., Barros, L., Barreira, J.C.M., Calhelha, R.C., Heleno, S.A., Sayer, C., Miranda, C.G., Leimann, F.V., Barreiro, M.F., Ferreira, I.C.F.R., Bioactive evaluation and application of different formulations of the natural colorant curcumin (E100) in a hydrophilic matrix (yogurt). *Food Chem.*, 261, 224–232, 2018.
136. Wu, Y.-C., Chen, C.-S., Chan, Y.-J., The outbreak of COVID-19: An overview. *J. Chin. Med. Assoc.*, 83, 3, 217, 2020.
137. Sengupta, A.M., Chatterjee, D., Ghosh, R., Role of probiotics in respiratory tract diseases with special reference to COVID-19: A review. *Asian J. Med. Sci.*, 11, 4, 64–70, 2020.

138. Lee, E.S., Shin, J.M., Son, S., Ko, H., Um, W., Song, S.H., Lee, J.A., Park, J.H., Recent advances in polymeric nanomedicines for cancer immunotherapy. *Adv. Healthcare Mater.*, 8, 4, 1801320, 2019.
139. Boretti, A. and Banik, B.K., Intravenous vitamin C for reduction of cytokines storm in acute respiratory distress syndrome. *PharmaNutrition*, 12, 100190, 2020.
140. Curley, P., Liptrott, N.J., Owen, A., Advances in nanomedicine drug delivery applications for HIV therapy. *Future Sci. OA*, 4, 1, FSO230, 2017.
141. Acevedo-Fani, A., Ochoa-Grimaldo, A., Loveday, S.M., Singh, H., Digestive dynamics of yoghurt structure impacting the release and bioaccessibility of the flavonoid rutin. *Food Hydrocolloids*, 111, 106215, 2021.

6

Therapeutic Suitability of Functional Phyto-Oils Against Oral and MDR-Pathogens Through Nanotechnology

Alaka Sahoo[1], Shasank S. Swain[2*], Ajaya Kumar Jena[1] and Maitreyee Panda[1†]

[1]Department of Skin & VD, Institute of Medical Sciences and SUM Hospital, Siksha 'O' Anusandhan Deemed to be University, Bhubaneswar, Odisha, India
[2]Division of Microbiology and NCDs, ICMR-Regional Medical Research Center, Chandrasekharpur, Bhubaneswar, Odisha, India

Abstract

In the crusade against emerged oral-pathogenic and multidrug-resistant (MDR) bacteria, exploring newer active antibacterial agents is continuously investigated with various natural sources. Preferably, emphasis on using the phyto-oils (POs)/essential oils (EOs) mined from various plants and generally used in health-ameliorating-cum-health-boosting regimens since the primitive era. From an ethnopharmacological point of view, POs are the most active antimicrobial, anti-inflammatory, and antioxidant properties with less toxicity in character. However, according to the drug development perspective, POs carry some adverse drug ability characteristics, such as poor solubility and low bioavailability, as a significant obstacle to converting an oral administrative mainstream antibacterial drug. Thus, POs are continuously less use in the therapeutic field, especially pathogenic bacteria. At present, the well-developed concept of nanoformulation/nanocarriers is a worthy concept for improving POs drug ability, biostability, bioactive, drug delivery toward potent antibacterial drug development modules. Thus, the article highlighted such exhaustive research of POs and nanoparticle stories-cum-facts to utilize POs toward the effective and safer antibacterial drugs against oral pathogens.

***Keywords*:** Multidrug-resistant oral pathogenic bacteria, natural antibacterial agent, phyto-oils and antibacterial activity, nanoformulation, bioavailability

**Corresponding author*: swain.shasanksekhar86@gmail.com
†Corresponding author: pandamaitreyee@gmail.com

Tanima Bhattacharya and Shakeel Ahmed (eds.) Nanotechnology in Functional Foods, (137–160)

6.1 Introduction

Currently, the emergence of multidrug-resistant (MDR) strains has created an obstacle in the primary health care management of several infectious diseases caused by number of Gram-positive (GP) and Gram-negative (GN) pathogenic bacteria [1, 2]. Alternatively, obsolete drugs and antibiotics have been structurally improved through the principle of medicinal chemistry to enhance the antibacterial potency against emerging MDR-bacteria [3, 4], which led to the generation of higher versions of semisynthetic antibiotics from parent structures [5–8]. Nevertheless, microorganisms even resistant to newer antibiotics as > 20,000 possible resistance genes from 400 types of bacterial communities have been reported [1, 9]. Thus, antimicrobial resistance (AMR) has gradually affected national healthcare policy and forecasting toward controlling MDR pathogens [1, 10]. Subsequently, medical specialists/physicians face many problems during the prescription of suitable antibiotics/antibacterial for any types of infectious diseases. Alternatively, potential natural remedies are essential to the human community, like the pre-antibiotic era of medical care, to avoid decreasing the use of more antibiotics and antimicrobial resistance [11–13].

Precisely, oral health illness is one of the highest burdens in both developing and developed nations. Statistically, 20% to 50% of the population affect dental, periodontal and other oral disorders globally. Approximately 700 microorganisms were isolated from the oral cavity, from which 50% of organisms are unculturable [14, 15]. Thus, oral health disorder now becomes chronic infective diseases caused by the rare and uncultivable anaerobic bacteria existing in the mouth cavity. Additionally, most dental surgeons frequently prescribed higher generations of antibiotics during instant treatment and healing; but these unnecessary prescriptions of antibiotics promote antibiotic resistance episodes later [16]. On the other hand, the floating oral bacterium also arranges a complex ecological community by attaching different cell surfaces in the mouth cavity known as biofilm, another multifaceted form of bacterial community to avoid applied antibiotic potency. Thus, the most prevalent oral pathogens, *Streptococcus mutans*, *S. salivarius*, *S. sanguinis* and opportunistic respiratory pathogens like *Pseudomonas aeruginosa*, *Klebsiella pneumoniae*, *Staphylococcus aureus*, *S. pneumoniae*, *Enterobacter cloacae*, and so on, along with biofilm-producing pathogens accelerate several local and systematic diseases including, metastatic inflammation/injury, cardiovascular disorder, neurodisorder, and so on [17–21].

As a possible solution, the vibrant classes of natural secondary metabolites (alkaloid, terpenoid, phenolic, flavonoids, steroids, etc.) from several plant sources are suitable for fighting against oral-cum-MDR pathogens with a nonhomologous pharmacological relation with newly emerged bacterial taxa. Indeed, emphasis on isolated phyto-oils (POs) from several plants as health-ameliorating and health-boosting agents has been used before nineteen centuries [22, 23, Figure 6.1]. With the continuous

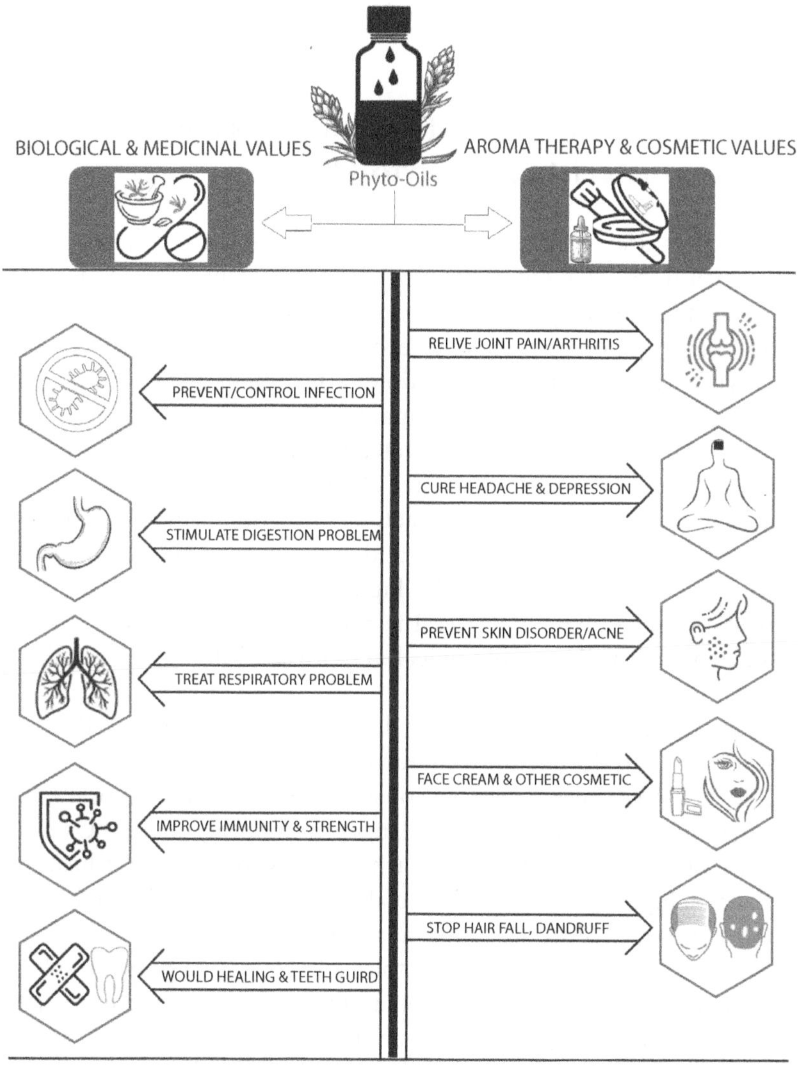

Figure 6.1 Graphical representation of biological-cum-medicinal and aroma therapy-cum-cosmetic values of POs reported from several aromatic and medicinal plants.

integration of modern chemistry-cum-instrumentation, several newer POs were isolated and used abundantly in pharmaceutical, food, and cosmetic industries with high economic value in the global market [24, 25]. From ethnopharmacological points of view, POs are the most active antimicrobial, antiinflammatory, and antioxidant properties with less-toxicity properties [26, 27]. Therefore, the discussion of drug chemistry and drug suitability and adopted nanoformulation toward enhancing POs drugability is the primary aim of this chapter.

6.2 Chemical Classification and Physicochemical Characteristics of Phyto-Oils

Usually, POs are volatile, natural, and aromatic in character, isolated from several aerial plant parts, leaves, flowers, and so on. Based on chemical structure and functional group, POs are classified into different classes, namely, polyphenols and terpenoids (carvacrol, chavicol, eugenol, thymol, etc.), alcohols (ascaridole, α-terpineol, β-santalol, borneol, etc.), aldehydes (benzaldehyde, cinnamaldehyde, citral, citronellal, etc.), esters (bornyl acetate, eugenol acetate, geraniol acetate, etc.), hydrocarbons (cymene, limonene, myrcene, phellandrene, etc.), ketones (camphor, fenchone, menthone, etc.), lactone and coumarin (alantrolactone, bergaptene, costuslactone, etc.), and oxides (ascaridole, bisabolone oxide, caryophyllene, etc.) [28–30, Figure 6.2]. Among the cited classes, terpenoids and polyphenols are the most abundant class of chemical constituents with diverse biological activities, while oxides are the less predominant class of POs.

However, according to the drug development perspective, POs bring some adverse physicochemical characteristics, such as poor solubility and low bioavailability as a significant obstacle to converting an oral administrative mainstream antibacterial drug. On the other hand, POs carbon chains with terpenoid and lactone rings are exposed to light/heat; thus, POs are at higher risk of commercial failure. Commonly, high volatility PO components are low bioavailability due to low aqueous solubility and direct use of such POs is intolerable [31]. As a possible solution, the well-developed nanoformulation and nanocarriers is a commendable concept for improving POs drug ability, drug availability, biostability, side specific drug delivery toward potent antibacterial drug development modules.

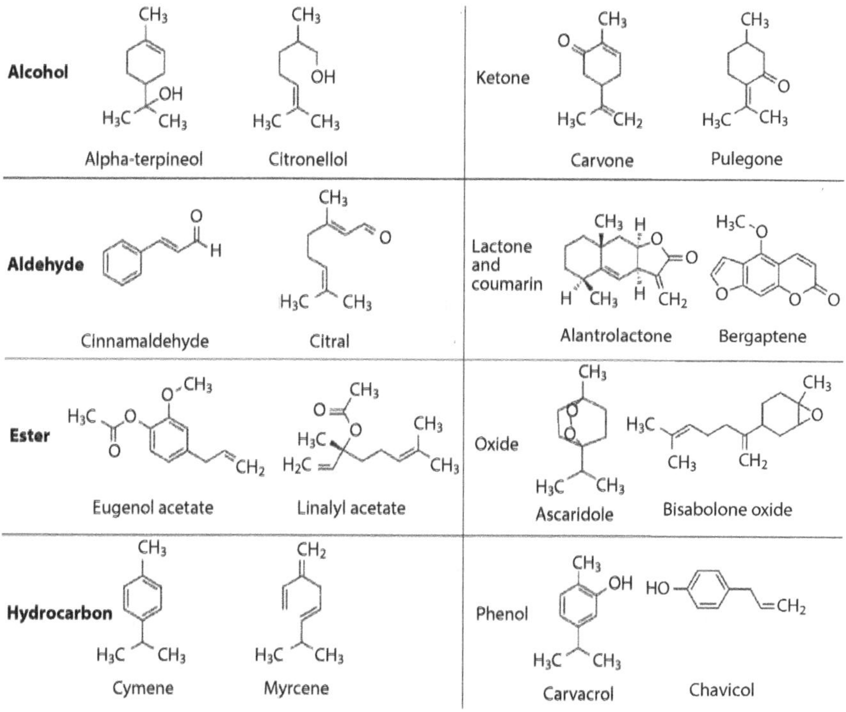

Figure 6.2 Chemical classification and composition of POs having potential antibacterial activities.

6.3 Therapeutic Action of Phyto-Oils Against Pathogenic Bacteria

A wide range of ethnomedicinal records and ethnopharmacological investigations show that POs are effective against various pathogens [24, 25, 28]. For example, phenols/monoterpenes class of POs can easily alter the bacterial cell wall components by interaction with lipophilic properties of bacterial cellular proteins, the glycoproteins, which have solid binding affinities to POs and a natural permeability ability to enter bacterial cell walls [32]. Furthermore, specific PO constituents have been used as antiseptics and dental root canal sealers [33, 34]. The tea tree oil from *Melaleuca alternifolia* leads to the loss of chemiosmosis control in GP and GN bacteria toward cell death [23, 35, 36, Table 6.1]. Additionally, fruit oils of *Eucalyptus globulus* had been active against the pathogenic strains of *P. aeruginosa*, *S. aureus*, *Bacillus subtilis*, and *E. coli*, with MIC values within 3 to 4 mg/mL and 3.6

Table 6.1 Antibacterial activity with MIC value of POs against several pathogenic bacteria.

Oil constituents	Pathogenic bacteria	MIC values	References
Anise	*B. cereus, E. coli, S. aureus*	≥ 1 % v/v	[37]
α-terpineol	*E. coli*	>2 mg/mL	[38]
Bergamot	*Streptococcus agalactiae*	0.0012–0.050 µL/mL	[39]
Carvacrol	*E. coli, S. aureus*	0.35–2.80 mg/mL	[40]
Cinnamon oil	*B. subtilis, E. coli, P. vulgaris; K. pneumoniae, S. aureus;*	>1.6 mg/mL 3.2 mg/mL	[41]
Citronello	*E. coli, S. aureus*	0.35–1.40 mg/mL	[40, 42]
Citrus limon	*Arcobacter butzleri*	> 4% v/v	[43]
Clove oil	*B. subtilis, P. vulgaris; E. coli, P. aeruginosa; K. pneumoniae, S. aureus*	>3.2 mg/mL, >1.6 mg/mL, >6.4 mg/mL	[41]
Eucalyptol	*E. coli, S. aureus*	2.80–5.60 mg/mL	[40, 42]
Eugenol	*Microsporum gypseum*	0.01–0.03 %	[44]
Geraniol	*E coli, S. aureus*	0.08–1.40 mg/mL	[40]
Geranium	*B. subtilis, E. coli; K. pneumoniae, P. vulgaris, P. aeruginosa, S. aureus*	>6.4 mg/mL, 12.8 to >12.8 mg/mL,	[41]
Lemon oil	*B. subtilis, K. pneumoniae P. aeruginosa, S. aureus; E. coli*	>12.8 mg/mL >6.4 mg/mL	[41]

(*Continued*)

Table 6.1 Antibacterial activity with MIC value of POs against several pathogenic bacteria. (*Continued*)

Oil constituents	Pathogenic bacteria	MIC values	References
Lime oil	*B. subtilis, E. coli, K. pneumonia; P. vulgaris; S. aureus*	>6.4 mg/mL >3.2 mg/mL 12.8 mg/mL	[41]
Limonene	*S. epidermidis; Arcobacter butzleri*	0.6–5 mg/mL 0.06, 0.125, 0.25% v/v	[45, 43]
Linalool	*S. aureus*	0.2–2.5 mg/mL	[45]
Marjoram	*B. cereus, S. aureus; Salmonella typhimuriu*	0.125 % v/v 0.062 % v/v	[37]
Orange oil	*B. subtilis, E. coli, K. pneumoniae, P. aeruginosa, S. aureus; P. vulgaris*	≥12.8 mg/mL >6.4 mg/mL	[41]
Pulegone	*S. aureus*	3.75 to >15 mg/mL	[45]
Rosemary oil	*B. subtilis, E. coli, K. pneumoniae, P. aeruginosa; P. vulgaris, S. aureus*	>6.4 mg/mL >12.8 mg/mL	[41]
Thyme	*S. agalactiae*	0.0012–0.050 μL/mL	[39]
Thymol	*E. coli, S. aureus*	0.7–1.40 mg/mL	[40, 42]
Triacetin	*E. coli, S. aureus*	22.40 mg/mL	[40, 42]
1,8-cineole	*B. subtilis, E. coli*	0.9–7.2, 1.8–7.2 mg/mL	[45]

and 9.0 mg/mL, respectively [46]. Individually, the phenolic monoterpene carvacrol (an isomeric form of thymol) was reported effectively against *S. aureus* and *E. coli* at 0.7 to 1.40 mg/mL *in vitro* [40, 47]. Furthermore, the MIC value of oregano and lavender had 1600 to 1800 and 2000 ppm against *E. coli*, whereas the MIC range had 800 to 900 and 1000 to 1200 ppm, against *S. aureus*, respectively [48].

6.4 POs and Antibiotics Combination Against Pathogenic Bacteria

Several promising antibacterial POs prompted scientists to combine a natural agent with an antibiotic to combat pathogenic bacteria [49, 50]. From previous records, carvacrol, cinnamaldehyde, cinnamic acid, eugenol and thymol combinedly with erythromycin, ciprofloxacin, amphotericin B exhibited more potential against pathogens [51, 52]. Indeed, combined drug formulation also targets multiple bacterial pathways at a time toward control of MDR bacteria [50, 52, 53]. PO with an antibiotic and even the combination of two individual POs were also comparatively effective against several oral pathogens and opportunistic respiratory bacteria. For example, the synergistic effect of basil and bergamot against *S. aureus*, oregano and bergamot against *B. subtilis* and oregano and perilla against *S. cerevisiae* were reported effective against pathogenic bacteria [54, Table 6.2].

6.5 Nanoformulation for Enhancement of Drug Ability/Bioavailability of POs

Today, nanoformulation with the desired chemical constituents or bioguided plant extract is an ultimate technology to improve drug ability, as well as active drug delivery-sustained release to the target location in application basis [68, 69]. From widely recorded research reports on optimization formulation of drugs, the particle sizes and morphology are the critical factors active release of bioactive agents in respect to "therapeutic rate and dose regimen" [68, 69]. Furthermore, nanocarriers protect thermal and photo degrading behaviors of POs, which eventually causes increased stability, function and flavor. These methods also protect POs from decay from oxidation or evaporation, enabling the antimicrobial activity through several diffusion properties [68, 69]. For example, thymol and other POs nanoemulsions exhibit potential antimicrobial activities [68–70]. Therefore, nanoparticle synthesis or nanoencapsulation are some promising strategies for overcoming the drug-likeness limitations associated with POs as antimicrobials [69, 70, Figure 6.3]. Comparatively, the biological action nanoparticle-PO/nano encapsulated-PO is higher than free PO in both consistent biological activity and target site drug delivery. Furthermore, nanodrug delivery systems also divided into polymerase based or lipid-based for specific feature, purpose and properties of POs using several mediating materials, such as

Table 6.2 Antibacterial activities of synergistic formulation of some PO-containing crude-phytoextract with antibiotics.

Oil-content-extracts/POs	Used antibiotics	Bacterial strains	References
Aniba rosaeodora	Gentamicin	*A. baumannii, B. cereus, B. subtilis, E. coli, S. aureus*	[40]
Cinnamomum zeylanicum	Amikacin	*Acinetobacter* sp.,	[55]
Coriandrum sativum	Chloramphenicol, ciprofloxacin, gentamicin	*A. baumannii*	[56]
Myrtus communis	Ciprofloxacine, polymixin B	*A. baumannii*	[57]
Pelargonium graveolens	Gentamicin	*A. baumannii, B. cereus, B. subtilis, E. coli, S. aureus*	[40]
Pelargonium graveolens	Norfloxacin	*B. cereus, S. aureus,*	[40]
Peucedanum officinale	Chloramphenicol, streptomycin, tetracycline	Gram-negative bacteria	[58]
Tetraclinis articlata	Amoxicilline	Pathogenic bacteria	[59]
Thymus broussonetii, T. maroccanus	Cefixime, ciprofloxacin gentamicin, pristinamycin	Pathogenic bacteria	[60]
Zataria multiflora	Vancomycin	*S. aureus*	[61]
Bergamottin	Norfloxacin	*S. aureus* (MRSA)	[62]
Carvacrol	Ampicillin	*E. coli, K. oxytoca, S. typhimurium, S. aureus*	[63]

(*Continued*)

Table 6.2 Antibacterial activities of synergistic formulation of some PO-containing crude-phytoextract with antibiotics. (*Continued*)

Oil-content-extracts/POs	Used antibiotics	Bacterial strains	References
Carvacrol	Bacitracin	*E. coli, S. aureus, S. typhimurium*	[63]
Carvacrol	Erythromycin	*S. typhimurium, S. pyogenes*	[63]
Carvacrol	Nalidixic acid	*S. derby, S. enteritidis, S. minnesota, S. typhimurium*	[64]
Carvacrol	Nitrofurantoin	*K. oxytoca*	[65]
Cinnamon bark	Piperacillin	*E. coli*	[66]
Eucalyptus	Chlorhexidine-digluconate	*S. epidermidis*	[67]
Eugenol	Vancomycin, β-lactams	*E. coli, E. aerogenes, P. vulgaris, P. aeruginosa*	[32]
Lavender	Piperacillin	*E. coli*	[66]
Oregano	Fluoroquinolones	*E. coli*	[46]
Peppermint	Meropenem	*E. coli*	[66]
Peppermint	Piperacillin	*E. coli*	[66]

chitosan, zein-sodium caseinate, alginate, phosphatidylcholine, polylactic-co-glycolic acid, and phosphatidylglycerol [71–73].

Niosomes represent an emerging class of new vascular nanocarrier drug delivery systems, getting great helpfulness as potential transdermal drug delivery systems. The bilayer structures assembled by nonionic surfactants in niosomes vesicular order can enhance a POs bioavailability to a specific drug target in time. Moreover, the amphiphilic nature promotes its efficiency in encapsulating lipophilic or hydrophilic POs with several medications sustained drug releases and systemic absorption [74]. The composition of nonionic surfactants is health-friendly, comparatively less toxic and an alternative technique of liposomes. Now it is widely used in

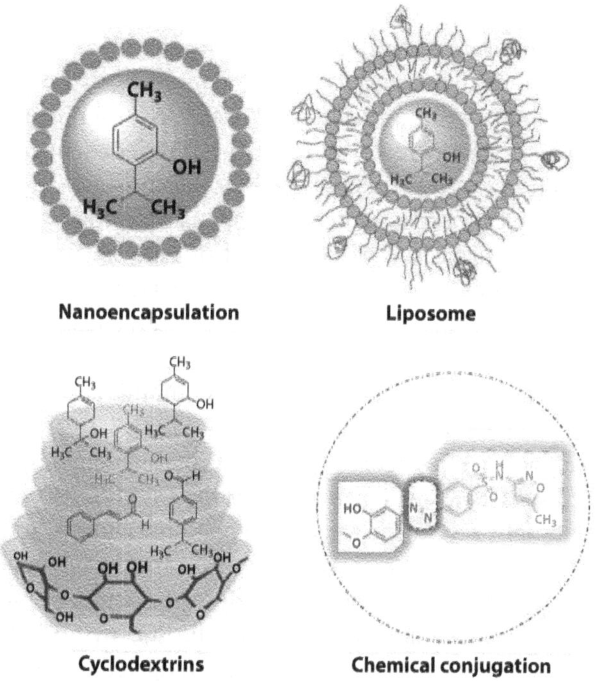

Figure 6.3 Schematic presentation of advanced techniques, such as nanoencapsulation, liposome, cyclodextrins, and chemical conjugation, for the use of active POs in antibacterial drug development modules.

conventional therapies against oral, ocular, transdermal, vaginal, and inhalation routes. For example, several active POs like eucalyptus and lemon oils have increased the transdermic vesicle delivery system [74, 75].

Similarly, "liposome" is another technique also used to improve the inherent volatility of POs through spherical lipid bilayer vesicles by intrinsic aqueous central with a hydrophobic outside coating [76, 77]. In this technique, the primary materials are lecithin, derived from egg yolk, and soybeans with cholesterol, which are safest constituents for our body. Notably, the colloid-carrier liposome has several benefits: solubilization of insoluble components, biocompatibility, bioavailability, and sustained-release [76, 78, 79]. The volatility and chemical instability of POs can reduce without changing the essential ingredients in the liposome encapsulation. Thus, liposomes directly improve the antimicrobial potency of POs with a passive absorption mechanism and concomitantly reduce the transport resistance. For example, by the liposome method, cinnamon oil and linolenic acids were reported more effectively against *S. aureus* biofilm [79, 80] and antibiotic-resistant *Helicobacter pylori* [81].

Recently, cyclodextrins and chemical conjugation are two different types of drug-ability refining techniques used as a significant add-on in drug development platform to increase the physicochemical properties, the low solubility of a PO [6, 82–84]. Cyclodextrins are unbending lipophilic hollows with decent aqueous suspension properties containing hydrophilic outer surfaces. It is composed of natural macrocyclic oligosaccharides with toroid-shaped structures for perfect molecular encapsulation. Cyclodextrins (α, β, and γ types) are a safeguard for most active POs from oxidation, light reactions, evaporation, thermal decomposition, reduction of desired tastes, microbiological contamination, ocular disturbances, drug-additive interactions, and so on. [83]. Using this technique, gallic acid, caffeic acid, pyrogallol, thymol, cinnamaldehyde-like POs were demonstrated activity against *S. aureus* and the *Candida tropicalis* [85, 86].

Correspondingly, several chemical conjugations, namely, antibody-drug, antibody-antibody, phytochemical-drug, and drug-drug concepts, were demonstrated with reliable outputs from the therapeutic point of view [6, 84]. Recently, chemical conjugation of a phenolic PO, thymol, eugenol, vanillin, and so on, with sulfa drugs revealed that conjugated products were more active against MRSA and *Mycobacterium leprosy* than individual [6, 8, Table 6.3; Figure 6.3]. The conjugation enhances the bioactivity and reduces the toxicity toward the control of bacterial pathogens. Nevertheless, chemical conjugation of an active PO with an existing drug is an emerging technique in recent drug development, but more studies should require standardization. Thus, the conjugation of POs with mainstream medicine could be a newly emerging technique in the development of health care products [6, 8].

6.6 Future Prospective of POs with Advanced Formulation for Antibacterial Drug Development

Still, approximately 80% of Indian aborigine tribes, hillside communities, and lower-class villagers use traditional plant extracts to treat several preliminary health disorders, communicable and noncommunicable diseases. Simultaneously, several medicinal and even common plants secondary metabolites exhibited potential controlling activity against several pathogenic bacteria and some are under investigation for appropriate utilization along with exploring other medicinal values [32, 42, 47, 98]. Today, the use of natural products/herbal therapy is approximately $62 billion economic value and as per the present calculation, it may grow up to $5 trillion in

Table 6.3 Currently used nanotechnology and conjugation techniques for enhancement of the bioactive drug delivery of POs.

Techniques	Formulation with POs	Advantages target bacteria	References
Nano-delivery technology for enhancement of bioavailability of POs			
Nanocapsulation, Nanoemulsion	Drug or chemical is dissolved, entrapped, encapsulated or attached a nanoparticle matrix to form nanocarrier	Target site and time specific with protection from enzymatic degradation, altering pharmaco-kinetics and reducing toxicity	[87–89]
Example	Thymol, menthol, clove oil, carvacrol, mustard oil, cinnamaldehyde, eugenol, oregano	*E. coli*, *B. cereus*, *Salmonella* sp., *P. aeruginosa*, *S. aureus*, *L. monocytogenes*, *S. typhi*, *S. epidermidis*	[90–92]
Other delivery technology for enhancement of bioavailability of POs			
Liposomes	Liposomes are nano size artificial vesicles produced from natural phospholipids and cholesterol, serve as immune-logical adjuvants and drug carriers	Its ability to encapsulate drugs with ideal lipophilicity in aqueous core of phosphor lipid bilayer and transport of drug without rapid degradation with minimum side effects	[93–95]
Example	*Thymus* sp. oils, carvacrol, menthol p-cymene, γ-terpinene, thymol, etc.,	*E. coli*, *S. aureus*, *L. monocytogenes* and pathogenic GN and GP bacterium	[92, 95]

(*Continued*)

Table 6.3 Currently used nanotechnology and conjugation techniques for enhancement of the bioactive drug delivery of POs. (*Continued*)

Techniques	**Formulation with POs**	**Advantages target bacteria**	**References**
Cyclodextrins	Pseudo amphiphilic molecules and several members, used industrially in pharmaceutical and allied applications.	With lipophilic inner cavities and hydrophilic outer surfaces, capable of interacting with any guest molecule to form noncovalent inclusion complexes.	[82, 96]
Example	Thymol, cinnamaldehyde, chamomile oil, bisabolol, garlic oil, linalool, etc.,	*E. coli* and *S. aureus.*	[83, 85, 96]
Chemical conjugation/ derivatization (not required)	Development of 'drug-phytochemical' or Phyto-chemical-phytochemical' conjugates through a suitable chemical reaction in principle of medicinal chemistry	Utilization of terpenes, phenolic and oil constituents used a particular therapeutic benefit with a mainstream drug for enhance bio-availability and permeability than individual POs	[6, 8]
Example	Thymol, eugenol, etc.; Using carbon dots with gum arabic.	Methicillin resistance *S. aureus*; Enhance drug delivery of ciprofloxacin hydrochloride.	[6, 8, 97]

the global market value in 2050 [54, 99, 100]. From which POs are such particular chemical group of constituents, nearly ~20 to 60 varieties of chemicals were reported by several hydro-distillation, solvent-solvent extraction along with advanced spectral techniques such as nuclear magnetic resonance (NMR), Fourier-transform infrared spectroscopy (FTIR), gas chromatography-mass spectrometry (GC-MS), high-Performance Liquid Chromatography (HPLC), etc., from several bio-assay guided plant fractions. The largely isolated common terpenoids class chemical PO-constituents are monoterpenes (C10) or sesquiterpenes (C15), or diterpenes (C20). A specific crude oil extract contains a variety of chemical constituents with different molecular weight as clove oil contains 85 % eugenol (164.2 g/mol) with ~10% to 12% eugenol acetate (206.24 g/mol), mint oil contains 59 % menthol (156.27 g/mol) and 19% menthone (154.25 g/mol), and those are separate in later stage from crude oil. The synergism with any other pharmaceutical agent, individual PO or crude PO, acts as a complementary biological activity enhancer of antibacterial, antifungal, antioxidant, anticancer properties has been recorded [22, 23, 101].

At present, >1000 thousands of individual PO-constituents were isolated from several medicinal and aromatic plants and recorded as potential antibacterial, wound healing, antibiofilm activities [39, 70]. From a long history, POs are the most potent remedy for aromatherapy, muscle pain, topical/ external application, and on the other hand, several POs have been used in agriculture for crop-food protection, fertilizer from insecticides, germ, weed killer in the form of bio-fertilizer, natural herbicide, pesticides, etc. [102]. Consequently, a watchful determination of inherent host cytotoxicity and genotoxicity profiles shows that most plant extracts/ POs are negligible/nontoxic side effects against mammalian cells and could be recommended for oral administration. However, several POs exhibited mutagenic and genotoxicity properties against the host in a higher dose [70, 103]. Thus, assessment of biological activity (antibacterial, antioxidant, antifungal, anticancer, antiviral, etc.) of isolated POs in suitable *in vitro* or *in vivo* models with proper toxicity profiles should be required toward utilization POs. Since out of 3,000 isolated POs, only 10% of POs were approved by the FDA for use in pharmacy, nutrition and cosmetic sectors based on the toxicity profile, from which carvacrol, one of the suitable POs, has been placed in 'generally recognized as safe (GRAS)' category food-grade chemical for use in pharmaceutic and cosmetic industries [47].

From the drug perspective, low water-solubility POs reduces the biological potency and limited specific antibacterial therapy. On the other hand, lipophilic classes of thymol contain physically and chemically unstable oxygen atoms, directly associated with the reduction of antibacterial potency. Indeed, these problems might be overcome after the preparation of thymol nanoemulsions or combined with a nanocarrier, as demonstrated previously [104]. Additionally, 'liposome and the most recently developed chemical conjugation' are the available other options for utilizing POs in therapeutic applications [6, 77]. Overall, highlighted advanced nano-techniques in terms of nanoemulsion, nanoencapsulation, are comparatively more effective techniques to enhance the biological activity, bioavailability, stability, effective drug delivery for most POs and directly increases the therapeutic use [31, 72, 102, Figure 6.4]. Thus,

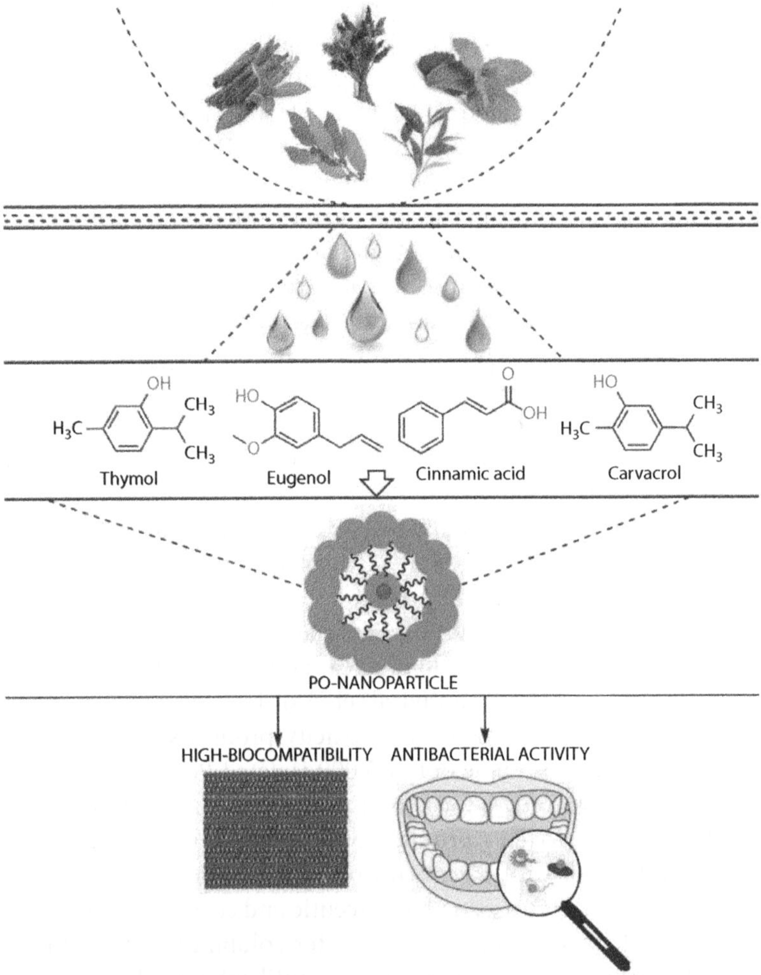

Figure 6.4 Schematic presentation of isolation, separation and nanoparticle formation of POs for use as ideal antibacterial agents against oral microbes.

highlighted such exhaustive research on POs and nanoparticle stories/facts again motivated the scientific community to utilize more POs in antibacterial drug development.

6.7 Conclusion

The collected scientific facts showed that POs could be used in current antibacterial therapy to control oral and other pathogens. Briefly, the physical combination of PO with other antibacterial or other PO, known as synergism formulation, is a dynamic approach against MDR pathogens. However, monocyclic terpenes with a volatile character of POs bearing a lack of ideal physiochemical property are significant obstacles toward the success of POs in antibacterial drug development. Purposefully, the advanced nanotechnology/nanocarrier and chemical conjugation techniques currently used to improve physicochemical, bioavailability, target drug delivery as thymol, carvacrol, gallic acid, eugenol like POs are successfully dementated. Moreover, POs may not resolve the current antibiotic unavailability or resistance problem entirely but could play substantial alternative antibacterial resources from drug development to microbe's free drug packeting. Thus, the highlighted experimental reports, chemical arrangement, mode of antibacterial action, and adopted techniques for POs help develop an alternative, potent, less toxic, and cost-effective antibacterial drug against emerged oral and MDR pathogens. Thus, the above methods, especially nanotechnology, offer a new direction/shape to PO-based therapy.

Acknowledgments

This work is a part of the Ph.D. thesis of A. Sahoo in Biotechnology, Siksha 'O' Anusandhan Deemed to be University, Bhubaneswar. Thanks to the Honorable President and Vice-chancellor, S'O'A Deemed to be University, Bhubaneswar, for the Ph.D. fellowship and research facilities.

References

1. Davies, J. and Davies, D., Origins and evolution of antibiotic resistance. *Microbiol. Mol. Biol. Rev.*, 74, 417, 2010.

2. Lerminiaux, N.A. and Cameron, A.D.S., Horizontal transfer of antibiotic resistance genes in clinical environments. *Can. J. Microbiol.*, 65, 34, 2019.
3. Finn, J., Application of SBDD to the discovery of new antibacterial drugs. *Methods Mol. Biol.*, 841, 291, 2012.
4. Wright, P.M., Seiple, I.B., Myers, A.G., The evolving role of chemical synthesis in antibacterial drug discovery. *Angew. Chem. Int. Ed. J.*, 53, 8840, 2014.
5. Hughes, D. and Karlén, A., Discovery and preclinical development of new antibiotics. *Ups. J. Med. Sci.*, 119, 162, 2014.
6. Swain, S.S., Paidesetty, S.K., Padhy, R.N., Development of antibacterial conjugates using sulfamethoxazole with monocyclic terpenes: A systematic medicinal chemistry based computational approach. *Comput. Methods Programs Biomed.*, 140, 185, 2017.
7. Jackson, N., Czaplewski, L., Piddock, L.J.V., Discovery and development of new antibacterial drugs: Learning from experience? *J. Antimicrob. Chemother.*, 73, 1452, 2018.
8. Swain, S.S., Paidesetty, S.K., Padhy, R.N., Hussain, T., Isoniazid-phytochemical conjugation: A new approach for potent and less toxic anti-TB drug development. *Chem. Biol. Drug Des.*, 96, 714, 2020.
9. Aslam, B., Wang, W., Arshad, M.I., Khurshid, M., Muzammil, S., Rasool, M.H., Nisar, M.A., Alvi, R.F., Aslam, M.A., Qamar, M.U., Salamat, M., Baloch, Z., Antibiotic resistance: A rundown of a global crisis. *Infect. Drug Resist.*, 11, 1645, 2018.
10. Dadgostar, P., Antimicrobial resistance: Implications and costs. *Infect. Drug Resist.*, 12, 3903, 2019.
11. Lee, C.R., Cho, I.H., Jeong, B.C., Lee, S.H., Strategies to minimize antibiotic resistance. *Int. J. Environ. Res. Public Health*, 10, 4274, 2013.
12. Rossiter, S.E., Fletcher, M.H., Wuest, W.M., Natural products as platforms to overcome antibiotic resistance. *Chem. Rev.*, 117, 12415, 2017.
13. Hall, T.J., Villapún, V.M., Addison, O., Webber, M.A., Lowther, M., Louth, S.E.T., Mountcastle, S.E., Brunet, M.Y., Cox, S.C., A call for action to the biomaterial community to tackle antimicrobial resistance. *Biomater. Sci.*, 8, 4951, 2020.
14. Nazir, M., Al-Ansari, A., Al-Khalifa, K., Alhareky, M., Gaffar, B., Almas, K., Global prevalence of periodontal disease and lack of its surveillance. *Sci. World J.*, 2020, 2146160, 2020.
15. Nazir, M.A., Prevalence of periodontal disease, its association with systemic diseases and prevention. *Int. J. Health Sci. (Qassim)*, 11, 72, 2017.
16. Fair, R.J. and Tor, Y., Antibiotics and bacterial resistance in the 21st century. *Perspect. Medicin. Chem.*, 6, 25, 2014.
17. Li, X., Kolltveit, K.M., Tronstad, L., Olsen, I., Systemic diseases caused by oral infection. *Clin. Microbiol. Rev.*, 13, 547, 2000.
18. Lamont, R.J., Koo, H., Hajishengallis, G., The oral microbiota: Dynamic communities and host interactions. *Nat. Rev. Microbiol.*, 16, 745, 2018.

19. Lu, M., Xuan, S., Wang, Z., Oral microbiota: A new view of body health. *Food Sci. Hum. Wellness*, 8, 8, 2019.
20. Mei, F., Xie, M., Huang, X., Long, Y., Lu, X., Wang, X., Chen, L., *Porphyromonas gingivalis* and its systemic impact: Current status. *Pathogens*, 9, 944, 2020.
21. Golub, L.M. and Lee, H.M., Periodontal therapeutics: Current host-modulation agents and future directions. *Periodontol. 2000*, 82, 186, 2020.
22. Cragg, G.M. and Newman, D.J., Natural products: A continuing source of novel drug leads. *Biochim. Biophys. Acta*, 1830, 3670, 2013.
23. Bhardwaj, K., Islam, M.T., Jayasena, V., Sharma, B., Sharma, S., Sharma, P., Kuča, K., Bhardwaj, P., Review on essential oils, chemical composition, extraction, and utilization of some conifers in Northwestern Himalayas. *Phytother. Res.*, 34, 2889, 2020.
24. Subramani, R., Narayanasamy, M., Feussner, K.D., Plant-derived antimicrobials to fight against multi-drug-resistant human pathogens. *3 Biotech.*, 7, 172, 2017.
25. Visioli, F., Davalos, A., López de Las Hazas, M.C., Crespo, M.C., Tomé-Carneiro, J., An overview of the pharmacology of olive oil and its active ingredients. *Br. J. Pharmacol.*, 177, 1316, 2020.
26. Nostro, A. and Papalia, T., Antimicrobial activity of carvacrol: Current progress and future prospectives. *Recent. Pat. Antiinfect. Drug Discovery*, 7, 28, 2012.
27. Madhumita, M., Guha, P., Nag, A., Bio-actives of betel leaf (*Piper betle* L.): A comprehensive review on extraction, isolation, characterization, and biological activity. *Phytother. Res.*, 34, 2609, 2020.
28. Bakkali, F., Averbeck, S., Averbeck, D., Idaomar, M., Biological effects of essential oils - a review. *Food Chem. Toxicol.*, 46, 446, 2008.
29. Seow, Y.X., Yeo, C.R., Chung, H.L., Yuk, H.G., Plant essential oils as active antimicrobial agents. *Crit. Rev. Food Sci. Nutr.*, 54, 625, 2014.
30. Chouhan, S., Sharma, K., Guleria, S., Antimicrobial activity of some essential oils-present status and future perspectives. *Medicines (Basel)*, 4, 58, 2017.
31. Samperio, C., Boyer, R., Eigel, W.N. 3rd, Holland, K.W., McKinney, J.S., O'Keefe, S.F., Smith, R., Marcy, J.E., Enhancement of plant essential oils' aqueous solubility and stability using alpha and beta cyclodextrin. *J. Agric. Food Chem.*, 58, 12950, 2010.
32. Hemaiswarya, S. and Doble, M., Synergistic interaction of eugenol with antibiotics against Gram negative bacteria. *Phytomedicine*, 16, 997, 2009.
33. Elshafie, H.S. and Camele, I., An Overview of the biological effects of some mediterranean essential oils on human health. *Biomed. Res. Int.*, 2017, 9268468, 2017.
34. Burčul, F., Blažević, I., Radan, M., Politeo, O., Terpenes, phenylpropanoids, sulfur and other essential oil constituents as inhibitors of cholinesterases. *Curr. Med. Chem.*, 27, 4297, 2020.
35. Burt, S., Essential oils: their antibacterial properties and potential applications in foods–a review. *Int. J. Food Microbiol.*, 94, 223, 2004.

36. Hamoud, R., Sporer, F., Reichling, J., Wink, M., Antimicrobial activity of a traditionally used complex essential oil distillate (Olbas(®) Tropfen) in comparison to its individual essential oil ingredients. *Phytomedicine*, 19, 969, 2012.
37. El-Shenawy, N.S., Oxidative stress responses of rats exposed to roundup and its active ingredient glyphosate. *Environ. Toxicol. Pharmacol.*, 28, 379, 2009.
38. Peana, A.T., D'Aquila, P.S., Panin, F., Serra, G., Pippia, P., Moretti, M.D., Anti-inflammatory activity of linalool and linalyl acetate constituents of essential oils. *Phytomedicine*, 9, 721, 2002.
39. Fabio, A., Cermelli, C., Fabio, G., Nicoletti, P., Quaglio, P., Screening of the antibacterial effects of a variety of essential oils on microorganisms responsible for respiratory infections. *Phytother. Res.*, 21, 374, 2007.
40. Rosato, A., Vitali, C., De Laurentis, N., Armenise, D., Antonietta Milillo, M., Antibacterial effect of some essential oils administered alone or in combination with Norfloxacin. *Phytomedicine*, 14, 727, 2007.
41. Prabuseenivasan, S., Jayakumar, M., Ignacimuthu, S., *In vitro* antibacterial activity of some plant essential oils. *BMC Complement. Altern. Med.*, 6, 39, 2006.
42. Rastogi, N., Domadia, P., Shetty, S., Dasgupta, D., Screening of natural phenolic compounds for potential to inhibit bacterial cell division protein FtsZ. Indian. *J. Exp. Biol.*, 46, 783, 2008.
43. Fisher, K. and Phillips, C., Potential antimicrobial uses of essential oils in food: is citrus the answer? *Trends Food Sci. Technol.*, 19, 156, 2008.
44. Lee, S.B., Cha, K.H., Kim, S.N., Altantsetseg, S., Shatar, S., Sarangerel, O., Nho, C.W., The antimicrobial activity of essential oil from *Dracocephalum foetidum* against pathogenic microorganisms. *J. Microbiol.*, 45, 53, 2007.
45. Sonboli, A., Eftekhar, F., Yousefzadi, M., Kanani, M.R., Antibacterial activity and chemical composition of the essential oil of *Grammosciadium platycarpum* Boiss. from Iran. *Z. Naturforsch. C. J. Biosci.*, 60, 30, 2005.
46. Si Saida, Z.B., Hayate Haddadi-Guemghara, H., Boulekbache-Makhloufa, L., Rigoub, P., Reminia, H., Adjaouda, A., Khoudjaa, N.K., Madani, K., Essential oils composition, antibacterial and antioxidant activities of hydro distillated extract of *Eucalyptus globulus* fruits. *Ind. Crops Prod.*, 89, 167, 2016.
47. Hyldgaard, M., Mygind, T., Meyer, R.L., Essential oils in food preservation: Mode of action, synergies, and interactions with food matrix components. *Front. Microbiol.*, 3, 12, 2012.
48. Martuccia, J.F., Gendeb, L.B., Neiraa, L.M., Ruseckaite, R.A., Oregano and lavender essential oils as antioxidant and antimicrobial additives of biogenic gelatin films. *Ind. Crops Prod.*, 71, 205, 2015.
49. Bush, K., Courvalin, P., Dantas, G., Davies, J., Eisenstein, B., Huovinen, P., Jacoby, G.A., Kishony, R., Kreiswirth, B.N., Kutter, E., Lerner, S.A., Levy, S., Lewis, K., Lomovskaya, O., Miller, J.H., Mobashery, S., Piddock, L.J., Projan, S., Thomas, C.M., Tomasz, A., Tulkens, P.M., Walsh, T.R., Watson, J.D.,

Witkowski, J., Witte, W., Wright, G., Yeh, P., Zgurskaya, H.I., Tackling antibiotic resistance. *Nat. Rev. Microbiol.*, 9, 894, 2011.
50. Langeveld, W.T., Veldhuizen, E.J., Burt, S.A., Synergy between essential oil components and antibiotics: a review. *Crit. Rev. Microbiol.*, 40, 76, 2014.
51. van Vuuren, S.F., Suliman, S., Viljoen, A.M., The antimicrobial activity of four commercial essential oils in combination with conventional antimicrobials. *Lett. Appl. Microbiol.*, 48, 440, 2009.
52. Magi, G., Marini, E., Facinelli, B., Antimicrobial activity of essential oils and carvacrol, and synergy of carvacrol and erythromycin, against clinical, erythromycin-resistant Group A *Streptococci*. *Front. Microbiol.*, 6, 165, 2015.
53. Mohamed, S.H., Mohamed, M.S.M., Khalil, M.S., Azmy, M., Mabrouk, M.I., Combination of essential oil and ciprofloxacin to inhibit/eradicate biofilms in multidrug-resistant *Klebsiella pneumoniae*. *J. Appl. Microbiol.*, 125, 84, 2018.
54. Swamy, M.K., Akhtar, M.S., Sinniah, U.R., Antimicrobial properties of plant essential oils against human pathogens and their mode of action: An updated review. *Evid. Based Complement. Alternat. Med.*, 2016, 3012462, 2016.
55. Guerra, F.Q., Mendes, J.M., Sousa, J.P., Morais-Braga, M.F., Santos, B.H., Melo Coutinho, H.D., Lima Ede, O., Increasing antibiotic activity against a multidrug-resistant *Acinetobacter* spp by essential oils of Citrus limon and Cinnamomum zeylanicum. *Nat. Prod. Res.*, 26, 2235, 2012.
56. Duarte, A., Ferreira, S., Silva, F., Domingues, F.C., Synergistic activity of coriander oil and conventional antibiotics against Acinetobacter baumannii. *Phytomedicine*, 19, 236, 2012.
57. Aleksic, V. and Knezevic, P., Antimicrobial and antioxidative activity of extracts and essential oils of Myrtus communis L. *Microbiol. Res.*, 169, 240, 2014.
58. Miladinović, D.L., Ilić, B.S., Kocić, B.D., Ćirić, V.M., Nikolić, D.M., Antibacterial investigation of thyme essential oil and its main constituents in combination with tetracycline. *J. Med. Food*, 18, 935, 2015.
59. Djouahri, A., Saka, B., Boudarene, L., Boudarene, L., Benseradj, F., Aberrane, S., Aitmoussa, S., *In vitro* synergistic/antagonistic antibacterial and anti-inflammatory effect of various extracts/essential oil from cones of *Tetraclinis articulata* (Vahl) Masters with antibiotic and anti-inflammatory agents. *Ind. Crops Prod.*, 56, 60, 2014.
60. Fadli, M., Chevalier, J., Bolla, J.M., Mezrioui, N.E., Hassani, L., Pages, J.M., Thymus maroccanus essential oil, a membranotropic compound active on Gram-negative bacteria and resistant isolates. *J. Appl. Microbiol.*, 113, 1120, 2012.
61. Mahboubi, M. and Bidgoli, F.G., Antistaphylococcal activity of Zataria multiflora essential oil and its synergy with vancomycin. *Phytomedicine*, 17, 548, 2010.

62. Abulrob, A.N., Suller, M.T., Gumbleton, M., Simons, C., Russell, A.D., Identification and biological evaluation of grapefruit oil components as potential novel efflux pump modulators in methicillin-resistant Staphylococcus aureus bacterial strains. *Phytochemistry*, 65, 3021, 2004.
63. Palaniappan, K. and Holley, R.A., Use of natural antimicrobials to increase antibiotic susceptibility of drug resistant bacteria. *Int. J. Food Microbiol.*, 14, 164, 2010.
64. Choi, J.G., Kang, O.H., Lee, Y.S., Oh, Y.C., Chae, H.S., Jang, H.J., Shin, D.W., Kwon, D.Y., Antibacterial activity of methyl gallate isolated from Galla Rhois or carvacrol combined with nalidixic acid against nalidixic acid resistant bacteria. *Molecules*, 14, 1773, 2009.
65. Zhang, D., Hu, H., Rao, Q., Zhao, Z., Synergistic effects and physiological responses of selected bacterial isolates from animal feed to four natural antimicrobials and two antibiotics. *Foodborne Pathog. Dis.*, 8, 1055, 2011.
66. Yap, P.S., Lim, S.H., Hu, C.P., Yiap, B.C., Combination of essential oils and antibiotics reduce antibiotic resistance in plasmid-conferred multidrug resistant bacteria. *Phytomedicine*, 20, 710, 2013.
67. Karpanen, T.J., Worthington, T., Conway, B.R., Hilton, A.C., Elliott, T.S., Lambert, P.A., Penetration of chlorhexidine into human skin. *Antimicrob. Agents Chemother.*, 52, 3633, 2008.
68. Donsì, F., Annunziata, M., Sessa, M., Ferrari, G., Nanoencapsulation of essential oils to enhance their antimicrobial activity in foods. *LWT-Food Sci. Technol.*, 44, 1908, 2011.
69. Kumari, S., Kumaraswamy, R.V., Choudhary, R.C., Sharma, S.S., Pal, A., Raliya, R., Biswas, P., Saharan, V., Thymol nanoemulsion exhibits potential antibacterial activity against bacterial pustule disease and growth promotory effect on soybean. *Sci. Rep.*, 8, 6650, 2018.
70. Puškárová, A., Bučková, M., Kraková, L., Pangallo, D., Kozics, K., The antibacterial and antifungal activity of six essential oils and their cyto/genotoxicity to human HEL 12469 cells. *Sci. Rep.*, 7, 8211, 2017.
71. Wattanasatcha, A., Rengpipat, S., Wanichwecharungruang, S., Thymol nanospheres as an effective anti-bacterial agent. *Int. J. Pharm.*, 434, 360, 2012.
72. Woranuch, S. and Yoksan, R., Eugenol-loaded chitosan nanoparticles: I. Thermal stability improvement of eugenol through encapsulation. *Carbohydr. Polym.*, 96, 578, 2013.
73. de Oliveira, E.F., Paula, H.C., de Paula, R.C., Alginate/cashew gum nanoparticles for essential oil encapsulation. *Colloids Surf. B. Biointerfaces*, 113, 146, 2014.
74. Eid, R.K., Essa, E.A., El Maghraby, G.M., Essential oils in niosomes for enhanced transdermal delivery of felodipine. *Pharm. Dev. Technol.*, 24, 157, 2019.
75. Azeem, A., Anwer, M.K., Talegaonkar, S., Niosomes in sustained and targeted drug delivery: Some recent advances. *J. Drug Targeting*, 17, 671, 2009.

76. Musthaba, S.M., Baboota, S., Ahmed, S., Ahuja, A., Ali, J., Status of novel drug delivery technology for phytotherapeutics. *Expert Opin. Drug Deliv.*, 6, 625, 2009.
77. Lu, Q., Lu, P.M., Piao, J.H., Xu, X.L., Chen, J., Zhu, L., Jiang, J.G., Preparation and physicochemical characteristics of an allicin nanoliposome and its release behavior. *LWT-Food Sci. Technol.*, 57, 686, 2014.
78. Neethirajan, S. and Jayas, D.S., Nanotechnology for the food and bioprocessing industries. *Food Bioprocess Tech.*, 4, 39, 2011.
79. Cui, H., Li, W., Li, C., Vittayapadung, S., Lin, L., Liposome containing cinnamon oil with antibacterial activity against methicillin-resistant *Staphylococcus aureus* biofilm. *Biofouling*, 32, 215, 2016.
80. Khorasani, S., Danaei, M., Mozafari, M.R., Nanoliposome technology for the food and nutraceutical industries. *Trends Food Sci. Technol.*, 79, 106, 2018.
81. Obonyo, M., Zhang, L., Thamphiwatana, S., Pornpattananangkul, D., Fu, V., Zhang, L., Antibacterial activities of liposomal linolenic acids against antibiotic-resistant *Helicobacter pylori. Mol. Pharm.*, 9, 2677, 2012.
82. Challa, R., Ahuja, A., Ali, J., Khar, R.K., Cyclodextrins in drug delivery: An updated review. *AAPS PharmaSciTech*, 6, 329, 2005.
83. Ciobanu, A., Mallard, I., Landy, D., Brabiec, G., Nistor, D., Fourmentin. Inclusion interactions of cyclodextrins and crosslinked cyclodextrin polymers with linalool and camphor in *Lavandula angustifolia* essential oil. *Carbohydr. Polym.*, 87, 1963, 2012.
84. de Goeij, B.E. and Lambert, J.M., New developments for antibody-drug conjugate-based therapeutic approaches. *Curr. Opin. Immunol.*, 40, 14, 2016.
85. Ponce, C.P.A., Buera, M.P., Elizalde, B.E., Encapsulation of cinnamon and thyme essential oils components (cinnamaldehyde and thymol) in β-cyclodextrin: effect of interactions with water on complex stability. *J. Food Eng.*, 99, 70, 2010.
86. Lima, V.N., Oliveira-Tintino, C.D., Santos, E.S., Morais, L.P., Tintino, S.R., Freitas, T.S., Geraldo, Y.S., Pereira, R.L., Cruz, R.P., Menezes, I.R., Coutinho, H.D., Antimicrobial and enhancement of the antibiotic activity by phenolic compounds: Gallic acid, caffeic acid and pyrogallol. *Microb. Pathog.*, 99, 56, 2016.
87. Jamil, B., Bokhari, H., Imran, M., Mechanism of Action: How Nano-Antimicrobials Act? *Curr. Drug Targets*, 18, 363, 2017.
88. Hosseini, S.F., Zandi, M., Rezaei, M., Farahmandghavi, F., Two-step method for encapsulation of oregano essential oil in chitosan nanoparticles: Preparation, characterization and in vitro release study. *Carbohydr. Polym.*, 95, 50, 2013.
89. Grabovac, V. and Bernkop-Schnürch, A., Development and *in vitro* evaluation of surface modified poly(lactide-co-glycolide) nanoparticles with chitosan-4-thiobutylamidine. *Drug Dev. Ind. Pharm.*, 33, 767, 2007.

90. Gomes, C., Moreira, R.G., Castell-Perez, E., Poly (DL-lactide-co-glycolide) (PLGA) nanoparticles with entrapped trans-cinnamaldehyde and eugenol for antimicrobial delivery applications. *J. Food Sci.*, 76, N16, 2011.
91. Hamed, S.F., Sadek, Z., Edris, A., Antioxidant and antimicrobial activities of clove bud essential oil and eugenol nanoparticles in alcohol-free microemulsion. *J. Oleo Sci.*, 61, 641, 2012.
92. Trombetta, D., Castelli, F., Sarpietro, M.G., Venuti, V., Cristani, M., Daniele, C., Saija, A., Mazzanti, G., Bisignano, G., Mechanisms of antibacterial action of three monoterpenes. *Antimicrob. Agents Chemother.*, 49, 2474, 2005.
93. Sharma, A., Mayhew, E., Bolcsak, L., Cavanaugh, C., Harmon, P., Janoff, A., Bernacki, R.J., Activity of paclitaxel liposome formulations against human ovarian tumor xenografts. *Int. J. Cancer*, 71, 103, 1997.
94. Sherry, M., Charcosset, C., Fessi, H., Greige-Gerges, H., Essential oils encapsulated in liposomes: a review. *J. Liposome Res.*, 23, 268, 2013.
95. Liolios, C.C., Gortzi, O., Lalas, S., Tsaknis, J., Chinou, I., Liposomal incorporation of carvacrol and thymol isolated from the essential oil of *Origanum dictamnus* L. and *in vitro* antimicrobial activity. *Food Chem.*, 112, 77, 2009.
96. Wang, J., Cao, Y., Sun, B., Wang, C., Physicochemical and release characterisation of garlic oil-β-cyclodextrin inclusion complexes. *Food Chem.*, 127, 1680, 2011.
97. Thakur, M., Pandey, S., Mewada, A., Patil, V., Khade, M., Goshi, E., Sharon, M., Antibiotic conjugated fluorescent carbon dots as a theranostic agent for controlled drug release, bioimaging, and enhanced antimicrobial activity. *J. Drug Deliv.*, 2014, 282193, 2014.
98. Shen, B., A new golden age of natural products drug discovery. *Cell*, 163, 1297, 2015.
99. Akhtar, M.S., Degaga, B., Azam, T., Antimicrobial activity of essential oils extracted from medicinal plants against the pathogenic microorganisms: A review. *Biol. Sci. Pharm. Res.*, 2, 1, 2014.
100. Katz, L. and Baltz, R.H., Natural product discovery: Past, present, and future. *J. Ind. Microbiol. Biotechnol.*, 43, 155, 2016.
101. Shakeri, A., Khakdan, F., Soheili, V., Sahebkar, A., Rezva Shaddel, R., Asili, J., Volatile composition, antimicrobial, cytotoxic and antioxidant evaluation of the essential oil from *Nepeta sintenisii* Bornm. *Ind. Crops Prod.*, 84, 224, 2016.
102. Zanetti, M., Carniel, T.K., Dalcanton, F., Anjos, R.S., Riella, H.G., de Araújo, P.H.H., Oliveira, D., Fiori, M., Use of encapsulated natural compounds as anti-microbial additives in food packaging: A brief review. *Trends Food Sci. Technol.*, 81, 51, 2018.
103. Sinha, S., Jothiramajayam, M., Ghosh, M., Mukherjee, A., Evaluation of toxicity of essential oils palmarosa, citronella, lemongrass and vetiver in human lymphocytes. *Food Chem. Toxicol.*, 68, 71, 2014.
104. Majeed, H., Bian, Y., Ali, B., Jamil, A., Majeed, U., Khan, Q.F., Iqbal, K.J., Shoemaker, C.F., Fang, Z., Essential oil encapsulations: uses, procedures, and trends. *RSC Adv.*, 5, 58449, 2015.

7

Lipid Nanoformulation of Nutraceuticals as Neurotherapeuticals in Neurological Disorders

Mohamad Taleuzzaman[1], Sanjay Chauhan[2], Devendra Singh Tomar[3], Pankaj Kumar Singh[3], Isha Talwar[4] and Md Noushad Javed[5*]

[1]*Faculty of Pharmacy, Department of Chemistry, Maulana Azad University, Jodhpur, Rajasthan, India*
[2]*Institute of Pharmacy, Ram-Eesh Institute of Vocational and Technical Education, Knowledge Park-1, Greater Noida, Uttar Pradesh, India*
[3]*Department of Pharmaceutics, National Institute of Pharmaceutical Education & Research (NIPER), Balanagar, Hyderabad, India*
[4]*Department of Pharmacology, Glocal School of Pharmacy, Glocal University, Mirzapure Pole, Saharanpur, U.P., India*
[5]*Department of Pharmaceutics, SMAS, KR Mangalam University, Gurugram, India*

Abstract

The central nervous system (CNS) is one of the most important organs that control the function of the body. So the impairment of its function indicates many disorders or diseases. The neurological disorder affects millions of people in the world. Recently, the role of different nutraceuticals, such as monoterpenes, was explored as an emerging neurotherapeuticals in various CNS disorders but still facing various challenges at clinical translation because traditional delivery approaches have several limitations which restrict drug availability in the brain, owing to the action of endothelial cell layers, poor water solubility, low bioavailability, poor intestinal absorption. To overcome the above limitations, however, scientists are struggling to approach nanoformulations-based dosage forms, preferably through lipid-based forms. These formulations succeed in giving the high therapeutic values by targeted delivery of drugs while nanosize bears advantages of quickly crossing the blood-brain barrier, minimizing efflux, controlled drug release characteristics, high bioavailability, and less adverse effects.

**Corresponding author*: rxnoushad@gmail.com

Tanima Bhattacharya and Shakeel Ahmed (eds.) Nanotechnology in Functional Foods, (161–192)

Hence, current chapters aim to integrate roles of different lipid nanoformulations, such as nanoemulsion, solid lipid nanoparticles (SLNs) and nanostructured lipid carriers (NLCs), which utilize physiological transport mechanisms across the blood-brain barrier to deliver the drug in targeted sites in CNS safely, so emerged as a potential solution to overcome delivery challenges in CNS disorders.

Keywords: CNS, nanoformulations, SLNs, NLCs, ethosome

7.1 Introduction

7.1.1 CNS Disorders

Neurological disorders are diseases related to the central and peripheral nervous systems. In other words, these are disordered states of the brain, spinal cord, cranial nerves, peripheral nerves, nerve roots, and autonomic nervous system, neuromuscular junction, and muscles [1, 2]. These diseases include epilepsy, Alzheimer's disease, other dementias, cerebrovascular disease, including stroke, migraine, and other headache disorders, multiple sclerosis [3, 4]. In addition, malnutrition, pathological stages of Parkinson's disease, neuroinfections, brain tumors, traumatic disorders of the nervous system due to head trauma, and neurological disorders further worsen the final conditions [5, 6].

Protection of the brain from contaminations of environmental toxins or foreign substances is essential because alteration in the concentration of critical components, which is required for homeostasis, inner and outer of neuronal cells, leads to impairments in nerve conduction and dysfunction of body control processes [7, 8]. There are also other reasons for CNS disorders, such as genetic makeup, environmental factors, and protein aggregation [9]. In neurological disorders, these abovementioned contributing factors lead to neuroinflammation and degeneration via dysregulation of the immune process or other abnormalities there [10, 11].

7.1.2 Epidemiology of Brain Diseases

A collaborative work between the World Health Organization (WHO), World Bank, and Harvard School of Public Health produce very detailed data on the Global Burden of Disease (GBD) [12]. In this data set, the burden of neurological disorders and many other chronic conditions attract the attention of international health communities [13]. This study revealed that previously used traditional epidemiological and health statistical methods used for the data collection on neurological disorders were undervalued because it only considered mortality but not disability parameters [14, 15]. Hence, the impacts of neurological disorders, including brain cancer cases, on the global health indexes are always predicted as underestimated values [16, 17]. Therefore to estimate several aspects of neurological care, availability of neurological services, and frequency of neurological disorders; WHO and the World Federation of Neurology (WFN) worked together in an international survey for neurological conditions to collect information from experts by involving 109 countries that represent almost 99 % of the world's population [14, 15]. Briefly, different types of neurological disorders affect hundreds of millions of people worldwide ranging from neurodevelopmental to neuroinflammation and neurodegenerative disorders [18]. Roughly 6.3% of the global disease burden is due to neurological disorders [19]. Furthermore, the report revealed that facilities are inadequate for patients, especially in low-income and developing countries. In contrast, there are massive inequalities in behavioral patterns with surveyed patients [20]. Individually, a neurological disorder of the global burden of disease in high, middle, middle, and low-income countries contributes 10.9%, 6.7%, 8.7%, and 4.5%, respectively [21].

More than 6 million patients die due to strokes every year, while low- and middle-income countries are the main contributors with a share of 80 % in such deaths [22]. Among neurological disorders, worldwide patients with epilepsy are more than 50 million, while 47.5 million people have dementia, with a reported 7.7 million new cases every year. Here, Alzheimer's disease contributes to 60% to 70% of cases of dementia. Furthermore, migraine shared more than 10 % of global neurological [23]. Some investigators reported that because of demographic changes in population and growth of average life expectancy, the occurrence of the central nervous system (CNS) associated diseases was tremendously increased [24]. In high-income countries, only 2.84 % of the total deaths are due to Alzheimer's and other dementias diseases [25]. While, with a rate of 16.8 %, neurological disorders, mediated deaths are significantly higher in lower middle-income countries than the 13.2 % death rate of middle-income nations [26].

Worldwide, 12% of total deaths are because of neurological disorders; 85% of deaths are because of cerebrovascular disease, one type of neurological disorder A data published in 2005 revealed at least 2% global burden of disease contributed by neuropsychiatric disorders while 4.3% of patients were suffering from cerebrovascular diseases and some neuroinfections, such as poliomyelitis, tetanus, meningitis, and Japanese encephalitis [27]. Cerebrovascular disease mediated deaths are 15.8 % in lower-middle countries, 9.6 % in upper middle-income countries, 9.5 % in high income countries while only 6.4 % in low-income countries [28].

7.2 Pharmaceutical Challenges in Brain Delivery

This section provides a brief review of some of the critical challenges, which are directly associated with the process of fabricating targeted delivery systems of neurotherapeuticals in CNS disorders (Figure 7.1).

7.2.1 Poor Solubility

Most neurotherapeuticals are highly lipophilic, so these tend to have very high solubility in the lipid phase but confer insolubility in aqueous media [29]. Hence, conventional pharmaceutical dosage forms of such drugs have minimal roles in clinics because of the inability of active loads effectively to distribute in the aqueous media of GIT [30]. However, the lipophilic nature of such drugs forms physical barriers to prevent the transfer of drug mass from one compartment to another one [31]. Nevertheless, due to their low solubility in water, such loads therapeuticals can be easily encapsulated forms within the lipid-based carrier systems, such as nanoemulsions, NLC, SLC, so on [32]. Here, the lipophilic nature of neurotherapeuticals is also advantageous because it helps formulators developing a suitable oil-based drug delivery system to felicitate targeted and controlled release profiles [33].

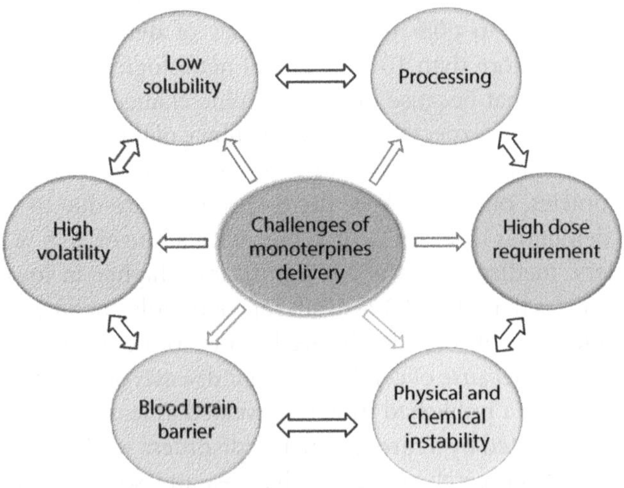

Figure 7.1 Schematic diagram of challenges associated to delivery of monoterpenes.

7.2.2 Physical and Chemical Instability

Most of the neurotherapeutics decompose fast when exposed to oxygen, light, high temperature, humidity, or oxygen during the phase of material processing, product development and storage [34, 35]. These abovementioned instability-related challenges pose significant challenges in pharmaceutical industries for practical use and translation of therapeutics in clinical usage [36]. During the preparation of dosage forms and final product storage; hence, strict control of these parameters to regulate quality and performances are critical requirements [37]. In pharmaceutical industries, QbD-enabled process offers some unique approach for facile control over the process to achieve desired goals [38]. Many successful attempts are made to effectively utilize emerging strategies of nanoencapsulation to improve the stability of loaded drugs and reduce sensitivity reactions [39].

7.2.3 Degradable Products

As mentioned above, many neurotherapeuticals are physically volatile, thermolabile and susceptible to environmental factors such as light, moisture and oxygen [40, 41]. While some drugs are chemically reactive or substrate of enzymes, such attributes processed these into some inert or toxic species [42]. Some neurotherapeuticals are the substrate of enzymes; hence, these drugs are enzymatically catalyzed after dose administration. While, in others, catalyzed products of therapeutics, especially peptides, lead not only to loss of therapeutical performances but also cause hyper-sensitivity, allergic, and other adverse reactions such as skin allergy or contact dermatitis [43–45]. For example, oxygenated byproducts of Linalool or caryophyllene induce allergenic and skin sensitization properties [46]. Similarly, many phenolic and flavonoids drugs such as eugenol and thymoqionone which bear significance in terms of their epigenetic mechanisms, owing to methyltransferase activities in deregulated cancer cells are very sensitive to oxygen, light, and heat-mediated degradation [47]. Similarly, limonene converts into harmful oxidized derivatives Pinene, while diepoxidation reaction produces carcinogenic diepoxylimonene species [48]. Hence, dosage form preparation techniques involve several steps such as heating, mixing, solvent evaporation, and high-pressure homogenization, which favor volatilization, instability and acceleration in incompatible reactions [49]. However, encapsulation of drugs within a secure system helps them to be protected from volatilization and degradation [50].

7.2.4 High-Dose Requirement

In an exception to a few potent drugs, the required clinical dose of neurotherapeuticals to fit within the therapeutic window zone is usually very high to treat CNS diseases [51]. Conventional approaches and traditional delivery systems are mainly based on the integration of monotherapy of neurotherapeuticals [52]. However, both these approaches and carriers cannot meet the requirements of loading such high doses of drugs, limiting the prospect of such therapies [52]. Subsequently, these carriers are also limited because of their inefficiency in the stability of neurotherapeutics loads, organ-specific delivery, and control over the release [53]. For example, due to poor stability-related issues of Linalool, conventional delivery systems require a relatively higher dose for biological activities [30]. However, incorporating such type of drugs into a suitable nanocarrier system to enable targeted and control release characteristics benefit in terms of the requirement for a low drug regimen for achieving desired pharmacological actions [54].

7.2.5 Volatility

Neurotherapeuticals with a high degree of volatilization are very short in half-life. Subsequently, conventional delivery systems exhibit loss of large fractions of dose during final product preparation and storage [55]. These pharmacotechnical concerns cause difficulties in determining the fate and monitoring of clinical performances from loads because of decreased pharmacologically active drugs from final dosage forms [56]. Further, these also cause lower patient compliance parameters, owing to damaged aesthetic appearance and smelling of products during storage. For example, volatile molecule linalool faces stability and short half-life-related issues; hence conventional dosage forms are retarding its therapeutical potentials [57]. In brief, due to short shelf life and loss of medicine from composition, fabrication of any efficient dosage forms and their clinical translation is challenging [58].

7.3 Physiological Barriers in Drug Delivery

7.3.1 Blood-Brain Barrier

Because of the complexity of the brain structures and functions, treatments of brain diseases are relatively more complicated than disorders of other organs [59]. The anatomical structure of the blood-brain barrier serves as a physiological barrier in the brain to limit the impacts of neurotherapeuticals in treating neurological diseases. It protects the brain and its role in transporting substances from the blood into the brain [60]. Our brains are a highly protected organ due to selective permeability and the gateway role of the blood-brain barrier (BBB) [61]. The blood-brain barrier is the least permeable capillaries due to tight junctions to protect the brain from an external bloodstream environment [62]. Hence, these systems should cross the BBB efficiently and safely for efficient actions and desired therapeutic effects [3].

Broadly, the region of BBB is divided into two parts; the first part is the neurovascular unit, while the second part is the junctional complexes. Here neurovascular have several components with specific functions [7]. For example, endothelial cells transport micronutrients and macronutrients and osmoregulatory inn functions while astrocytes regulate metabolism, modulation of neuronal transmissions and brain development. Similarly, pericytes regulate cell proliferation endothelial cells, neurons modulate the permeability of blood-brain barriers through neuronal-microvascular communications while the extracellular matrix maintains tight junctions. Second junctional complexes are divided into four parts where occluding ensures rigidity of the tight junction. Claudins impart primary function as a barrier, junctional adhesion molecules (involved in the development process) membrane-associated guanylate kinase-type proteins modulate the permeability of BBB [63, 64]. So, such structural units of BBB, which are made up of endothelial cells and tight junctions from a continuous cell membrane, control the transport of molecules. Apart from tight junctions, the presence of efflux transporters, such as the permeability-glycoprotein (P-gp) efflux pump on the blood-brain barrier also possess additional protections as a barrier in the transportation of lipophilic compound [65]. Hence, most CNS-acting neuropsychiatric drugs are facing a high rate of P-gp pumps mediated efflux during the treatment of brain disorders. The dose of the most conventional P-gp efflux inhibitors is very high, leading to toxicity events [66, 67]. However, now, attempts are also being made toward discovering novel P-gp inhibitors, which are potent enough to improve permeability toward BBB and inhibit the efflux action of P-gp pumps. Many researchers addressed this challenge by integrating some viable strategies, such as incorporating some natural phenolic components in any nanoformulation, as a part of codelivery approaches so that systems may efficiently deliver loads to the brain for expressing pharmacological actions [68].

However, to develop novel drug delivery systems for the treatment of brain diseases, it is critical to understand the anatomical structures and physiological functioning of BBB [69]. Targeted drug delivery to the brain involves some non-invasive novel formulation-based approaches because invasive techniques for site-specific drug delivery are very complicated and challenging in nature [70, 71]. To develop non-invasive novel drug delivery-based strategies, some planned integration of nanotechnology-based approaches is usually required so that loads may efficiently cross BBB [72]. This nanotechnology-based drug delivery may also enable specific But development of new drugs and novel formulations for brain diseases, requiring more time [71]. For example, owing to the anti-oxidant properties of essential oils, these exhibit potent protective actions towards the isolated neuronal cells [73, 74]. However, in *in vivo* systems, for efficient action of any neurotherapeuticals, they must cross the blood-brain barrier before reaching the brain.

7.3.2 The Blood-Cerebrospinal Fluid (CSF) Barrier

Cerebrospinal fluid (CSF) barrier explains anatomical structures, which impart restriction between the systemic circulation and the cerebrospinal fluid, i.e., function as a gateway to prevent the entry of outside chemicals into the CNS through filtration. Here, in the choroid plexus, epithelial cells form a tight junction as a mechanical barrier and bear fenestration between the capillaries to supply within the region. The surface area of CSF is around 1000 times lesser than BBB, so that even a reason for the role of the blood-CSF barrier is a hurdle in delivering drugs. Due to the presence of epithelial tight junctions, drugs and other molecules that, although managed to enter within the CSF peripheral area through these openings, are strictly regulated before entering CSF [75].

7.4 Approach for Nanotechnology-Enabled Delivery Systems and Their Prospects

Therapeutic impacts of bioactive neurotherapeuticals face several issues due to low permeability, restricted passage through biological barriers such as the blood-brain barrier (BBB), short-range systemic circulation time, and low bioavailability [76]. Therefore, the fabrication of these compounds for CNS targeted delivery also faces challenges due to low solubility, instability and biotransformation-related concerns with active loads [69]. However, such challenges in treating CNS disorders can be addressed by employing nanotechnology-based approaches to design suitable and efficient nanocarriers for brain-targeted drug delivery systems [63, 64]. Hence, in the last few decades, many nanocarriers are being rationalized and clinically used to treat CNS disorders because nanotherapeutics enable drug-targeting approaches to enhance both the efficacy and safety profile of loaded drugs. [77]. This field is emerging over the conventional system because it encapsulates drug molecules in nanocarriers such as liposomes, nanoparticles, nanoemulsion, polymeric nanomicelles, complex polymers nanocrystal, and nanofibers reach to the desired location from these nano-carriers. Reduction in the dimension of carriers loading neurotherapeuticals, usually some binding polymers or lipids are used, which alter the surface properties, increase solubility, enhance permeability, and improve overall bioavailability performances [77]. Here, the dimension of particles, composition of formulation and surface properties are critical factors to decide the fate of loads regarding their ability to cross the blood-brain barrier. However, the most significant impacts of this technique are to increase bioavailability, site targeted actions and reduction in side effects.

For treating several brain disorders such as Alzheimer's disease and brain cancer, poly (lactide-co-glycolic) acid, alginate, and other polymers are extensively used as a fabricating material to prepare nanoparticles of desired attributes as well as efficient encapsulation of therapeutic drugs [78]. While, several distinct advantages of NCs such as their high stability, enhanced capacity to encapsulate both hydrophilic and hydrophobic drugs, biological (enzymes, peptides or nucleic acids), and other bioactive substances as well as their rationalized formulation for the transdermal, parental, oral, and inhalational type of delivery routes. To deliver the neurotherapeuticals in CNS, there are always requirements and strategies to conceptualize NCs as penetrating approaches against impermeable BBB [79]. For example, with targeted drug delivery systems of CNS, the fabrication of NCs to deliver bioactive substances directly to the brain is challenging. For example, poorly water-soluble phenols were successfully encapsulated into the novel colloid-based delivery system and polymeric nano-micellar systems [80]. For example, nanoformulation of curcumin increased bioavailability and neurophysiological performances owing to a significant reduction in oxidative stress and inflammatory markers. These carriers also favor a greater extent of doxorubicin internalization with improved cytotoxic effects on cancer cells [63, 64].

7.5 Lipid Nanocarriers

Lipid-based drug delivery systems are in use since the 19th century because these overcome some limitations of conventional formulations. Three different lipid nanoparticles are available- homogenous drug-lipid matrix, drug enriched core, and drug enriched shell [81]. Hence, these different generations of lipid nanoparticles offer approaches to load both lipophilic and hydrophilic drugs in the matrix. Generally, biocompatible and tolerable lipids from classes of triglycerides, fatty acids, steroids, and waxes are usually preferred to fabricate such dosage forms [32]. These lipid-based carriers are evaluated for parameter particle size and size distribution, zeta potential, polymorphism, degree of crystalline, drug loading, entrapment efficiency, and drug release type of parameters [31]. Nanoparticles easily bypass the reticuloendothelial system (RES) because of their small size and high drug encapsulation efficiency.

Thus, these are effectively used to deliver the active loads within the CNS region to target specific brain tissues. Other benefits of lipid nanoparticles are the characteristic establishment of gradient flows from the blood to brain tissue, opening tight junctions to ease passage through BBB, favoring transcytosis of lipid nanoparticles through the endothelium layer, and increased duration of retention time in brain capillaries (Figure 7.2).

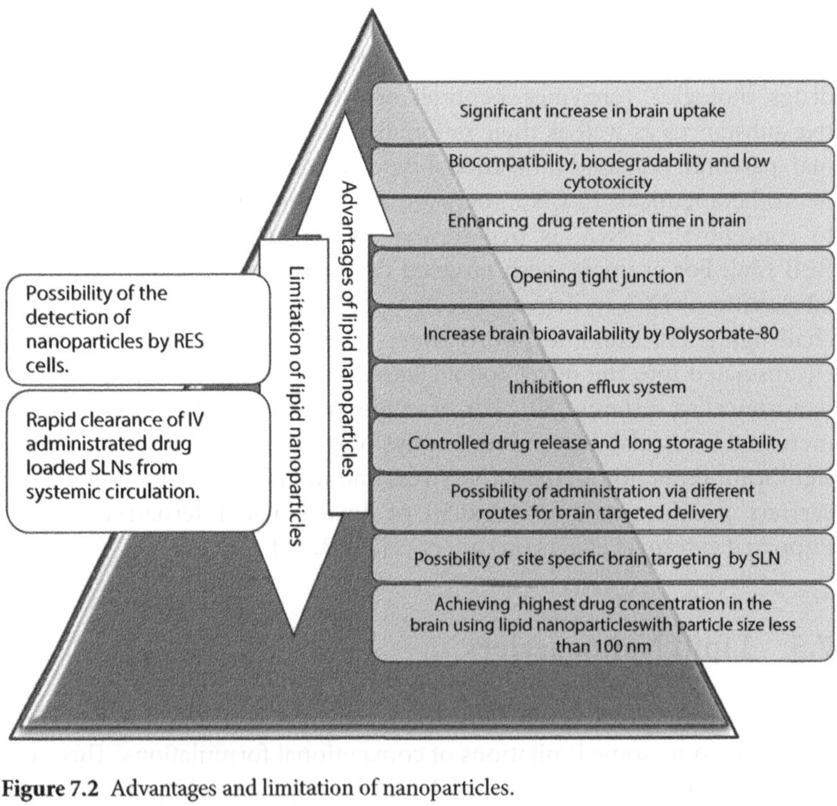

Figure 7.2 Advantages and limitation of nanoparticles.

Subsequently, integration of judiciously selected and appropriate surfactants also offers higher stabilities, desired particle size, enhanced accumulation drugs in the brain and better therapeutic responses [81]. Many lipid nanocarriers are also used to encapsulate various natural compounds to treat brain diseases, so these carriers promise hope in drug discovery against brain disease.

7.5.1 Solid Lipid Nanoparticles (SLN) and Nanostructured Lipid Carriers (NLC)

Among lipid-based delivery systems, solid lipid nanoparticles (SLNs) and nanostructured lipid carriers (NLCs) are viral and draw many interests. Compared to the liposome, lipid NC with solid matrix preferred for low-aqueous soluble drugs, owing to the advantages of enhanced protections for the incorporated active compounds and a good range of stability. SLN comprises saturated monoacid triglycerides and hard fats, solid at the body and room temperature. The solid lipid matrix prevents the mobility of the incorporated active compounds and controls the release from the carrier. While in comparison to SLN, NLCs have less ordered solid matrix, but higher fractions of active compounds can be loaded, require lesser possessing efforts and lesser incidence of undesirable expulsion of drugs from the matrix [82] (Table 7.1).

7.5.2 Ethosomes

Ethosomes are a new generation derivative of generic liposomes, which are frequently used to promote transdermal permeation of lipophilic drugs for their effectiveness in the percutaneous drug delivery route, from topical application of the formulation. Ethosomes are highly stable at room temperature, bearing good entrapment efficiency and improved compatibility with stratum corneum. These are mainly composed of phospholipids, water and higher quantities of ethanol (20–45%). From *in vitro* and *in vivo* studies based on observation, ethosomal systems are well established to improve drug delivery via transdermal routes. These ethosomal carriers are exciting and efficient to ensure lipophilic and hydrophilic drugs within deep layers of the skin. Although the exact mechanism of interactions between the ethosomes and the skin is still unclear, ethanol interacts with lipid molecules of cell membranes to enhance the membrane's fluidity. Briefly, due to high solubilization efficiency, ethanol favors drugs in solution state within the carriers. It distorts cell membranes' lipid structures so

Table 7.1 SLNs and NLCs for brain disorders.

S. no.	Nano-products	Route of administration	Facts	References
1	Resveratrol and Grape Extract-loaded Solid Lipid Nanoparticles SLNs).	Intravenous	Brain disorder improved more efficiently by resveratrol and grape extract delivery into the brain using SLNs. In addition, high cellular uptake of the OX26 monoclonal antibody in endothelial cells of the brain had shown.	[83]
2	Curcumin Solid Lipid Nanoparticles and Nanostructured Lipid Carriers.	Intravenous	Curcumin bioavailability in the brain founded high with entrapment efficiency in SLNs, and NLCs were found 82% ± 0.49 and 94% ± 0.74, respectively. Curcumin-loaded NLCs (AUC0-t = 505.76 ng/g h) and curcumin-loaded SLNs (AUC0-t = 116.31 ng/g h) that is very high as compare to curcumin dosage form.	[84]
3	Solid lipid nanoparticles hydroxy safflor yellow A (SLNs-HSYA).	Oral	The absorption of HSYA increased by 3.97 fold, and less cytotoxicity was found in Caco-2 and Hela cells in the brain with SLNs-HSYA formulation usage. Thus, overall therapeutic effects were high.	[85]
4	Nicotinamide encapsulated solid lipid nanoparticles (NTM-SLNs).	Intraperitoneal injection	For treating Alzheimer's disease, this formulation had shown very effectively that preserve the neuronal cells and decline the tau hyperphosphorylation in rat animal model.	[86]

(*Continued*)

Table 7.1 SLNs and NLCs for brain disorders. (*Continued*)

S. no.	Nano-products	Route of administration	Facts	References
5	Rizatriptan Benzoate-encapsulated Solid Lipid Nanoparticles (RB-SLNs)	Intravesnous	The RB-SLNs optimized by the Box-Behnken design (BBD) suggested particle size 71.8 ± 1.9% and entrapment efficiency 71.8±1.9%. 18.43 folds higher brain uptake of these nanoformulations compare to the pure drug in its free form.	[87]
6	Baicalin-loaded cationic solid lipid nanoparticles.	-	PEG-CSLNs are an excellent carrier that crosses the BBB and delivers a drug into the brain. The high therapeutic result had shown when baicalin-loaded cationic solid lipid nanoparticles were used to treat brain disorders.	[88]
7	Thymoquinone-loaded solid lipid nanoparticles (TQ-SLNs).	Neurons	The brain disorder Huntington's treatment with TQ-SLNs gives high therapeutic value. TQ optimum level counteracts 3-Nitropropionic acid-induced motor impairments, improves tyrosine hydroxylase (TH) immune reactive neurons.	[89]

(*Continued*)

Table 7.1 SLNs and NLCs for brain disorders. (*Continued*)

S. no.	Nano-products	Route of administration	Facts	References
8	Sumatriptan succinate loaded chitosan solid lipid nanoparticles (STS-CT-SLNs).	Oral	The drug brain/blood ratio increases 4.54 fold, the optimized formulation has a particle size, zeta potential and entrapment efficiency of 192-301.4nm, 30.2-51.4mV and 76.3-91.1%, respectively. This formulation is very effective for the management of migraines.	[90]
9	Dual-Modified Novel Biomimetic Nanocarriers.	Cells	To treat brain disorders such as glioma, formulated a red blood cell membrane coated solid lipid nanoparticles (RB-CS-SLNs) based nanocarriers used, deliver vinca alkaloid vincristine as a better therapeutic alternative.	[91]
10	Tamoxifen-conjugated solid lipid nanoparticles (TM-SLNs)	Cells	The BBB permeability coefficient for etoposide enhanced by Melanotransferrin antibody (MA) and Tamoxifen (TX) conjugated on etoposide (ETP)- entrapped solid lipid nanoparticles formulation target the human brain microvascular endothelial and U87MG cells.	[92]

Table 7.2 Ethosomes for CNS diseases.

S. no.	Nano-product	Route of administration	Effects	References
1.	Deliver vinpocetine and piracetam by ethosome.	Intranasal	Two cognitive enhancers encapsulated and delivered in the brain with percentages of 29.99 and 57.78 % vinpocetine and piracetam, respectively, which effectively decreased acetylcholinesterase activity and increased lipid peroxidation by 32-42%.	[94]
2.	Ethosomes-Ligustrazine phosphate (ETH-LP)	Transdermal	The therapeutic effect of ETH-LP had founded significantly higher than the aqueous one for the treatment of Alzheimer's disease (AD). Using this formulation could be recovered antioxidant enzyme activities and an optimum level of MDA.	[57]
3.	Novel nanovesicular drug delivery systems like ethosomes	---------------	To cure CNS diseases like Alzheimer's and Parkinson's, the ethosomes are the best carrier. The advantage of this carrier is biocompatible, biodegradable and less immunogenic and capable of carrying high drug payloads to deliver both hydrophilic and lipophilic drugs.	[95]

(*Continued*)

Table 7.2 Ethosomes for CNS diseases. (*Continued*)

S. no.	Nano-product	Route of administration	Effects	References
4.	Griseofulvin-Ethosomal vesicular carrier	Transdermal	Griseofulvin delivers through ethosomal vesicular carrier observed a better result, developed VCs containing varying amounts of ethanol, average spherical vesicles of an average size of 137.70 ± 51.62 nm and $68.0 \pm 5.6\%$ entrapment efficiency (EE).	[96]
5.	Ethosomes as Vesicles	Transdermal	Many CNS drugs belonging to different therapeutic classes deliver through ethosomes give better permeability and overall better therapeutic effects than traditional dosage forms.	[97]

that higher retention and permeation of drugs felicitate higher fractions to reach within systemic circulation via deeper layers of skin. Therefore, this approach is a reasonable means for the efficient delivery of bio-active loads within the CNS via non-invasive and non-oral routes. Some studies also suggested that ethosomes might also promote the permeability of drugs across the BBB [93] [Table 7.2].

7.5.3 Nanoemulsions

To overcome the major drawbacks associated with conventional drug delivery, various types of nano-sized emulsions (10–100 nm) are nowadays being conceptualized as an advanced mode of delivery to improve the performance of active pharmaceutical ingredients. Tiny droplets of nanoemulsion impart higher surface area for absorption, enhanced bioavailability due to felicitated dermal and mucosal transportation, and unique transparent optical properties [98]. The main components of nanoemulsion are oil, water, and emulsifying/stabilizing. Here, judicious selection of surfactants and their concentration is critical in nanoemulsion because surfactants reduce interfacial tension and impart stability. Such a surfactant role is a primary requirement concentration for turning micron-sized lipid globules into nano-sized droplets of emulsions [99].

Furthermore, many lipophilic drugs are being used as a neurotherapeutic, so nanoemulsion-based approaches help efficiently load such drugs within the oil phase, favoring higher penetration through biological membranes and barriers. Here, drug-loaded lipid droplets of nanoemulsion interact with skin lipids interact and are solubilized within lipid bilayers of cells, as one of the possible mechanisms of action [100]. For example, terpenes loaded (limonene, terpinolene, menthol, eucalyptol and nerolidol) nano-micelles exhibit significant colloidal stability and favor more bioavailability by enhanced permeation and accumulation in the skin [101,102]. Similarly, delivering p-cymene via transdermal routes is efficient as permeation enhancers. It enhances the fluidity of intercellular lipids in the stratum corneum to modulate intracellular pathways [103]. Ahad *et al.* research study focused upon anethole, menthone, and Eugenol as natural skin enhancers [104]. Many essential oils and their components enhance transdermal permeation and decrease resistance through BBB [105]. Kinds of the literature suggest that in comparison to synthetic permeation enhancers, natural agents should be preferred as a barrier-crossing activity of both lipophilic and hydrophilic compounds because of their neuroprotective actions and higher efficiency [106] [Table 7.3].

Table 7.3 Nanoemulsions for CNS disorders.

S. no.	Nano-emulsion	Route of administration	Particle size	Effects	References
1	Cefuroxime-loaded nanoemulsion (CLN)	Parenteral	100.32 ± 0.75 nm	This nanoemulsion optimized and found high penetration into the brain that enhanced the bioavailability overall high therapeutic value.	[107]
2	Lactoferrin Loaded Huperzine A Nanoemulsio (Lf-Hup A-NE)	Intranasal	15.24 ± 0.67 nm	Lf-HupA-NE formulation optimized with response methodology, it is an excellent selective acetylcholinesterase inhibitor. Drug availability increases in the central nervous system.	[108]
3	Kaempferol-loaded mucoadhesive nanoemulsion (KPF-MNE)	Intranasal	180.53 ± 4.90 nm	Free drug availability of KPF increases largely in CNS that induces apoptosis of glioma cells. Nanoemulsion prepared high-pressure homogenization technique.	[109]

(*Continued*)

Table 7.3 Nanoemulsions for CNS disorders. (*Continued*)

S. no.	Nano-emulsion	Route of administration	Particle size	Effects	References
4	Vitamin E Loaded Naringenin Nanoemulsion	Intranasal	38.70 ± 3.11 nm	In Parkinson's disease (PD), the concentration of 6-OHDA increases, its level would be down when used intranasal delivered Vitamin E loaded NRG NE along with levodopa.	[110]
5	Antimigraine drug loaded mucoadhesive nanoemulsion.	Intranasal	54.63 ± 3.24 nm	Zolmitriptan (ZT) is used for the treatment of migraines. The residence time increases in CNS by 0.3% concentration of chitosan as mucoadhesive in nanoemulsion.	[111]
6	Amiloride-nanoemulsion	Intranasal	89.36 ± 11.18 nm	For the treatment of seizures, the better bioavailability of drug amiloride in CNS is shown in Nanoemulsion-nanoformulation.	[79]
7	Vitamin E loaded resveratrol nanoemulsion	Intranasal	102 ± 1.46 nm	To treat Parkinson's disease, this nanoemulsion was founded to be effective because the pharmacology of the drug suggested reducing oxidative stress, i.e., antioxidant activity.	[112]

(*Continued*)

Table 7.3 Nanoemulsions for CNS disorders. (*Continued*)

S. no.	Nano-emulsion	Route of administration	Particle size	Effects	References
8	Nanoemulsions Riluzole	Intranasal	23.92 ± 0.52 nm	The CNS disorder Amyotrophic Lateral Sclerosis (ALS) a motor neuron disease (MND). The drug used for the treatment is riluzole, O/W nanoemulsion gives a high level of bioavailability in the brain.	[113]
9	Tanshinone I (TSI) nanoemulsion modified with lactoferrin.	----	20.96 ± 0.15 nm	TSI is a bioactive compound used to treat brain disorders. Still, limitations like poor water solubility, blood-brain barrier permeability, and low bioavailability can be overcome by TSI nanoemulsion modified with lactoferrin.	[114]
10	Selegiline Nanoformu-lation	Intranasal	61.43 ± 4.10 nm	The antioxidant properties of selegiline cure brain disorders. Its traditional dose has significantly less bioavailability that is enhanced in nanoformulation.	[70]

7.6 Conclusion and Prospects

Disturbances in the central or peripheral nervous system may be because of a nutrient-deficient diet. Ancient people rely on spices and natural products to treat different ailments, shown a remarkable result. For CNS disorders, many bioactive compounds are used to cure the disease. The advancement of technology creates an opportunity to examine and utilize phytochemicals that have therapeutic properties, led to focus on research in the nutrition of human health to dietary substances, with this innovation a term called nutraceuticals a combination of nutrition and pharmaceuticals [115]. The low pharmacokinetic properties of herbal medicine/natural compounds inhibit its therapeutic potential. Such limitations can be overcome by incorporating herbal medicine into various drug delivery formulations. Nanotechnology provides nanoencapsulation, a prominent new area of research for drug delivery in the last few years. Sometimes, nutraceuticals' role strengthens the therapeutic effects of medicine by joining a mechanism or several pathways such as increased reuptake of prevented monoamines and special neurobiological effects [116]. Using nutraceuticals has several benefits like a healthy diet, provides longer life, assists in psychological benefits, and is frequently used to treat neurological disease; fewer side effects, more population significantly older prefer nutrient-rich foods for the lifestyle-related disease [117]. Furthermore, neutraceuticals curative effects impute their neuroprotective, anti-inflammatory, anti-oxidant, hypolipidemic, and healing properties by targeting various ligands and receptors that enhance protein synthesis. Finally, it leads to neuroprotection. Monoterpenes can be inhibited protein folding, and their degradation leads to a healthy nervous system. Experiment research on monoterpenes nutraceuticals revealed a new era for treating, preventing, and curative effects of CNS disorder.

The body's defense mechanism deteriorated because of lifestyle changes; scavenging free oxygen radicals suppressed by antioxidants leads to overloaded oxidative stress. With the age level of antioxidants decrease in the body. Nanotechnologies application in clinical neuroscience is still in the infancy stage; scientists have a great interest in this area, on the potential of this novel delivery for a better therapeutic effect. Various nanoscale delivery systems are available among these lipid nanocarriers are one of the efficient delivery process systems. The primary function of nanocarriers (NC) is to give a platform where bioactive compounds shields from destruction in an early stage with better pharmacokinetic properties. For brain treatment, disorder monoterpenes are alternatives but need nanocarrier systems to deliver the bioactive compounds to the central nervous systems in optimum concentrations. Moreover, there are particular challenges associated with monoterpenes' conventional delivery: high volatility, short shelf life, solubility as monoterpenes are lipophilic, physical and chemical stability, high dose requirement, and blood-brain barrier.

Addressing these challenges by incorporating monoterpenes in nanoformulations, especially targeting transdermal nanoparticles, can help overcome this blood-brain barrier issue. Various drug targeting and carrier systems (liposomes, nanoparticles, nanoemulsion, polymeric nanomicelles, ethosomes, complex polymers nanocrystal, and nanofibers) and different routes of drug administration intranasal, buccal, and transdermal. Among these routes of administration approaching transdermal nanoparticles, ethosomes can be seen as very specific to overcome this blood-brain barrier issue, which is a significant hurdle in brain delivery. Future challenges lie in a more in-depth characterization of their therapeutic mechanisms and developing effective, safe, and brain-targeted delivery systems for their transdermal administration. Even so, nutraceuticals exhibit properties that vary from person to person. Regularly consuming a recommended dosage encourages good neurological health.

References

1. Pottoo, F.H., Tabassum, N., Javed, M.N., Nigar, S., Sharma, S., Barkat, M.A., Alam, M.S., Ansari, M.A., Barreto, G.E., Ashraf, G.M., Raloxifene potentiates the effect of fluoxetine against maximal electroshock induced seizures in mice. *Eur. J. Pharm. Sci.*, 146, 105261, 2020.
2. Javed, M.N., Ashraf, G.M., Barreto, G.E., Naim, M., Current trends for rationalizing brain targeting nanoparticles in neurological disorders (Part-I). *Curr. Drug Metab.*, 21, 9, 648, 2020.
3. Pottoo, F.H., Sharma, S., Javed, M.N., Barkat, M.A., Harshita, Alam, M.S., Naim, M.J., Alam, O., Ansari, M.A., Barreto, G.E., Ashraf, G.M., Lipid-based nanoformulations in the treatment of neurological disorders. *Drug Metab. Rev.*, 52, 1, 185–204, 2020.
4. Pottoo, F.H., Tabassum, N., Javed, M.N., Nigar, S., Rasheed, R., Khan, A., Barkat, M.A., Alam, M.S., Maqbool, A., Ansari, M.A., Barreto, G.E., The synergistic effect of raloxifene, fluoxetine, and bromocriptine protects against pilocarpine-induced status epilepticus and temporal lobe epilepsy. *Mol. Neurobiol.*, 56, 1233–1247, 2019.
5. Kumar, A., Walia, H., Pottoo, F.H., Javed, M.N., Insights of nanophytomedicines as a combinatorial therapy in disease diagnosis and treatment, in: *Nanophytomedicine*, S. Beg, M. Barkat, F. Ahmad (Eds.), pp. 113–132, 2020.

6. Verma, R., Vyas, P., Kaur, J., Javed, M., Sarafroz, M., Ahmad, M., Gilani, S.J., Taleuzzaman, M., Approaches for ear-targeted delivery systems in neurosensory disorders to avoid chronic hearing loss mediated neurological diseases. *CNS Neurol. Disord.-Drug Targets (Formerly Curr. Drug Targets-CNS & Neurological Disorders)*, 6, 22, 479–491, 2022.
7. Shatzmiller, S., Lapidot, I., Zats, G., Blood brain barrier crossing for therapeutic and diagnostic agents. *SM J. Neurol. Disord. Stroke*, 2, 1012, 2016.
8. Alam, M., Javed, M.N., Najmi, A.K., Ahmad, F.J., Imam, S.S., Akhtar, M., Thymoquinone lipid nanoparticles cut the Gordian knots of depression via neuroprotective BDNF and downregulation of neuro-inflammatory NF-κB, IL-6, and TNF-α in LPS treated rats. *Curr. Drug Metab.*, 22, 12, 978–988, 2021.
9. Aslam, M., Javed, M., Deeb, H.H., Nicola, M.K., Mirza, M., Alam, M., Akhtar, M., Waziri, A., Lipid nanocarriers for neurotherapeutics: Introduction, challenges, blood-brain barrier, and promises of delivery approaches. *CNS Neurol. Disord.-Drug Targets (Formerly Curr. Drug Targets-CNS & Neurological Disorders)*, 2021.
10. Javed, N., Pottoo, F.H., Baretto, G.E., Nano-nutraceuticals dietary supplements for athletes: An eminent approach for managing traumatic brain injuries. *Curr. Mol. Pharmacol.*, 15, 1, 1–2, 2022.
11. Pottoo, F.H., Javed, M., Barkat, M., Alam, M., Nowshehri, J.A., Alshayban, D.M., Ansari, M.A., Estrogen and serotonin: Complexity of interactions and implications for epileptic seizures and epileptogenesis. *Curr. Neuropharmacol.*, 17, 3, 214–231, 2019.
12. Murray, C.J., Lopez, A.D. and World Health Organization, *The Global burden of disease: A comprehensive assessment of mortality and disability from diseases, injuries, and risk factors in 1990 and projected to 2020*, 1, 1996. https://apps.who.int/iris/handle/10665/41864.
13. Pandey, M., Saleem, S., Nautiyal, H., Pottoo, F.H., Javed, M.N., PINK1/Parkin in neurodegenerative disorders: Crosstalk between mitochondrial stress and neurodegeneration, in: *Quality Control of Cellular Protein in Neurodegenerative Disorders*, M. Uddin, and G.M. Ashraf, (Eds.), pp. 282–301, IGI Global, Hershey, USA, 2020.
14. Janca, A., Parkinson's disease from WHO perspective and a public health point of view. *Parkinsonism Relat. Disord.*, 9, 1, 3–6, 2002.
15. Janca, A., Aarli, J.A., Prilipko, L., Dua, T., Saxena, S., Saraceno, B., WHO/WFN survey of neurological services: A worldwide perspective. *J. Neurol. Sci.*, 247, 1, 29–34, 2006.
16. Pottoo, F.H., Barkat, M.A., Ansari, M.A., Javed, M.N., Jamal, Q.M., Kamal, M.A., Nanotechnological based miRNA intervention in the therapeutic management of neuroblastoma. *Semin. Cancer Biol.*, 69, 100–108, 2021.
17. Pottoo, F.H., Javed, M.N., Rahman, J.U., Abu-Izneid, T., Khan, F.A., Targeted delivery of miRNA based therapeuticals in the clinical management of Glioblastoma Multiforme. *Semin. Cancer Biol.*, 69, 391–398, 2021.

18. Nigar, S., Pottoo, F.H., Tabassum, N., Verma, S.K., Javed, M.N., Molecular insights into the role of inflammation and oxidative stress in epilepsy. *J. Adv. Med. Pharm. Sci.*, 10, 1, 1–9, 2016.
19. Wittchen, H.U., Jacobi, F., Rehm, J., Gustavsson, A., Svensson, M., Jönsson, B., Olesen, J., Allgulander, C., Alonso, J., Faravelli, C., Fratiglioni, L., The size and burden of mental disorders and other disorders of the brain in Europe. *Eur. Neuropsychopharmacol.*, 21, 655–679, 2011.
20. Khan, F., Owolabi, M.O., Amatya, B., Hamzat, T.K., Ogunniyi, A., Oshinowo, H., Elmalik, A., Galea, M.P., Challenges and barriers for implementation of the World Health Organization global disability action plan in low- and middle-income countries. *J. Rehabil. Med.*, 50, 367–376, 2018.
21. Ahmad, N., Ahmad, R., Abbas Naqvi, A., Ashafaq, M., Alam, M.A., Ahmad, F.J., Al-Ghamdi, M.S., The effect of safranal loaded mucoadhesive nanoemulsion on oxidative stress markers in cerebral ischemia. *Artif. Cells Nanomed. Biotechnol.*, 45, 4, 775–787, 2017.
22. Gaziano, T.A., Bitton, A., Anand, S., Abrahams-Gessel, S., Murphy, A., Growing epidemic of coronary heart disease in low- and middle-income countries. *Curr. Probl. Cardiol.*, 35, 2, 72–115, 2010.
23. Mehndiratta, M.M. and Wadhai, S.A., International Epilepsy Day - A day notified for global public education & awareness. *Indian J. Med. Res.*, 141, 2, 143–144, 2015.
24. Christensen, K., Doblhammer, G., Rau, R., Vaupel, J.W., Ageing populations: The challenges ahead. *Lancet*, 374, 9696, 1196–1208, 2009.
25. Masserini, M., Nanoparticles for brain drug delivery. *Int. Sch. Res. Notices*, 2013, 1–18, 2013, Article ID 238428. https://doi.org/10.1155/2013/238428.
26. Miranda, J.J., Kinra, S., Casas, J.P., Davey Smith, G., Ebrahim, S., Non-communicable diseases in low- and middle-income countries: Context, determinants and health policy. *Trop. Med. Int. Health*, 13, 10, 1225–1234, 2008.
27. GBD 2017 Mortality Collaborators, Global, regional, and national age-sex-specific mortality and life expectancy, 1950-2017: A systematic analysis for the Global Burden of Disease Study 2017. *Lancet*, 392, 10159, 1684–1735, 2018.
28. Khatib, R., Arevalo, Y.A., Berendsen, M.A., Prabhakaran, S., Huffman, M.D., Presentation, evaluation, management, and outcomes of acute stroke in low- and middle-income countries: A systematic review and meta-analysis. *Neuroepidemiology*, 51, 1-2, 104–112, 2018.
29. Raman, S., Mahmood, S., Hilles, A.R., Javed, M.N., Azmana, M., Al-Japairai, K.H.S., Polymeric nanoparticles for brain drug delivery - A review. *Curr. Drug Metab.*, 21, 1, 2020, https://doi.org/10.2174/138920022166620050807 4348.
30. Vergallo, C., Nutraceutical vegetable oil nanoformulations for prevention and management of diseases. *Nanomaterials*, 10, 6, 1232, 2020.

31. Javed, M.N., Alam, M.S., Waziri, A., Pottoo, F.H., Yadav, A.K., Hasnain, M.S., Almalki, F.A., QbD applications for the development of nanopharmaceutical products, in: *Pharmaceutical quality by design*, pp. 229–253, Academic Press, London, UK, 2019.
32. Javed, M.N., Kohli, K., Amin, S., Risk assessment integrated QbD approach for development of optimized bicontinuous mucoadhesive limicubes for oral delivery of rosuvastatin. *AAPS PharmSciTech*, 19, 1377–1391, 2018, https://doi.org/10.1208/s12249-018-0951-1.
33. Javed, M.N., Dahiya, E.S., Ibrahim, A.M., Alam, M.S., Khan, F.A., Pottoo, F.H., Recent advancement in clinical application of nanotechnological approached targeted delivery of herbal drugs, in: *Nanophytomedicine*, S. Beg, M. Barkat, F. Ahmad (Eds.), Springer, Singapore, 2020, https://doi.org/10.1007/978-981-15-4909-0_9.
34. Turek, C. and Stintzing, F.C., Stability of essential oils: A review. *Compr. Rev. Food Sci. Food Saf.*, 12, 1, 40–53, 2013.
35. Hădărugă, D.I., Hădărugă, N.G., Costescu, C.I., David, I., Gruia, A.T., Thermal and oxidative stability of the Ocimum basilicum L. essential oil/Îβ-cyclodextrin supramolecular system. *Beilstein J. Org. Chem.*, 10, 1, 2809–2820, 2014.
36. Mohanta, B.C., Javed, M.N., Hasnain, M.S., Nayak, A.K., Polyelectrolyte complexes of alginate for controlling drug release, in: *Alginates in Drug Delivery*, Chapter-12, pp. 297–321, 2020.
37. Hasnain, M.S., Javed, M.N., Alam, M.S., Rishishwar, P., Rishishwar, S., Ali, S., Nayak, A.K., Beg, S., Purple heart plant leaves extract-mediated silver nanoparticle synthesis: Optimization by Box-Behnken design. *Mater. Sci. Eng.: C*, 99, 1105–1114, 2019.
38. Alam, M.S., Javed, M.N., Pottoo, F.H., Waziri, A., Almalki, F.A., Hasnain, M.S., Garg, A., Saifullah, M.K., QbD approached comparison of reaction mechanism in microwave synthesized gold nanoparticles and their superior catalytic role against hazardous nirto-dye. *Appl. Organomet. Chem.*, 33, 9, e5071, 2019.
39. Singhal, S., Gupta, M., Alam, M.S., Javed, M.N., Ansari, J.R., Carbon allotropes-based nanodevices: Graphene in biomedical applications, in: *Nanotechnology Device Design and Applications*, pp. 241–269, CRC Press, USA, 2022.
40. Kumar, R., Dhamija, G., Ansari, J.R., Javed, M.N., Alam, M.S., C-dot nanoparticulated devices for biomedical applications, in: *Nanotechnology Device Design and Applications*, pp. 271–299, CRC Press, USA, 2022.
41. Akhter, M., Ahmad, J., Javed, M., Haque, R., Khalilullah, H., Gupta, M., Ali, J., Porphyrin-based nanomaterials for cancer nanotheranostics, in: *Cancer Nanotheranostics*, pp. 275–295, Springer, Cham, 2021.
42. Schweiggert, U.R., Carle, Schieber, A., Conventional and alternative processes for spice production – A review. *Trends Food Sci. Technol.*, 18, 5, 260–268, 2007.

43. Pottoo, F.H., Abu-Izneid, T., Ibrahim, A.M., Javed, M.N., AlHajri, N., Hamrouni, A.M., Immune system response during viral infections: Immunomodulators, cytokine storm (CS) and immunotherapeutics in COVID-19. *Saudi Pharm. J.: SPJ*, 29, 2, 173–187, 2021.
44. Christensson, J.B., Forsström, P., Wennberg, A.M., Karlberg, A.T., Matura, M., Air oxidation increases skin irritation from fragrance terpenes. *Contact Derm.*, 60, 1, 32–40, 2009.
45. Divkovic, M., Pease, C.K., Gerberick, G.F., Basketter, D.A., Hapten–protein binding: From theory to practical application in the *in vitro* prediction of skin sensitization. *Contact Derm.*, 53, 4, 189–200, 2005.
46. Sköld, M., Karlberg, A.T., Matura, M., Börje, A., The fragrance chemical β-caryophyllene—air oxidation and skin sensitization. *Food Chem. Toxicol.*, 44, 4, 538–45, 2006.
47. Kaleem, M., Alhosin, M., Khan, K., Ahmad, W., Hosawi, S., Nur, S.M., Choudhry, H., Zamzami, M.A., Al-Abbasi, F.A., Javed, M.D., Epigenetic basis of polyphenols in cancer prevention and therapy, in: *Polyphenols-Based Nanotherapeutics for Cancer Management*, pp. 189–238, Springer, Singapore, 2021.
48. Sarigiannis, D.A., Karakitsios, S.P., Gotti, A., Liakos, I.L., Katsoyiannis, A., Exposure to major volatile organic compounds and carbonyls in European indoor environments and associated health risk. *Environ. Int.*, 37, 4, 743–65, 2011.
49. Javed, M.N., Pottoo, F.H., Shamim, A., Hasnain, M.S., Alam, M.S., Design of experiments for the development of nanoparticles, nanomaterials, and nanocomposites, in: *Design of Experiments for Pharmaceutical Product Development*, pp. 151–169, Springer, Singapore, 2021.
50. Bharti, C., Alam, M.S., Javed, M.N., Khalid, M., Saifullah, F.A., Manchanda, R., Silica based nanomaterial for drug delivery, in: *Nanomaterials: Evolution and Advance Towards Therapeutic Drug Delivery (Part 2)*, pp. 57–89, Bentham Science Publishers, Sharjah, UAE, 2021.
51. Rodenak-Kladniew, B., Islan, G.A., de Bravo, M.G., Durán, N., Castro, G.R., Design, characterization and *in vitro* evaluation of linalool-loaded solid lipid nanoparticles as potent tool in cancer therapy. *Colloids Surf. B: Biointerfaces*, 154, 123–32, 2017.
52. Javed, M.N., Ashraf, G.M., Barreto, G.E., Naim, M., Trends in rationalized brain targeted nanoparticles in neurological disorders (Part-III). *Curr. Drug Metab.*, 22, 4, 250, 2021.
53. Waziri, A., Bharti, C., Aslam, M., Jamil, P., Mirza, A., Javed, M.N., Pottoo, U., Ahmadi, A., Alam, M.S., Probiotics for the Chemoprotective role against the toxic effect of cancer chemotherapy. *Anti-Cancer Agents Med. Chem.*, 22, 4, 654-667, 2022.

54. Alam, M.S., Naseh, M.F., Ansari, J.R., Ahmadi, A., Saifullah, M.K., Garg, A., Approaches of synthesis for higher yields of nanoparticles, in: *Nanomaterials in the Battle Against Pathogens and Disease Vectors*, CRC Press, USA, 2022.
55. Zhu, G.Y., Xiao, Z.B., Zhou, R.J., Yi, F.P., Fragrance and flavor microencapsulation technology, in: *Advanced Materials Research*, vol. 535, pp. 440–445, Trans Tech Publications Ltd, Freienbach, Switzerland, 2012.
56. Pandit, J., Alam, M.S., Ansari, J.R., Singhal, M., Gupta, N., Waziri, A., Sharma, K., Potto, F.H., Multifaced applications of nanoparticles in biological science, in: *Nanomaterials in the Battle Against Pathogens and Disease Vectors*, CRC Press, USA, 2022.
57. Shi, J., Wang, Y., Luo, G., Ligustrazine phosphate ethosomes for treatment of Alzheimer's disease, *in vitro* and in animal model studies. *AAPS PharmSciTech*, 13, 2, 485–92, 2012.
58. Soottitantawat, A., Takayama, K., Okamura, K., Muranaka, D., Yoshii, H., Furuta, T., Ohkawara, M., Linko, P., Microencapsulation of l-menthol by spray drying and its release characteristics. *Innovative Food Sci. Emerg. Technol.*, 6, 2, 163–170, 2005.
59. Javed, M.N., Akhter, M.H., Taleuzzaman, M., Faiyazudin, M., Alam, M.S., Cationic nanoparticles for treatment of neurological diseases, in: *Fundamentals of Bionanomaterials*, pp. 273–292, Elsevier, Netherlands, 2022.
60. Li, J., Wei, X., Wang, Q., Chen, J., Chang, G., Kong, L., Su, J., Liu, Y., Homogeneous isolation of nanocellulose from sugarcane bagasse by high pressure homogenization. *Carbohydr. Polym.*, 90, 4, 1609–1613, 2012.
61. Mishra, S., Sharma, S., Javed, M.N., Pottoo, F.H., Barkat, M.A., Alam, M.S., Amir, M., Sarafroz, M., Bioinspired nanocomposites: Applications in disease diagnosis and treatment. *Pharm. Nanotechnol.*, 7, 3, 206–219, 2019.
62. Sharma, S., Javed, M.N., Pottoo, F.H., Rabbani, S.A., Barkat, M.A., Harshita, Sarafroz, M., Amir, M., Bioresponse inspired nanomaterials for targeted drug and gene delivery. *Pharm. Nanotechnol.*, 7, 3, 220–233, 2019.
63. Teleanu, D.M., Chircov, C., Grumezescu, A.M., Volceanov, A., Teleanu, R.I., Blood-brain delivery methods using nanotechnology. *Pharmaceutics*, 10, 4, 269, 2018.
64. Teleanu, D.M., Negut, I., Grumezescu, V., Grumezescu, A.M., Teleanu, R.I., Nanomaterials for drug delivery to the central nervous system. *Nanomaterials (Basel)*, 9, 3, 371, 2019.
65. Subramaniyan Parimalam, S., Badilescu, S., Sonenberg, N., Bhat, R., Packirisamy, M., Lab-on-a-chip for the development of pro-/anti-angiogenic nanomedicines to treat brain diseases. *Int. J. Mol. Sci.*, 20, 24, 6126, 2019.
66. Saraiva, C., Praça, C., Ferreira, R., Santos, T., Ferreira, L., Bernardino, L., Nanoparticle-mediated brain drug delivery: Overcoming blood-brain barrier to treat neurodegenerative diseases. *J. Control. Release*, 235, 34–47, Aug 10, 2016.

67. He, Q., Liu, J., Liang, J., Liu, X., Li, W., Liu, Z., Ding, Z., Tuo, D., Towards improvements for penetrating the blood-brain barrier-recent progress from a material and pharmaceutical perspective. *Cells*, 7, 4, 24, 23, 2018.
68. Fahmy, H.M., Fathy, M.M., Abd-Elbadia, R.A., Elshemey, W.M., Targeting of thymoquinone-loaded mesoporous silica nanoparticles to different brain areas: *In vivo* study. *Life Sci.*, 1, 222, 94–102, 2019.
69. Patel, M.M. and Patel, B.M., Crossing the blood-brain barrier: Recent advances in drug delivery to the brain. *CNS Drugs*, 31, 109–133.6, 2017.
70. Kumar, S., Dang, S., Nigam, K., Ali, J., Baboota, S., Selegiline nanoformulation in attenuation of oxidative stress and upregulation of dopamine in the brain for the treatment of Parkinson's Disease. *Rejuvenation Res.*, 21, 5, 464–476, 2018.
71. Kumar, M., Sharma, P., Maheshwari, R., Tekade, M., Shrivastava, S.K., Tekade, R.K., Beyond the blood-brain barrier: Facing new challenges and prospects of nanotechnology-mediated targeted delivery to the brain, in: *Nanotechnology-Based Targeted Drug Delivery Systems for Brain Tumors*, pp. 397–437, Academic Press, London, UK, 2018.
72. Dong, X., Current strategies for brain drug delivery. *Theranostics*, 8, 1481–1493, 20182018.
73. Dobetsberger, C. and Buchbauer, G., Actions of essential oils on the central nervous system: An updated review. *Flavour Fragr. J.*, 26, 5, 300–316, 2014.
74. López, V., Nielsen, B., Solas, M., Ramírez, M.J., Jäger, A.K., Exploring Pharmacological Mechanisms of Lavender (Lavandula angustifolia) Essential Oil on Central Nervous System Targets. *Front. Pharmacol.*, 8, 280, 2017.
75. Barchet, T.M. and Amiji, M.M., Challenges and opportunities in CNS delivery of therapeutics for neurodegenerative diseases. *Expert Opin. Drug Deliv.*, 6, 3, 211–25, 2009.
76. Upadhyay, R.K., Drug delivery systems, CNS protection, and the blood brain barrier. *Biomed. Res. Int.*, 2014, 869269, 2014.
77. Jahangir, M.A., Anand, C., Muheem, A., Gilani, S.J., Taleuzzaman, M., Zafar, A., Jaffer, M., Verma, S., Barkat, M.A., Nano phytomedicine based delivery system for CNS disease. *Curr. Drug Metab.*, 21, 9, 661–673, 2020.
78. Patra, J.K., Das, G., Fraceto, L.F., Campos, E.V., del Pilar Rodriguez-Torres, M., Acosta-Torres, L.S., Diaz-Torres, L.A., Grillo, R., Swamy, M.K., Sharma, S., Habtemariam, S., Nano based drug delivery systems: Recent developments and future prospects. *J. Nanobiotechnol.*, 16, 1, 71, 2018.
79. Ahmad, N., Ahmad, R., Alam, M.A., Ahmad, F.J., Amir, M., Impact of ultrasonication techniques on the preparation of novel amiloride-nanoemulsion used for intranasal delivery in the treatment of epilepsy. *Artif. Cells Nanomed. Biotechnol.*, 46, sup3, S192–S207, 2018.
80. Ratheesh, G., Tian, L., Venugopal, J.R., Ezhilarasu, H., Sadiq, A., Fan, T.P., Ramakrishna, S., Role of medicinal plants in neurodegenerative diseases. *Biomanuf. Rev.*, 2, 1, 2, 2017.

81. Ghasemiyeh, P. and Mohammadi-Samani, S., Solid lipid nanoparticles and nanostructured lipid carriers as novel drug delivery systems: Applications, advantages and disadvantages. *Res. Pharm. Sci.*, 13, 4, 288–303, 2018.
82. Naseri, N., Valizadeh, H., Zakeri-Milani, P., Solid lipid nanoparticles and nanostructured lipid carriers: Structure, preparation and application. *Adv. Pharm. Bull.*, 5, 3, 305–313, 2015.
83. Loureiro, J.A., Andrade, S., Duarte, A., Neves, A.R., Queiroz, J.F., Nunes, C., Sevin, E., Fenart, L., Gosselet, F., Coelho, M.A., Pereira, M.C., Resveratrol and grape extract-loaded solid lipid nanoparticles for the treatment of Alzheimer's Disease. *Molecules*, 22, 2, 277, 2017.
84. Sadegh Malvajerd, S., Azadi, A., Izadi, Z., Kurd, M., Dara, T., Dibaei, M., Sharif Zadeh, M., Akbari Javar, H., Hamidi, M., Brain delivery of curcumin using solid lipid nanoparticles and nanostructured lipid carriers: Preparation, optimization, and pharmacokinetic evaluation. *ACS Chem. Neurosci.*, 10, 1, 728–739, 2019.
85. Zhao, B., Gu, S., Du, Y., Shen, M., Liu, X., Shen, Y., Solid lipid nanoparticles as carriers for oral delivery of hydroxysafflor yellow A. *Int. J. Pharm.*, 535, 1-2, 164–171, 15, 2018.
86. Vakilinezhad, M.A., Amini, A., Javar, H.A., Zarandi, B.F., Montaseri, H., Dinarvand, R., Nicotinamide loaded functionalized solid lipid nanoparticles improves cognition in Alzheimer's disease animal model by reducing Tau hyperphosphorylation. *Daru*, 26, 2, 165–177, 2018.
87. Girotra, P. and Singh, S.K., Multivariate optimization of Rizatriptan Benzoate-loaded solid lipid nanoparticles for brain targeting and migraine management. *AAPS PharmSciTech*, 18, 2, 517–528, 2017.
88. Liu, Z., Zhao, H., Shu, L., Zhang, Y., Okeke, C., Zhang, L., Li, J., Li, N., Preparation and evaluation of Baicalin-loaded cationic solid lipid nanoparticles conjugated with OX26 for improved delivery across the BBB. *Drug Dev. Ind. Pharm.*, 41, 3, 353–61, 2015.
89. Ramachandran, S. and Thangarajan, S., Thymoquinone loaded solid lipid nanoparticles counteracts 3-nitropropionic acid induced motor impairments and neuroinflammation in rat model of Huntington's disease. *Metab. Brain Dis.*, 33, 5, 1459–1470, 2018.
90. Hansraj, G.P., Singh, S.K., Kumar, P., Sumatriptan succinate loaded chitosan solid lipid nanoparticles for enhanced anti-migraine potential. *Int. J. Biol. Macromol.*, 81, 467–76, 2015.
91. Fu, S., Liang, M., Wang, Y., Cui, L., Gao, C., Chu, X. *et al.*, Dual-modified novel biomimetic nanocarriers improve targeting and therapeutic efficacy in glioma. *ACS Appl. Mater. Interfaces*, 11, 2, 1841–1854, 2019.
92. Kuo, Y.C. and Wang, I.H., Enhanced delivery of etoposide across the blood-brain barrier to restrain brain tumor growth using melanotransferrin antibody- and tamoxifen-conjugated solid lipid nanoparticles. *J. Drug Targeting*, 24, 7, 645–54, 2016.

93. Soares, T.B., Loureiro, L., Carvalho, A., Oliveira, M.E.C.R., Dias, A., Sarmento, B., Lúcio, M., Lipid nanocarriers loaded with natural compounds: Potential new therapies for age related neurodegenerative diseases. *Prog. Neurobiol.*, 168, 21–41, 2018.
94. Nasr, M. and Wahdan, S.A., Neuroprotective effects of novel nanosystems simultaneously loaded with vinpocetine and piracetam after intranasal administration. *Life Sci.*, 226, 117–129, 2019.
95. Neupane, Y.R., Mahtab, A., Siddiqui, L., Singh, A., Gautam, N., Rabbani, S.A. *et al.*, Biocompatible nanovesicular drug delivery systems with targeting potential for autoimmune diseases. *Curr. Pharm. Des.*, 26, 42, 5488–5502, 2020.
96. Mbah, C.C., Builders, P.F., Agubata, C.O. *et al.*, Development of ethosomal vesicular carrier for topical application of griseofulvin: Effect of ethanol concentration. *J. Pharm. Investig.*, 49, 27–36, 2019.
97. Akhtar, N., Varma, A., Pathak, K., Ethosomes as vesicles for effective transdermal delivery: From bench to clinical implementation. *Curr. Clin. Pharmacol.*, 11, 168, 2016, https://doi.org/10.2174/1574884711666160813231352.
98. Jaiswal, M., Dudhe, R., Sharma, P.K., Nanoemulsion: An advanced mode of drug delivery system. *3 Biotech.*, 5, 2, 123–12.
99. Azeem, A., Rizwan, M., Ahmad, F.J., Iqbal, Z., Khar, R.K., Aqil, M., Talegaonkar, S., Nanoemulsion components screening and selection: A technical note. *AAPS PharmSciTech*, 10, 1, 69–76, 2009.
100. Wojtunik Kulesza, K.A., Kasprzak, K., Oniszczuk, T., Oniszczuk, A., Natural monoterpenes: Much more than only a scent. *Chem. Biodivers.*, 16, 12, e1900434, 2019.
101. Kahraman, E., Neşetoğlu, N., Güngör, S., Ünal, D Ş., Özsoy, Y., The combination of nanomicelles with terpenes for enhancement of skin drug delivery. *Int. J. Pharm.*, 551, 1-2, 133–140, 2018.
102. Rangsimawong, W., Obata, Y., Opanasopit, P., Ngawhirunpat, T., Takayama, K., Enhancement of galantamine HBr skin permeation using sonophoresis and limonene-containing PEGylated liposomes. *AAPS PharmSciTech*, 19, 3, 1093–1104, 2018.
103. Nan, L., Liu, C., Li, Q., Wan, X., Guo, J., Quan, P., Fang, L., Investigation of the enhancement effect of the natural transdermal permeation enhancers from Ledum palustre L. var. angustum N. Busch: Mechanistic insight based on interaction among drug, enhancers and skin. *Eur. J. Pharm. Sci.*, 124, 105–113, 2018.
104. Ahad, A., Aqil, M., Ali, A., The application of anethole, menthone, and Eugenol in transdermal penetration of valsartan: Enhancement and mechanistic investigation. *Pharm. Biol.*, 54, 6, 1042–1051, 2016.
105. Adorjan, B. and Buchbauer, G., Biological properties of essential oils: An updated review. *Flavour Fragr. J.*, 25, 6, 407–426, 2010.

106. Chen, J., Jiang, Q.D., Chai, Y.P., Zhang, H., Peng, P., Yang, X.X., Natural terpenes as penetration enhancers for transdermal drug delivery. *Molecules*, 21, 12, 1709, 2016.
107. Harun, S.N., Nordin, S.A., Gani, S.S.A., Shamsuddin, A.F., Basri, M., Basri, H.B., Development of nanoemulsion for efficient brain parenteral delivery of cefuroxime: Designs, characterizations, and pharmacokinetics. *Int. J. Nanomedicine*, 13, 2571–2584, 2018.
108. Jiang, Y., Liu, C., Zhai, W., Zhuang, N., Han, T., Ding, Z., The optimization design of lactoferrin loaded HupA nanoemulsion for targeted drug transport via intranasal route. *Int. J. Nanomedicine*, 14, 9217–9234, 2019.
109. Colombo, M., Figueiró, F., de Fraga Dias, A., Teixeira, H.F., Battastini, A.M.O., Kaempferol-loaded mucoadhesive nanoemulsion for intranasal administration reduces glioma growth *in vitro*. *Int. J. Pharm.*, 543, 1-2, 214–223, 2018.
110. Gaba, B., Khan, T., Haider, M.F., Alam, T., Baboota, S., Parvez, S., Ali, J., Vitamin E loaded naringenin nanoemulsion via intranasal delivery for the management of oxidative stress in a 6-OHDA Parkinson's disease model. *Biomed. Res. Int.*, 2019, 2382563, 2019.
111. Abdou, E.M., Kandil, S.M., Miniawy, H.M.F.E., Brain targeting efficiency of anti migraine drug loaded mucoadhesive intranasal nanoemulsion. *Int. J. Pharm.*, 529, 1-2, 667–677, 2017.
112. Pangeni, R., Sharma, S., Mustafa, G., Ali, J., Baboota, S., Vitamin E loaded resveratrol nanoemulsion for brain targeting for the treatment of Parkinson's disease by reducing oxidative stress. *Nanotechnology*, 25, 48, 485102, 2014.
113. Parikh, R.H. and Patel, R.J., Nanoemulsions for intranasal delivery of riluzole to improve brain bioavailability: Formulation development and pharmacokinetic studies. *Curr. Drug Deliv.*, 13, 7, 1130–1143, 2016.
114. Wu, Y., Zhang, B., Kebebe, D., Guo, L., Guo, H., Li, *N.et al.*, Preparation, optimization and cellular uptake study of tanshinone I nanoemulsion modified with lactoferrin for brain drug delivery. *Pharm. Dev. Technol.*, 24, 8, 982–991, 2019.
115. Bungau, S.G. and Popa, V.C., Between religion and science some aspects concerning illness and healing in antiquity. *Transylv. Rev.*, 24, 3–18, 2015.
116. Van der Burg, K.P., Cribb, L., Firth, J., Karmacoska, D., Sarris, J., Nutrient and genetic biomarkers of nutraceutical treatment response in mood and psychotic disorders: A systematic review. *Nutr. Neurosci.*, 24, 4, 279–295, 2021.
117. Makkar, R., Behl, T., Bungau, S., Zengin, G., Mehta, V., Kumar, A., Uddin, M., Ashraf, G.M., Abdel-Daim, M.M., Arora, S., Oancea, R., Nutraceuticals in neurological disorders. *Int. J. Mol. Sci.*, 21, 12, 4424, 2020.

8

Nanofunctional Foods as Immunity Booster in COVID-19

Muhammad Qasim Barkat[1], Ahsan Nawaz[2,3*], Sana Asghar[4], Farwa Basit[5], Musaddique Hussain[6], Liaqat Hussain[7] and Ximei Wu[1]

[1]Department of Pharmacology and Key Laboratory of CFDA for Respiratory Drug Research, Zhejiang University, School of Medicine, Hangzhou, China
[2]College of Civil Engineering & Architecture, Zhejiang University, Hangzhou, China
[3]Honorary Faculty Member, Novel Global Community Foundation, NSW, Australia
[4]Collage of Life Science, Zhejiang University, Hangzhou, China
[5]School of Agriculture & Biotechnology, Zhejiang University, Hangzhou, China
[6]Department of Pharmacology, Faculty of Pharmacy, The Islamia University of Bahawalpur, Bahawalpur, Pakistan
[7]Department of Pharmacology, Faculty of Pharmaceutical Sciences, Government College University Faisalabad, Faisalabad, Pakistan

Abstract
COVID-19 has become a major pandemic disease in recent days since December 2019. It has affected and killed many people globally without differentiating kids, younger children, and adults. The WHO has estimated that approximately 148,329,348 people were affected and 3,128,962 were killed by COVID-19 in 223 countries. COVID mostly targets the immune system of patients and leads to a more lethal condition in a few days without considering potent treatment. Recently, there is no specific treatment available but only supportive, protective, and natural treatments are using in the form of functional foods ingredients, such as vitamins, natural plants, probiotics, and flavonoids, all they enhance the immunity level of immuno-compromised COVID-19 patients. These functional foods are fortified with bioactive compounds and provide antioxidants to boosting immunity by application of nanotechnology. Nanotechnology is a targeted drug delivery system that has reduced size (nm) and increased surface area as well as provides more effectiveness and safety. Moreover, it also improves the color, taste, flavor, bioavailability, and absorption of foodstuff into the body. In this chapter, we

**Corresponding author*: ahsanklasra@zju.edu.cn; ahsanklasra@gmail.com

Tanima Bhattacharya and Shakeel Ahmed (eds.) Nanotechnology in Functional Foods, (193–222)

have summarized the immune effects of functional foods on COVID-19 patients through influencing nanotechnology.

***Keywords*:** Functional foods, COVID-19, immunity, nanotechnology

8.1 Introduction

Coronavirus was first identified in the Wuhan city of China since December 31, 2019. It was found in the wholesale market where snakes, fish, bat, and poultry were common businesses and eventually spread to other places. First, it was given the name as Severe Acute Respiratory Syndrome Corona Virus 2 (SARS-COV-2) through the Chinese Center for Disease Control (CCDC) when CCDC collected the virus sample on January 7 through cotton swab from infected people but later World Health Organization (WHO) titled as COVID-19 [1]. The WHO declared on 30 January 2020 that the epidemic COVID-19 to be a worldwide public health emergency [2]. Coronavirus name indicates that it has a crown-like appearance on its surface. Coronaviridae is a family of coronavirus. It has a 65- to 125-mm diameter covered by single-stranded ribonucleic acid (RNA) with a length of 26 to 32 kbs. It has less lethal (229e and NL63-α virus or OC43 and HKU1-β virus) and more lethal viruses like middle east respiratory syndrome (MERS) or severe acute respiratory syndrome (SARS) [3]. However, the existing pandemic COVID-19 is also a β-type virus (more lethal), and their structure has been represented in Figure 8.1.

COVID-19 is based on the immune system of the individual, and if someone has a strong immune system, then he can easily manage this

Figure 8.1 Configuration of COVID-19 structure.

disease. Because the immune system protects the body against infectious agents like viruses, bacteria, parasites, and fungi. It is always active, and provides protection, but its functioning to be enhanced when a subject becomes infected. Such type of immunity is obtained through a higher metabolic rate, renewable sources of energy, and substrate molecules resulting from a healthy diet or functional foods. Thus, enough supply of natural foods, diet, or supplements is essential to provide more support for the immune system's proper functioning [4] through utilizing nanotechnology for improving food safety status and shelf life. Recently, a wide range of functional foods or supplements, such as vitamins, trace elements, herbs, and natural plants, are used to improve the immune system through the application of nanotechnology.

The nanotechnology term was first used by Norio in 1974, and their concepts were presented by Richard in 1959. The nano term is originated from the Greek word that means dwarf. The study of nanoscience deals with atomic, molecular, macromolecular phenomena, and management of the material on the nanometer scale that differ from a larger scale. Nanotechnology has described by the initiative of national nanotechnology as the regulation and understanding of the material at a scale of about 1 to 100 nm or 1 to 1000 nm, where specific phenomenon makes enable novel application: including nanoscale science and technology, measuring, evaluating, and management of the material at this length scale. The food and drug administration (FDA) states that the size of a nanoparticle is less than a micrometer and having unique properties. Now, about 8879 products are registered in the nanotechnology database and 2467 companies are manufacturing nanotechnology products in 62 countries of the world [5–8].

These nanomaterials have unique characteristics such as novel physicochemical properties color, taste, strength, optical, solubility, toxicity, thermodynamic, and magnetic and also exhibit high surface/volume ratio [9, 10]. Thus, nanotechnology offers many opportunities with novel properties in the field of food, medicine, and others. The demand for nanoparticles in the food industry has been tremendously increased even though all of them comprise vital elements and having non-toxic properties as well as remain stable when exposed to high temperature and pressure [6]. A complete procedure for food manufacturing, processing, and packaging is given by nanotechnology. Nanomaterials bring a significant change not only inconsistency and preservation of the food but also in the health advantage offered by the food. To apply nanotechnology in manufacturing of food, many industries, organizations, and researchers are using novel methods and techniques [11].

The major aim of this chapter is to emphasize the role of monofunctional foods, supplements, and herbal plants in supporting or improving the immune system through the application of nanotechnology, mainly, but not solely regarding the strong defense against the pandemic coronavirus disease. Important functions of functional foods in mitigating gut microbial dysbiosis and cytokine storms have been described. Moreover, the COVID-19 key characteristics and their effects on the immune system have been revealed.

8.2 Pathophysiology and Clinical Manifestation

The pathogenesis of the virus could be described into two asymptomatic and symptomatic phases. In asymptomatic or without detection of the virus, the entry of the virus is initiated through angiotensin-converting enzyme 2 receptor (ACE2). Subsequently, it delays the interferon immune response directly through the production of ROS (reactive oxygen species) and replicates them into the lungs, blood vessels, and other vital organs. Monocytes, macrophages, and granulocytes are activated or even release TNF-α, IL-1, 6, and 8 (proinflammatory cytokines) that lead to inflamed lungs, as well as influence other organs. Consequently, the greater release of cytokines stimulates the production of fibroblasts, and HA-synthase-2 generated by endothelium in alveolar epithelial cells as well as elevates the hyaluronan (HA) that has good water-holding potential. Meanwhile, fluid to be filled in the lungs of neumonia patients. These factors promote the fibroblast and dysfunction of multi-organ that ultimately causes death [12]. The pathophysiological representation of the coronavirus has been depicted in Figure 8.2.

In COVID-19, the pneumonic condition might produce complexity and viral infection can have the capacity to initiate an immune reaction in the host [13]. In certain conditions, cytokine storm type reaction takes place, resulting significant damage to the tissue. The interleukin-6 performs a key role in this reaction. The development of IL-6 is primarily done through activated leukocytes and performs its action on various cells and tissues [14]. It can promote the B lymphocytes differentiation and some other cells but also prevents the other cells and tissue from growing. It plays a vital role for bone health, thermoregulation, and boosts the function of the CNS as well as encourages the production of acute-phase protein [15].

The covid-19 virus can enter through nasal and larynx mucosal membranes then moves toward the lungs. Meanwhile, it will immediately target the specific organs or systems, such as the respiratory, gastrointestinal,

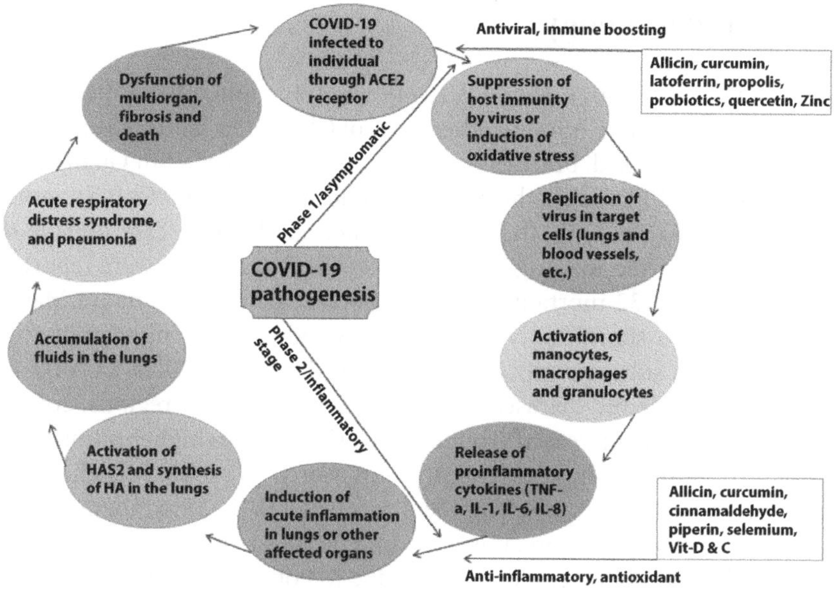

Figure 8.2 Representation of pathogenesis of the COVID-19. It includes two phases like asymptomatic (phase 1) or symptomatic (phase 2). Food supplements exert their actions, such as antiviral and immunity boosting in phase 1, whereas antiinflammatory and antioxidant in phase 2 to counteract the pathogenesis of COVID-19.

cardiovascular, and renal systems where they express their receptor-like ACE2 (18-20). During the 2nd attack, the virus initiates to worsen the patient's condition for 7 to 14 days after onset. When the disease is in its early stage, the reduction of B lymphocyte may take place which can influence the production of antibodies in the patient. Further, the IL-6 is a major factor involved in inflammatory diseases, which greatly increased and led to the worsening of the disease about 2 to 10 days after infection [15–17].

The most common clinical presentation of corona virus are fatigue, joint and chest pain, dyspnea, cough, rhinitis, headache, sore throat, sputum production, anosmia, lack of appetite, conjunctivitis, sicca syndrome, dysgeusia, myalgia, vertigo, and diarrhea. Besides this, diabetic, lung, and heart disease patients could be more affected by COVID-19 [18, 19]. The normal incubation period of COVID-19 is 2-10 days, but it can be prolonged. It can transmit through sneezing, coughing, handshaking, and hugging. The droplets of an infected person can spread up to one meter. Thus, the droplets of an infected person would be transmitted from one infected person to another in a one-meter area through breathing, coughing, and sneezing [18, 19].

8.3 Epidemiology

In early December 2019, the COVID-19 was transmitted from the Wuhan (7th popular City of China) to other parts of China. On January 10, 2020, it was rapidly increased, but 1st confirmed case was reported in Guangdong on January 19, 2020. Soon, the virus was rapidly spread from China to other parts of the World [20]. The first confirmed COVID-19 case outside of China was reported in Bangkok on January 13, 2020. There were reported 8565 cases and 132 mortalities in 67 regions other than China on March 2, 2020. It also significantly transmitted worldwide but mostly affected Italy and Iran, subsequently, WHO declared 11, March 2020; COVID-19 is a global pandemic disease. After European and Asian regions, corona cases were rapidly increased around the world, but also higher numbers of cases were observed in poor countries on March 31, 2020. It is challenging to calculate the accurate numbers for the COVID-19 pandemic virus across the world. Therefore, it is very difficult to quantify how much population affected by the epidemic virus due to complex criteria for testing, diagnosing, hospitalization, and estimation of cases [21–23]. According to the WHO, COVID-19 has been spread to 223 countries and infected 148,329,348 people, as well as killed 3,128,962 globally till April 28, 2021. In the past, there were also other pandemic diseases such as SARS and the MERS that have been spread around the world, and their comparisons with COVID-19 were given in Table 8.1 [19, 24, 25].

8.4 Role of Immune System

The immune system plays a major role when a human is exposed to an infectious agent. The immune system adopts different approaches to invade the infection according to its nature. These specific approaches follow the same general principle to tackle the infection, but specific immune mechanisms might be different. Mostly, bacteria do not attack the host cells but persist to the immune system of the host. Then such bacteria will be phagocytosed by dendritic cells, macrophages, monocytes, and neutrophils (innatte phagocytic cells), destroyed within phagocytic intracellular vacuoles and finally to be digested. Fragments of antigens (digested bacteria) can be demonstrated on the phagocyte surface through the major histocompatibility class II (MHC II). The antigen-specific T lymphocyte (CD4+) recognize these antigens and activates the adaptive immunity towards the bacteria, which contains the composition of B and T lymphocytes as well as further

Table 8.1 Comparison of COVID-19 with previous pandemic diseases.

Characteristics	SARS-CoV	MERS-CoV	SARS-CoV-2 or COVID-19
Clinical Features	Cough, dyspnea, fever, muscle pain, sore throat, headache, lethargy, diarrhea, shivering	Cough, fever, dyspnea, muscle pain, abdominal pain, vomiting,	Cough, fever, dyspnea, sore throat, fatigue, joint and chest pain, myalgia, vertigo, headache, lack of appetite, pneumonia, conjunctivitis, rhinitis, sicca syndrome, diarrhea
Incubation period (days)	2-7 (virus spread after the appearance of symptoms)	2-14 (virus might be spread before the appearance of symptoms)	1-14 (virus might be spread before the appearance of symptoms)
1st Origin	China (Guangdong)	Saudi Arabia	China (Wuhan)
Reported Year	2002 (July)	2012 (Dec)	2019 (Dec)
Period	2002-2003	2012-2020	Dec 2019-till date
Source	Bats, civet cats, raccoon dogs	Dromedary camels	Bats, pangolin
Receptor	AC2	DPP4 (CD26)	ACE2
Globally spread	29	26	223
Fatality rate	10%	35%	2.1%
Population affected	8098	2506	148,329,348
Confirmed death	774	862	3,128,962

Abbreviation: SARS-COV (Severe Acute Respiratory Syndrome Coronavirus), MERS (Middle East Respiratory Syndrome), SARS-COV-2 (Severe Acute Respiratory Syndrome Coronavirus-2 or COVID-19).

stimulation of innate immune cells. Consequently, this response targets and kills the bacteria. But viruses attack the host cells instead of remaining extracellular and can activate the antigen on the infected cell's surface through MHCI. These antigen recognitions are made possible by cytotoxic T lymphocytes (CD8+) that lead to destroying the antigen-presenting host cells. Natural killer cells (NKC) detect the infected cells with virus and act as an analog to cytotoxic T-lymphocytes via destroying the diseased cells. Hence, this approach targets the destruction of host cells through virally infected cells that harbor the viruses. Indeed, the destruction of host cells releases the viruses that trigger the fight between the infected virus and host immune cells [4]. Here, the major roles of the immune system that enable potential host defense are given below:

A) Developing a barrier to restrain the pathogen entry into the body
B) Recognition of pathogen in case of entrance
C) Killing or eliminating the pathogen
D) Producing immunological memory

8.4.1 Developing a Barrier

The immune system protects to counter the pathogen entry from the external environment into the body through creating a barrier function. The barriers might be different according to their location and conditions like a physical barrier (skin and the mucosal layer of respiratory, gastrointestinal, and genitourinary tract), chemical barrier (acidic stomach-pH), and biological barrier (living organisms in the intestinal tract or on the skin, antimicrobial proteins, and immunoglobulin A secretion in tears and saliva) [4].

8.4.2 Identification of Pathogens

The innate or nonspecific immune cells (dendritic cells, monocytes, and macrophages) recognize the pathogen. This is accomplished through pattern recognition receptor (PRR) that leads to the recognition of microbe-associated molecular patterns (MAMPs) which are common among pathogens. However, the response of frontline-host defense is stimulated when MAMPs are recognized by PRRs and PRR consists of toll-like receptors (RLRs). Now, humans have more than 10 TLRs, each of them identifying MAMPs from viruses, bacteria, parasites, and fungi. Many TLRs are shown their expression on the innate immune cells like bacteria due to the presence of extracellular and TLR4 recognizes gram +Ve bacteria while TLR2 recognize -Ve bacteria. Whereas viruses enter the host cells and intracellular TLRs identify viral DNA, viral single- and double-stranded RNA. Besides, TLR7 and 8 are mostly identified in dendritic, macrophages or monocytes cells, and others but probably be essential in the detection of single-stranded RNA of coronaviruses, although both extracellular and intracellular PRRs are expected to identify the coronavirus protein coat (glycoproteins) [26–29].

8.4.3 Elimination of Pathogens

The pathogen-like extracellular bacteria can be phagocytosed through phagocytic cells and activates the CD4+ helper T lymphocytes by MHCII that lead to the development of IL-2 and interferon-γ (INFγ). This INFγ stimulates the antibody production through B-lymphocytes and in resulting neutralizing the bacteria and the process of phagocytosis become more effective. Like the phagocytosis, the recognition of innate immune cells by PRRs that activates the nuclear factor kappa β cells (NFκB) and initiated cytokines productions like IL-12, IL-1β, and tumor necrosis factor (TNF) [30, 31]. Whereas the infection of viral cell promotes the release of IFN-α and β, as well as initiates the NKC activation that leads to antiviral resistance. Likewise, NKC is activated by virally infected cells that kill the infected cells. The development of dendritic cells is induced by PRR signaling that is necessary for the exhibition of viral antigen and initiating the adaptive immunity. The viral antigen demonstration to cytotoxic T-lymphocytes (CD8+) through MHCI upregulation on infected virus cells (dendritic and respiratory epithelial cells). The activated process ultimately kills the infected cells with virus via release of perforins (pore-forming proteins). The demonstration of viral antigens through MCHII and cytokine contribute to the stimulation of T-lymphocytes (CD4+) by moving to T-helper 1 phenotype. Subsequently, they develop IL-2 which initiates the activity of T-lymphocyte, as well as IFN-γ that lead to B lymphocytes differentiation to plasma cells and ultimately induce the antiviral antibodies. The antibodies produced may bind to free-moving viruses and deactivate them [4]. So, the process of antiviral immunity is depicted in Figure 8.3.

8.4.4 Immunological Memory

The immune system can specifically and rapidly identify an antigen that the body has earlier faced and activate a subsequent immune response. Immunological memory can be described in two phases. Initially, antibodies can stay in the blood circulation for several months or even years

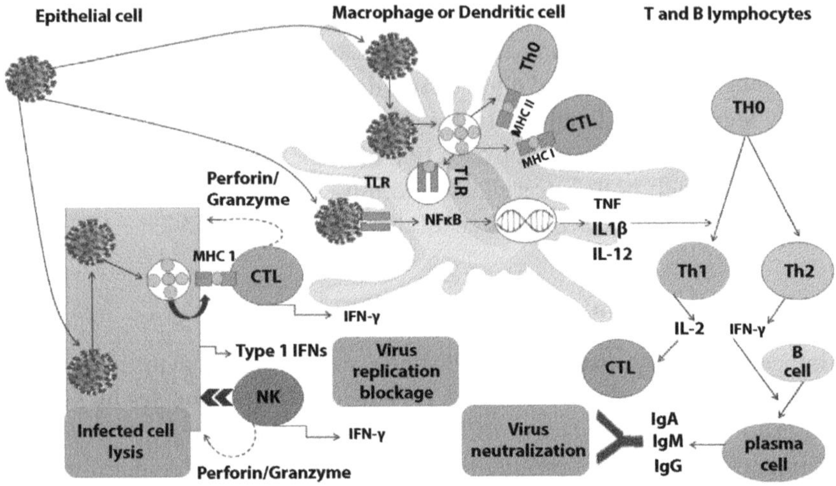

Figure 8.3 Anti-viral immunity. The infected cell directly activates the NKC to kill the virally infected cell. Maturation of dendritic cells are initiated by PPR for detection of virus as well as induce immunity. Meanwhile, CTL are activated via upregulation of MHC1 and kill virus through activated T lymphocytes or release of perforin granzyme on epithelial cells as well dendritic cells. Whereas virus presentation by MHCII and cytokines activate the T lymphocytes. These cells like Th1 release IL-2 in the presence of TNF, IL1β, IL-12 (NFκB) and stimulate CTL while Th2 releases IFN-γ and promotes the B cell to the plasma cell to produce multiple antibodies. These antibodies detect the antigen or viruses and inhibit their replication and ultimately neutralize them. CTL, cytotoxic T lymphocyte; IFN, interferon; NFκB, nuclear factor kappa β cells; NK, natural killer cell; TLR, Toll-like receptor; Th, helper T lymphocyte; MHC, major histocompatibility class; B cell, B lymphocyte; Ig, immunoglobulin; TNF, tumor necrosis factor; IL, interleukin

and delivering protection to the body. Then a limited number of memory cells like T cells and B lymphocytes persist after the termination of the strong immune response. If these cells detect the same antigen, then they prompted their development and have the potential to respond immediately, as well as eliminate the antigen. Immunological memory cells remain for a longer time up too many decades and these memory cells are the basis of vaccination [4].

8.5 Diagnosis

The main test for using the diagnosis of the COVID-19 worldwide is the reverse transcriptase-polymerase chain reaction (RT-PCR). But the WHO also recommended some other tests because of observed higher levels in infected patients, such as higher levels of creatinine, C-reactive protein, erythrocyte sedimentation rate, ferritin, elevated lactate dehydrogenase, and prolong prothrombin time. Elevated levels of aminotransferase, leukocytosis, leukopenia, and lymphopenia were reported in suspected patients. Oropharyngeal, nasopharyngeal, and expectorated sputum were also considered. Moreover, antibody tests are also recommended in COVID-19–infected patients because they express the immunity level of the patients. Antibodies are immunoglobulin (Ig) protein that protects the body against COVID-19 or another upcoming disease. It can detect in infected patients after 1-3 weeks of the infection [2, 23].

8.6 Management and Treatment

Yet not even a single specific anti-viral or monoclonal antibody has been recommended as a potential treatment for the COVID-19 victims and more than a hundred clinical trials of the antiviral drug are being under investigation. Many drugs are using worldwide to tackle this coronavirus pandemic disease like anti-malarial, anti-HIV, and anti-swine flu drugs. But all these drugs can reduce the symptoms or infection rate for a limited time and only manage till the arrival of upcoming symptoms. Many efforts were being made to manage this viral disease through public health interventions such as isolating, quarantine, tracing, and collecting the information of infected persons to prevent the viral transmission from the infected to the healthy individual [32]. Meanwhile, preventive measures were recommended by WHO to counter this epidemic viral respiratory syndrome and strictly followed by the law enforcement authority. Preventive measures such as wear a face mask, social distancing, frequently washing or disinfectant hands with soap or hand sanitizer, avoid social gatherings and public places, stay home-stay safe, and avoid unnecessarily going out. But lockdown, smart or mini lockdown, and curfew were also implemented to curtail the spread of the virus through restraining regular activities [33]. COVID-19 mostly targets the immune system of the patients if we improve the immunity then this disease would be easily overcome. Immunity can be boosted through taking healthy diets such as functional foods, herbal products, and supplements. Many natural immunity boosters are available in the markets that can manage this disease by providing more strength to the immune system of the patients. Thus, these functional foods are playing a vital role in immunity through the application of nanotechnology.

8.7 Nanotechnological Application

The application of nanotechnology can be briefed into two major categories like food nanostructured and nanosensing. A category for food nanostructured consists of processing and packaging foods. These substances might be utilized as food additives, antimicrobial, and anticaking agents, carriers for smart nutrients delivery, mechanical strength improving fillers and, packaging materials durability. But food nanosensing can be used for achieving the best food quality as well as safety [34]. Here, we are discussing how nanotechnology is improving the manufacturing, processing, and packaging of food items as well as nanotechnological based application has been summarized in Figure 8.4.

8.7.1 Food Processing and Quality or Taste of Nanofunctional Foods

Nanotechnology is enhancing the shelf life of foodstuff and protected the food wastage against microbial attack [35]. Nano polymer has replaced conventional packaging material, whereas nanoparticles have improved the encapsulation and make effective delivery than conventional encapsulation [36]. Moreover, these delivery systems having deep penetration capacity into tissues because of nano size and therefore allow the transport of active biological compounds towards the target sites within the body [37]. Nanotechnology offers many opportunities to improve the food quality as well as boosting their taste. Hence, nanoencapsulation have been utilized to enhance the release of flavor preservation and to provide

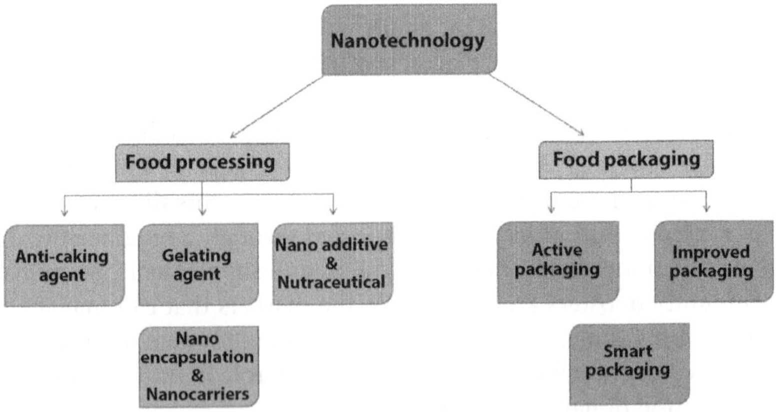

Figure 8.4 Application of nanotechnology in food sectors.

food balance [38]. Nanoparticles have the potential to enhance the bioavailability of nutraceutical products because of subcellular size resulting in improved drug bioavailability than larger particles that consume more time to release encapsulated compounds [6, 39].

8.7.2 Improved Nutrient Content, Packing Material, and Shelf Life of Functional Foods

Many bioactive molecules including vitamins, carbohydrates, lipids, and proteins are degraded due to increased pH and enzyme activity of stomach or duodenum. The encapsulation technique prevents these active molecules from vulnerable conditions but also increases their assimilation into foodstuff as compared to non-encapsulated products. To deliver vital health benefits, nanoparticle-based edible capsules are being manufactured for improving the provision of micronutrients (medicine or vitamins) in regular foods [40]. In functional foods, the nanoencapsulation have longer shelf life of food stuff through slowing the breakdown process and protect until the product reached to target site. Besides, the digestive nanocoating on different food products might deliver resistance against moisture or gas exchange, and provide flavors, colors, enzymes, as well as antibrowning, and antioxidants agents and might improve the stability of the foods after the packaging wrapper has been opened [6]. A suitable product packaging should have durability and biodegradability along with moisture and gas permeability [41]. Active and smart food packaging based on nanomaterials has many benefits over existing methods including better materials with improved barrier properties, mechanical strength, nano sensing antimicrobial films for the identification of the pathogen, and alerting users to the health condition of the foodstuff [42]. Furthermore, the application of nanoparticles improves the quality and extends the shelf life of the packaging food materials. Many organic or inorganic nanoparticles are using to enhance polymer composites and developing more cost-effective resistant packaging material [43].

8.7.3 Pathogen Detection by Nanosensors

In food microbiology, the nanosensors can be used for the pathogen detection in food material of processing plants to ensure the availability or evaluation of food constituents, inform customers and retailers on the safety level of food [6]. As an indicator, the nanosensor contributes to variations in environmental conditions, like moisture or temperature, degradation of products, and contamination of microbes [44]. The nanotechnology could

Table 8.2 To improve shelf life and food safety status by active food packaging.

Type	Properties	Agents	References
Carbon dioxide emitters	Spoilage inhibition through microbial action	Citric acid, Sodium bicarbonate or Ascorbate	[51, 52]
Ethylene scavengers	Reduced ripening of vegetables and fruits	KMno4, SiO2, Ag, Pdcl2, TiO2, Pd-Impregnated zeolite, Polyvinyl chloride (PVC) film including zinc oxide nanoparticles	[53, 54]
Moisture absorbers	Reduced microbial growth	Chlorides (Al, Ca, K, Mg, Na), Inorganic (montmorillonite, natural clay, silica gel, zeolite), Oxides (Ba, Ca), Bentonite, Organic (cellulose and derivatives, fructose, sorbitol, xylitol), Polymer-based (absorbent resin, starch copolymers, polyvinyl alcohol)	[55]
Oxygen scavengers	Fat oxidation protection	Enzyme-based (glucose oxidase, lactase), Inorganic (sulfite, thiosulfate, zinc oxide), Metallic (activated iron, iron powder, zinc), Organic (ascorbic, acid, catechol, tocopherol), Polymer-based (polymer metallic complex)	[56]

also lead to the identification of pathogens, pesticides, and toxins that deliver their function in the food monitoring or tracking chain [45–47]. Biosensors carbon-based nanotubules have been effectively used for the screening of toxins, microorganisms, and further damaged products in foods because of their fast identification and economic viability [48]. Several researchers have used quartz crystal microbalance (QCM) surfaces on small molecule detection that have been altered by various functional groups or biomolecules, like many polymers, enzymes, amines, and lipids [49].

8.7.4 Safety Issues

The nanomaterials have some safety issues despite many beneficial effects. Several investigators have raised the safety issues related to nanomaterial and emphasis on the leakage or migration possibilities of nanoparticles from packaging to the food as well as their effects on human health [6]. However, the material is known to be generally considered as a safe (GRAS) substance, further investigations should be conducted to examine the risk of nanomaterial because nanostates physiochemical properties are distinct from macrostates. The nanosize of these materials may rise the possibility of bio-concentration within body tissues or organs [50]. Even so, the experts need to set criteria for consumer's products to ensure health, quality, and safety of the products and environmental protection. Food packaging can improve the shelf life and safety of food items that have been given in Table 8.2.

8.8 Functional Foods

The word "functional food" was first utilized in the Japan. They were the pioneer who observed that food could provide energy and nutrient to the body and many functional foods are using now in Japan. After Japan, the USA and Europe have utilized foods and assigned different names like nutraceutical, pharma foods, and designer foods. Meanwhile, Western civilizations are experiencing immune-mediated and stomach problems so; they are trying to improve these diseases and improving quality of life through analyzing the effects of functional foods on the immune system. Functional foods are only foods, not a medicine, and can only get beneficial effects by consuming the appropriate amount of these foods in a normal diet. Importantly, these foods are only utilized to improve the physiological functions and immune system to protect against diseases. Here, some functional foods are describing to mitigate the SARS-CoV-2 disease through strengthening and boosting the immune system [57, 58].

8.8.1 Vitamin D

It is a fat-soluble vitamin and has antiviral, antioxidant, and immunomodulatory functions [59, 60]. Vitamin D (Vit-D) receptor is mainly expressed by the human airway epithelium, thereby promoting the defensive effects of Vit-D toward respiratory infections [61]. Vit-D also reduces the type T1 cytokines-proinflammatory expression such as TNF-α, IFN-γ, interleukin-8, -12, -16, while type T2 cytokines like interleukin-4, 5, 10, and regulatory or suppressor T cells are increasing [62, 63]. According to the epidemiological data, deficiency of Vit-D increases the susceptibility to acute viral infection in the respiratory system [64] while its supplementation increases the response of innate immunity to such type of infections even those caused by hepatitis or influenza viruses [12]. The Vit-D mostly founds in eggs, liver, oily fish, and fortified foods [4]. It could act a supportive role in viral infection through the reduction of inflammation caused by the virus instead of directly affects the viral load. In COVID-19 infection, it might suppress the cytokine storm. The standard dose of Vit-D is 12,000 IU/month that facilitates in decreasing the occurrence of lungs diseases or other chronic illness particularly in older or all ages groups [65, 66]. Vit-D deficiency is mostly observed in older people if they can take these supplements on daily basis then would get significant results [67].

8.8.2 Vitamin C

Vitamin C (Vit-C) plays an essential role to tackle the infection because of possessing vital effects on immune health [68] and supports, as well as enhances the ability of immune cells against infection. Vit-C supplementation has reduced the severity and incidence of respiratory tract infection, including viral infection or the common cold. It has 1 to 3 g/day dose range and 60 mg recommended daily allowance (RDA). Various, fruits, vegetables, herbs, and spices have the best source of Vit-C likewise they contain different RDA like coriander (35%), cardamom (35%), turmeric (43%), and thyme fresh (267%) [12]. Many studies indicated that Vit-C delivers synergistic effects like antioxidant, antiviral, and immunomodulatory when administered with quercetin. According to current clinical data, administration of Vit-C (500 mg) with quercetin (250–500 mg) twice daily for 7 days in higher or mild symptomatic patients while Vit-C (up to 3 g) with quercetin (500mg) twice daily for 7 days in acute respiratory distress syndrome (ARDS), it enhances the recovery in COVID-19 subjects [69]. Thus, Vit-C as a food supplement can assist in alleviating the respiratory tract infection or enhancing the anti-inflammatory, antioxidant, and immunity booster functions against COVID-19 syndrome.

8.8.3 Zinc

Zinc (Zn) is the main metal implicated in several biochemical mechanisms because of its features as signaling molecules, structural elements, and cofactor. It modulates inflammatory activity and exhibit antioxidant and antiviral activities [70]. It can suppress the entry, synthesis, proliferation, protein translation, and budding of the respiratory viruses [70, 71] and exposure of Zn (100 μM) has been demonstrated to constrain the recombinant ACE-2 receptors in the rat lungs [72]. The major sources of zinc are meat, shellfish, cheese, seeds, grains, cereals, and wholegrain breads [4]. A combination of Zn2+ cations with Zn ionophore pyrithone has inhibited the RNA polymerase of COVID-19 by blocking its replication. Acute respiratory infections were reduced by approximately 35% through oral Zn supplements and reduce the length of flu-like symptoms in 2 days, resulting in improving the recovery rate. Various studies recommended dose ranges of Zn from 20-92 mg/week and regarded it as promising supportive therapy for COVID-19 disease because of anti-viral, antioxidants, as well as anti-inflammatory effects [12].

8.8.4 Selenium

It is an important trace element that can play a role in the immune system and several physiological processes. It's abundantly present in broccoli, corn, cabbage, garlic, and onion. It produces its effects through the induction of selenoproteins into the body. The recommended dose of selenium is 100 μg/day that improved innate immunity, T cell proliferation, and NK cell function. It strengthens the vaccine response as well as stronger immunity to the pathogen and suppresses the hyper inflammation in lungs or intestinal tissues. The stronger antioxidant activity of selenium suppresses the inflammatory response in corona virus patients by inhibiting the IL-6 and IL-1β [73, 74]. Therefore, selenium supplementation indicates enhanced the counts of CD+ T cells and augments the glutathione peroxidase, catalase, and antioxidants from several selenoenzymes [12]. Selenium boosts immunity by serving as an enzyme precursor responsible for the modification of essential post-translational proteins. Due to its antioxidant and anti-inflammatory effects, selenium may be utilized as an immunity enhancer against the epidemic COVID-19 infection.

8.8.5 Curcumin

Curcumin (curcuma) has many health benefits, such as antiviral, anti-inflammatory, antibacterial, antioxidant, and antifungal [75]. It prevents the production of TNF-alpha and IL-6 in lipopolysaccharide-stimulated microglial BV2 cells by negatively affecting the MAPK signaling or NF-κB activation [76, 77]. The recommended daily dose of curcumin is 500 mg/kg [78]. It has a broad range of antiviral activity including COVID-19 and others through several mechanisms including inhibiting the virus entry, encapsulation, protease, replication, and modulating many signaling pathways. It potentially inhibits the ACE2 receptors and stimulates the production of host interferon to activate the host innate immune response (54, 55). Moreover, it exhibits antioxidant activity through induction of antioxidants enzymes and neutralization of free radicals [12, 79]. These findings reveal that the potential effects of curcumin act as immunity booster and antioxidants in COVID-19. Thus, curcumin might be essential supplements in tackling the pathogenesis of respiratory coronavirus disease.

8.8.6 Cinnamaldehyde

Cinnamaldehyde is an organic compound that naturally presents in essential oils in cinnamon. Its flavor or smells due to trans-isomer form [80] and possess anti-inflammatory properties. It reduces the inflammation stimulated by TNF-α via overwhelming the stimulation of NF-κB [81]. It is often suggested to reduce the output of the prostaglandins (PGEs) through reducing the action of IL-1β caused by COX-2 thus reducing the risk of dose-dependent hyper inflammation [82]. The above data reveal that cinnamaldehyde has a potential antiinflammatory activity, which might be beneficial for mitigating the hyper inflammation caused by SARS-CoV-2 in the lungs.

8.8.7 Allicin

Garlic belongs to the allium family and has many nutraceutical properties. A major thiosulfinate was identified as allicin in freshly garlic extract that has several therapeutic effects due to its antioxidant, antiviral, and anti-inflammatory properties. It reduces the inflammation via restricting the IL-8, 1L-1β, IFN-γ, and IP-10 expression caused by TNF-α and reducing the deprivation of NF-κB inhibitory protein (IκB) in the intestinal epithelium [83]. Many garlic derivatives such as methyl thiosulfinate, ajoene, allyl, allicin, and methyl allyl thiosulfinate have potent activity against several viruses, and their effects were reported at 200 ng/ml concentration. Research findings have investigated that fresh natural garlic possesses many biological activities instead of heating or drying processing and could be useful as a prophylactic agent against current coronavirus disease [12, 84].

8.8.8 Piperine

Piperine is the main alkaloid in the cinnamates group and is derived from black pepper (ethanolic extract). It has many medicinal properties like potential anti-inflammatory activity and thus could be appropriated to inhibit the hyper inflammation caused by SARS-CoV-19 [85]. It also enhances innate immunity through promoting phagocytic activity and suppresses LPS-induced interleukin-6, 1β, TNF-α, and the development of PGE2 in BV2 cells [86, 87]. Moreover, it has strong antioxidant activity with IC50 of 1.82 mM for superoxide radicals as well as 1.23mM for lipid peroxidation [88, 89]. Due to such effects, piperine could be used as a therapeutic or prophylactic compound to curb the COVID-19–induced oxidative stress and inflammation.

8.8.9 Propolis

Propolis is formed from honeybees and has potential biological activities like anti-diabetic, anti-inflammatory, laxative, anti-microbial, immunomodulatory, anti-tumor, and hepatoprotective. The immunomodulatory property is associated with flavonoids along with phenolic acids artepillin C and caffeic acid phenethyl esters. Propolis exerts immunomodulatory effects on immune cells activated by the regulation of MAPK signaling and extracellular signal mediated kinase 2 [90, 91]. It inhibits lymphoproliferation at higher concentration while their effects are reversed at lower concentration and exhibits stronger activity against a wide range of viruses. It enhances antibody production and could be used in vaccines as an adjuvant [12, 92]. Altogether, propolis has antiviral and immunomodulatory effects and can be used as supportive therapy in COVID-19.

8.8.10 Probiotics

The *Bifidobacterium* and *Lactobacillus* species are commonly used as probiotics followed by *Bacillus*, *Enterococcus*, *Escherichia coli*, and *Streptococcus*. Probiotics (fermented milk, nondairy foods, and desserts) do not only boost up gut health but also enhances the functioning and regulation of the system. Generally, it is investigated that the gut microbiome at distal mucosal sites influences the local and systemic immune responses, including lungs [93, 94]. Probiotics improve the antigen-presenting cells, NK cells, mucosal antibodies, and interferons. The Lactobacillus strain-Plantarum DR7 improves the IL-4, and IL-10 (anti-inflammatory) [95–97]. Administration of *Lactobacillus acidophilus* CMCC878 has decreased the lung injury in mice infected with *Pseudomonas aeruginosa*, and *Staphylococcus aureus* by suppressing the inflammation and bacterial growth [98, 99]. Probiotics induce anti-inflammatory and immunomodulatory effects through modification of PRR, MAPK, and NF-κB mechanisms that reduce the Th2-mediated responses and activate Th1 responses [100, 101]. Probiotics are playing their role as an antiinflammatory, immunity enhancer, as well as enterocytes and gut can be considered for corona virus infection [94, 102] that is why probiotics could be indicated as prophylactic agents to tackle the etiology of corona virus infection.

8.8.11 Lactoferrin

It is a biologically active and non-lethal glycoprotein that has been examined for various viral infections including COVID-19. Lactoferrin (Lf) inhibits entry and replication of the virus as well as reduces the oxidative stress induced by H2O2 in human endothelial cells (umbilical vein) [103, 104]. Studies have revealed that it possesses antioxidant and immunomodulatory effects through stimulating the T-cell activation, cytokines suppression (TNF-α and IL-6), and ferritin. Moreover, Lf has a stronger antiviral effect due to zinc saturation [105, 106]. It is primarily included in infant formulas and clinical development as a nutritional additive, with 100 mg to 4.5 g/day daily doses for multiple purposes despite significant side effects. Hence, it could be included as a protective/supportive agent in the COVID-19 treatment [106].

8.8.12 Quercetin

It is a popular bioactive compound having anti-inflammatory, antiviral, and antioxidant activities. It blocks IL-8 (lungs A549 cells), IL-1α (glial cells), TNF-α (LPS induced macrophages), and TNF-α (mRNA). It possesses antiviral activity on both DNA and RNA viruses and removes the entry or replication of virus. In mice, proinflammatory cytokines were expressed and rhinovirus-mediated pulmonary inflammation were observed [12]. In the epidemic corona virus situation, Vit-D, quercetin, and estradiol influence the activity of 61%, 70%, and 85% of COVID-19 proteins in human cells according to the gene set enrichment analyses. Meanwhile, a combination of Vit-D, quercetin, and estradiol may alter 73% of human genes encoding COVID-19 targets as compared to Vit-D, and quercetin indicating a potential agent against respiratory viral disease. Besides, a combination of quercetin with Vit-C produces a synergistic effect like antiviral and immunomodulatory targeting COVID-19 [69, 107]. Collectively, these findings propose that quercetin exerts stronger effects and could be remodeled as a prophylactic candidate to mitigate the COVID-19 disease. A summary of clinical trials of functional foods against COVID-19 and their biological actions have been given in Tables 8.3 and 8.4.

Table 8.3 Registered clinical trials of functional foods against COVID-19 (Source: ClinicalTrials.gov).

Serial #	Functional foods	Title	Phase	Clinical trial
1	Vit-D	Vitamin D testing and treatment for Covid-19	I	NCT04407286
2	Vit-C	Administration of intravenous Vit C in novel coronavirus infection and decreased oxygenation	I	NCT04357782
3	Zinc	Placebo controlled trial to evaluate zinc for the treatment of Covid-19 in the outpatient setting	IV	NCT04621461
4	Curcumin	A phase II, controlled clinical study designed to evaluate the effect of Artemic in patients diagnosed with Covid-19	II	NCT04382040
5	Propolis	Brazilian green propolis extract as an adjunct treatment for hospitalized Covid-19 patients	II,III	NCT04800224
6	Probiotics	Oxygen-zone as adjuvant treatment in early control of Covid-19 progression and modulation of the gut microbial flora	II	NCT04366089
7	Lactoferrin	Efficacy and safety of liposomal lactoferrin in Covid-19 patients with mild to moderate disease and in Covid-19 asymptomatic	II, III	NCT04475120
8	Quercetin	Efficacy of Psidii Guava's extract for Covid-19	III	NCT04810728

Table 8.4 Summarized biological actions of functional foods.

Serial #	Functional foods	Biological actions	References
1	Vit-D	Anti-inflammatory, immunity boosting	[59, 60, 62, 63]
2	Vit-C	Antioxidant, immunity boosting	[69]
3	Zinc	Anti-viral	[70]
4	Selenium	Antioxidant, immunity boosting	[73, 74]
5	Curcumin	Anti-viral, antioxidant, anti-inflammatory, immunity boosting	[12, 75, 79]
6	Cinnamaldehyde	Anti-inflammatory	[12, 80]
7	Allicin	Anti-viral, anti-inflammatory	[12, 83, 84]
8	Piperine	Anti-inflammatory, antioxidant	[85, 88, 89]
9	Propolis	Anti-inflammatory, immunity boosting	[90, 91]
10	Probiotics	Anti-inflammatory, immunity boosting	[100, 101]
11	Lactoferrin	Anti-viral	[103, 104]
12	Quercetin	Anti-viral	[12]

8.9 Conclusion

The severe COVID-19 respiratory disease has been transmitted to more than 200 countries. The WHO reported that approximately 148 million people were affected and 3.1 million killed worldwide. It is rapidly spreading through droplets to one another. Unfortunately, the world has remained to fail to develop specific medicines against this pandemic disease, and some of the countries have launched vaccines on a trial basis that would not be available to every individual due to limited supply. Meanwhile, the world health authorities have introduced some functional foods or supplements such as allicin, cinnamaldehyde, curcumin, lactoferrin, piperine, probiotics, propolis, quercetin, selenium, Vit-C, and D, and zinc as well as observing their effects on the immune system against COVID-19. These functional foods have not only improved the immune system of the infected individual but also exerted antiviral, anti-inflammatory, and antioxidants effects through the application of nanotechnology. Moreover, nanotechnology is a target delivery system at the nanoscale to provide more protection, attractive packaging, color, texture, and good taste to functional foods through the application of different nanomaterials of different sizes. Thus, monofunctional foods with recommended doses may act as immunity booster as well as antiviral medicines in pandemic COVID-19 disease.

References

1. Ha, B.T.T. *et al.*, Combating the COVID-19 Epidemic: Experiences from Vietnam. *Int. J. Environ. Res. Public Health*, 17, 9, 3125, 2020.
2. Sohrabi, C. *et al.*, World Health Organization declares global emergency: A review of the 2019 novel coronavirus (COVID-19). *Int. J. Surg.*, 76, 71–76, 2020.
3. Yang, Y. *et al.*, The deadly coronaviruses: The 2003 SARS pandemic and the 2020 novel coronavirus epidemic in China. *J. Autoimmun.*, 109, 102434, 2020.
4. Calder, P.C., Nutrition, immunity and COVID-19. *BMJ Nutr. Prev. Health*, 3, 1, 74, 2020.
5. Primožič, M., Knez, Ž., Leitgeb, M., (Bio) nanotechnology in food science—Food packaging. *Nanomaterials*, 11, 2, 292, 2021.
6. Singh, T. *et al.*, Application of nanotechnology in food science: Perception and overview. *Front. Microbiol.*, 8, 1501, 2017.
7. He, X., Deng, H., Hwang, H.-m., The current application of nanotechnology in food and agriculture. *J. Food Drug Anal.*, 27, 1, 1–21, 2019.

8. Cavalcanti, I.D.L. and Nogueira, M.C.d.B.L., Pharmaceutical nanotechnology: Which products are been designed against COVID-19? *J. Nanopart. Res.*, 22, 9, 1–11, 2020.
9. Rai, M., Yadav, A., Gade, A., Silver nanoparticles as a new generation of antimicrobials. *Biotechnol. Adv.*, 27, 1, 76–83, 2009.
10. Gupta, A. *et al.*, Nanoemulsions: Formation, properties and applications. *Soft Matter*, 12, 11, 2826–2841, 2016.
11. Dasgupta, N. *et al.*, Nanotechnology in agro-food: From field to plate. *Food Res. Int.*, 69, 381–400, 2015.
12. Mrityunjaya, M. *et al.*, Immune-boosting, antioxidant and anti-inflammatory food supplements targeting pathogenesis of COVID-19. *Front. Immunol.*, 11, 2337, 2020.
13. Letko, M., Marzi, A., Munster, V., Functional assessment of cell entry and receptor usage for SARS-CoV-2 and other lineage B betacoronaviruses. *Nat. Microbiol.*, 5, 4, 562–569, 2020.
14. Pyle, C.J. *et al.*, Early IL-6 signalling promotes IL-27 dependent maturation of regulatory T cells in the lungs and resolution of viral immunopathology. *PLoS Pathog.*, 13, 9, e1006640, 2017.
15. Chen, C. *et al.*, Advances in the research of cytokine storm mechanism induced by Corona Virus Disease 2019 and the corresponding immunotherapies. *Zhonghua shao shang za zhi= Zhonghua shaoshang zazhi= Chin. J. Burns*, 36, E005–E005, 2020.
16. Bennardo, F., Buffone, C., Giudice, A., New therapeutic opportunities for COVID-19 patients with Tocilizumab: Possible correlation of interleukin-6 receptor inhibitors with osteonecrosis of the jaws. *Oral. Oncol.*, 106, 104659, 2020.
17. Rose-John, S., Interleukin-6 family cytokines. *Cold Spring Harb. Perspect. Biol.*, 10, 2, a028415, 2018.
18. Carfì, A., Bernabei, R., Landi, F., Persistent symptoms in patients after acute COVID-19. *JAMA*, 324, 6, 603–605, 2020.
19. CDC, Symptoms of Cornavirus, 2021, [cited 2021 05 February]; Available from: https://www.cdc.gov/coronavirus/2019-ncov/symptoms-testing/symptoms.html.
20. Jin, Y. *et al.*, Virology, epidemiology, pathogenesis, and control of COVID-19. *Viruses*, 12, 4, 372, 2020.
21. Hsu, L.Y., Chia, P.Y., Lim, J., The Novel Coronavirus (SARS-CoV-2) Pandemic. *Ann. Acad. Med. Singapore*, 49, 105, 105–107, 2020.
22. Lipsitch, M., Swerdlow, D.L., Finelli, L., Defining the epidemiology of Covid-19—studies needed. *N. Engl. J. Med.*, 382, 13, 1194–1196, 2020.
23. Di Gennaro, F. *et al.*, Coronavirus diseases (COVID-19) current status and future perspectives: a narrative review. *Int. J. Environ. Res. Public Health*, 17, 8, 2690, 2020.
24. Who, WHO Coronavirus Disease (COVID-19) Dashboard, 2021, [cited 2021 08 February]; Available from: https://covid19.who.int.

25. Hao, W. *et al.*, COVID-19 epidemic spread and the impact on public health & safety policy: An analysis of the adoption of preventive measures and effective management: Evidence from Pakistan. *Rev. Argent. Clín. Psicol.*, 29, 4, 722–736, 2020.
26. Dosch, S.F., Mahajan, S.D., Collins, A.R., SARS coronavirus spike protein-induced innate immune response occurs via activation of the NF-κB pathway in human monocyte macrophages *in vitro*. *Virus Res.*, 142, 1-2, 19–27, 2009.
27. Hu, W. *et al.*, SARS-CoV regulates immune function-related gene expression in human monocytic cells. *Viral Immunol.*, 25, 4, 277–288, 2012.
28. Wang, Y. and Liu, L., The membrane protein of severe acute respiratory syndrome coronavirus functions as a novel cytosolic pathogen-associated molecular pattern to promote beta interferon induction via a Toll-like-receptor-related TRAF3-independent mechanism. *MBio*, 7, 1, e01872-15, 2016.
29. Al-Qahtani, A.A. *et al.*, Middle east respiratory syndrome corona virus spike glycoprotein suppresses macrophage responses via DPP4-mediated induction of IRAK-M and PPARγ. *Oncotarget*, 8, 6, 9053, 2017.
30. García-Sastre, A. and Biron, C.A., Type 1 interferons and the virus-host relationship: A lesson in detente. *Science*, 312, 5775, 879–882, 2006.
31. Sallard, E. *et al.*, Type 1 interferons as a potential treatment against COVID-19. *Antiviral Res.*, 178, 104791, 2020.
32. Singh, P. *et al.*, Potential inhibitors for SARS-CoV-2 and functional food components as nutritional supplement for COVID-19: A Review. *Plant Foods Hum. Nutr.*, 75, 1–9, 2020.
33. Nawaz, A. *et al.*, Epidemic spread and its management through governance and leadership response influencing the arising challenges around COVID-19 in Pakistan—A lesson learnt for low income countries with limited resource. *Front. Public Health*, 8, 817, 2020.
34. Ezhilarasi, P. *et al.*, Nanoencapsulation techniques for food bioactive components: A review. *Food Bioprocess Technol.*, 6, 3, 628–647, 2013.
35. Pradhan, N. *et al.*, Facets of nanotechnology as seen in food processing, packaging, and preservation industry. *BioMed. Res. Int.*, 2015, 1–17, 2015.
36. Bratovčić, A. *et al.*, Application of polymer nanocomposite materials in food packaging. *Croat. J. Food Sci. Technol.*, 7, 2, 86–94, 2015.
37. Lamprecht, A. *et al.*, Lipid nanocarriers as drug delivery system for ibuprofen in pain treatment. *Int. J. Pharm.*, 278, 2, 407–414, 2004.
38. Nakagawa, K., Nano-and microencapsulation of flavor in food systems, in: *Nano and microencapsulation for foods*, pp. 249–272, 2014.
39. Dekkers, S. *et al.*, Presence and risks of nanosilica in food products. *Nanotoxicology*, 5, 3, 393–405, 2011.
40. Koo, O.M., Rubinstein, I., Onyuksel, H., Role of nanotechnology in targeted drug delivery and imaging: A concise review. *Nanomed.: Nanotechnol. Biol. Med.*, 1, 3, 193–212, 2005.

41. Couch, L.M. *et al.*, Food nanotechnology: Proposed uses, safety concerns and regulations. *Agro Food Ind. Hi-Tech*, 27, 1, 36–39, 2016.
42. Mihindukulasuriya, S. and Lim, L.-T., Nanotechnology development in food packaging: A review. *Trends Food Sci. Technol.*, 40, 2, 149–167, 2014.
43. Sorrentino, A., Gorrasi, G., Vittoria, V., Potential perspectives of bio-nanocomposites for food packaging applications. *Trends Food Sci. Technol.*, 18, 2, 84–95, 2007.
44. Bouwmeester, H. *et al.*, Review of health safety aspects of nanotechnologies in food production. *Regul. Toxicol. Pharm.*, 53, 1, 52–62, 2009.
45. Liu, S. *et al.*, Recent advances in nanosensors for organophosphate pesticide detection. *Adv. Powder Technol.*, 19, 5, 419–441, 2008.
46. Inbaraj, B.S. and Chen, B., Nanomaterial-based sensors for detection of foodborne bacterial pathogens and toxins as well as pork adulteration in meat products. *J. Food Drug Anal.*, 24, 1, 15–28, 2016.
47. Palchetti, I. and Mascini, M., Electroanalytical biosensors and their potential for food pathogen and toxin detection. *Anal. Bioanal. Chem.*, 391, 2, 455–471, 2008.
48. Nachay, K., Analyzing nanotechnology. *Food Technol. (Chicago)*, 61, 1, 34–36, 2007.
49. Kanazawa, K. and Cho, N.-J., Quartz crystal microbalance as a sensor to characterize macromolecular assembly dynamics. *J. Sens.*, 2009, 1–17, 2009.
50. Savolainen, K. *et al.*, Nanotechnologies, engineered nanomaterials and occupational health and safety–A review. *Saf. Sci.*, 48, 8, 957–963, 2010.
51. Han, J.W. *et al.*, Food packaging: A comprehensive review and future trends. *Compr. Rev. Food Sci. Food Saf.*, 17, 4, 860–877, 2018.
52. Vilela, C. *et al.*, A concise guide to active agents for active food packaging. *Trends Food Sci. Technol.*, 80, 212–222, 2018.
53. Sadeghi, K., Lee, Y., Seo, J., Ethylene scavenging systems in packaging of fresh produce: A review. *Food Rev. Int.*, 37, 2, 155–176, 2021.
54. Zhou, L. *et al.*, Effect of PE/AG2O Nano-packaging on the quality of apple slices. *J. Food Qual.*, 34, 3, 171–176, 2011.
55. Gaikwad, K.K., Singh, S., Ajji, A., Moisture absorbers for food packaging applications. *Environ. Chem. Lett.*, 17, 2, 609–628, 2019.
56. Gaikwad, K.K., Singh, S., Lee, Y.S., Oxygen scavenging films in food packaging. *Environ. Chem. Lett.*, 16, 2, 523–538, 2018.
57. López-Varela, S., González-Gross, M., Marcos, A., Functional foods and the immune system: A review. *Eur. J. Clin. Nutr.*, 56, 3, S29–S33, 2002.
58. Granato, D. *et al.*, Functional foods: Product development, technological trends, efficacy testing, and safety. *Annu. Rev. Food Sci. Technol.*, 11, 93–118, 2020.
59. Prietl, B. *et al.*, Vitamin D and immune function. *Nutrients*, 5, 7, 2502–2521, 2013.
60. Wimalawansa, S.J., Vitamin D deficiency: Effects on oxidative stress, epigenetics, gene regulation, and aging. *Biology*, 8, 2, 30, 2019.

61. Chen, Y. *et al.*, Vitamin D receptor inhibits nuclear factor κB activation by interacting with IκB kinase β protein. *J. Biol. Chem.*, 288, 27, 19450–19458, 2013.
62. Lemire, J.M. *et al.*, Immunosuppressive actions of 1, 25-dihydroxyvitamin D3: Preferential inhibition of Th1 functions. *J. Nutr.*, 125, suppl_6, 1704S–1708S, 1995.
63. Jeffery, L.E. *et al.*, 1, 25-Dihydroxyvitamin D3 and IL-2 combine to inhibit T cell production of inflammatory cytokines and promote development of regulatory T cells expressing CTLA-4 and FoxP3. *J. Immunol.*, 183, 9, 5458–5467, 2009.
64. Monlezun, D.J. *et al.*, Vitamin D status and acute respiratory infection: Cross sectional results from the United States National Health and Nutrition Examination Survey, 2001–2006. *Nutrients*, 7, 3, 1933–1944, 2015.
65. Ginde, A.A. *et al.*, High-dose monthly vitamin D for prevention of acute respiratory infection in older long-term care residents: A randomized clinical trial. *J. Am. Geriatr. Soc.*, 65, 3, 496–503, 2017.
66. Charan, J. *et al.*, Vitamin D for prevention of respiratory tract infections: A systematic review and meta-analysis. *J. Pharmacol. Pharmacother.*, 3, 4, 300, 2012.
67. Goncalves-Mendes, N. *et al.*, Impact of vitamin D supplementation on influenza vaccine response and immune functions in deficient elderly persons: a randomized placebo-controlled trial. *Front. Immunol.*, 10, 65, 2019.
68. Carr, A.C. and Maggini, S., Vitamin C and immune function. *Nutrients*, 9, 11, 1211, 2017.
69. Colunga Biancatelli, R.M.L. *et al.*, Quercetin and vitamin C: An experimental, synergistic therapy for the prevention and treatment of SARS-CoV-2 related disease (COVID-19). *Front. Immunol.*, 11, 1451, 2020.
70. Read, S.A. *et al.*, The role of zinc in antiviral immunity. *Adv. Nutr.*, 10, 4, 696–710, 2019.
71. Ishida, T., Review on the role of Zn2+ ions in viral pathogenesis and the effect of Zn2+ ions for host cell-virus growth inhibition. *Am. J. Biomed. Sci. Res.*, 2, 1, 28–37, 2019.
72. Speth, R. *et al.*, Concentration-dependent effects of zinc on angiotensin-converting enzyme-2 activity (1067.4). *FASEB J.*, 28, 1067.4, 2014.
73. Avery, J.C. and Hoffmann, P.R., Selenium, selenoproteins, and immunity. *Nutrients*, 10, 9, 1203, 2018.
74. Mahmoodpoor, A. *et al.*, The effect of intravenous selenium on oxidative stress in critically ill patients with acute respiratory distress syndrome. *Immunol. Invest.*, 48, 2, 147–159, 2019.
75. Catanzaro, M. *et al.*, Immunomodulators inspired by nature: A review on curcumin and echinacea. *Molecules*, 23, 11, 2778, 2018.
76. Jin, C.y. *et al.*, Curcumin attenuates the release of pro-inflammatory cytokines in lipopolysaccharide-stimulated BV2 microglia 1. *Acta Pharmacol. Sin.*, 28, 10, 1645–1651, 2007.

77. Cho, J.-W., Lee, K.-S., Kim, C.-W., Curcumin attenuates the expression of IL-1β, IL-6, and TNF-α as well as cyclin E in TNF-α-treated HaCaT cells; NF-κB and MAPKs as potential upstream targets. *Int. J. Mol. Med.*, 19, 3, 469–474, 2007.
78. Zahedipour, F. *et al.*, Potential effects of curcumin in the treatment of COVID-19 infection. *Phytother. Res.*, 34, 11, 2911–2920, 2020.
79. Chen, T.-Y. *et al.*, Inhibition of enveloped viruses infectivity by curcumin. *PLoS One*, 8, 5, e62482, 2013.
80. Rao, P.V. and Gan, S.H., Cinnamon: A multifaceted medicinal plant. *Evid.-Based Complement. Alternat. Med.*, 2014, 1–12, 2014.
81. Liao, B.-C. *et al.*, Cinnamaldehyde inhibits the tumor necrosis factor-α-induced expression of cell adhesion molecules in endothelial cells by suppressing NF-κB activation: Effects upon IκB and Nrf2. *Toxicol. Appl. Pharmacol.*, 229, 2, 161–171, 2008.
82. Guo, J.-Y. *et al.*, Cinnamaldehyde reduces IL-1β-induced cyclooxygenase-2 activity in rat cerebral microvascular endothelial cells. *Eur. J. Pharmacol.*, 537, 1-3, 174–180, 2006.
83. Lang, A. *et al.*, Allicin inhibits spontaneous and TNF-α induced secretion of proinflammatory cytokines and chemokines from intestinal epithelial cells. *Clin. Nutr.*, 23, 5, 1199–1208, 2004.
84. Shin, J.-H. *et al.*, Short-term heating reduces the anti-inflammatory effects of fresh raw garlic extracts on the LPS-induced production of NO and pro-inflammatory cytokines by downregulating allicin activity in RAW 264.7 macrophages. *Food Chem. Toxicol.*, 58, 545–551, 2013.
85. Bang, J.S. *et al.*, Anti-inflammatory and antiarthritic effects of piperine in human interleukin 1β-stimulated fibroblast-like synoviocytes and in rat arthritis models. *Arthritis Res. Ther.*, 11, 2, 1–9, 2009.
86. Bae, G.-S. *et al.*, Inhibition of lipopolysaccharide-induced inflammatory responses by piperine. *Eur. J. Pharmacol.*, 642, 1-3, 154–162, 2010.
87. Wang-Sheng, C. *et al.*, Piperine attenuates lipopolysaccharide (LPS)-induced inflammatory responses in BV2 microglia. *Int. Immunopharmacol.*, 42, 44–48, 2017.
88. Vaibhav, K. *et al.*, Piperine suppresses cerebral ischemia–reperfusion-induced inflammation through the repression of COX-2, NOS-2, and NF-κB in middle cerebral artery occlusion rat model. *Mol. Cell. Biochem.*, 367, 1, 73–84, 2012.
89. Mittal, R. and Gupta, R., *In vitro* antioxidant activity of piperine. *Methods Find. Exp. Clin. Pharmacol.*, 22, 5, 271–274, 2000.
90. Wolska, K. *et al.*, Immunomodulatory effects of propolis and its components on basic immune cell functions. *Indian J. Pharm. Sci.*, 81, 4, 575–588, 2019.
91. Sforcin, J.M. and Bankova, V., Propolis: is there a potential for the development of new drugs? *J. Ethnopharmacol.*, 133, 2, 253–260, 2011.
92. Sforcin, J., Propolis and the immune system: a review. *J. Ethnopharmacol.*, 113, 1, 1–14, 2007.

93. Schreck Bird, A. *et al.*, Probiotics for the treatment of infantile colic: A systematic review. *J. Pharm. Pract.*, 30, 3, 366–374, 2017.
94. Zelaya, H. *et al.*, Respiratory antiviral immunity and immunobiotics: Beneficial effects on inflammation-coagulation interaction during influenza virus infection. *Front. Immunol.*, 7, 633, 2016.
95. Baud, D. *et al.*, Using probiotics to flatten the curve of coronavirus disease COVID-2019 pandemic. *Front. Public Health*, 8, 186, 2020.
96. Mortaz, E. *et al.*, Probiotics in the management of lung diseases. *Mediators Inflammation*, 2013, 1–10, 2013.
97. Chong, H.-X. *et al.*, Lactobacillus plantarum DR7 improved upper respiratory tract infections via enhancing immune and inflammatory parameters: A randomized, double-blind, placebo-controlled study. *J. Dairy Sci.*, 102, 6, 4783–4797, 2019.
98. Shoaib, A., Xin, L., Xin, Y., Oral administration of Lactobacillus acidophilus alleviates exacerbations in Pseudomonas aeruginosa and Staphylococcus aureus pulmonary infections. *Pak. J. Pharm. Sci.*, 32, 1621–30, 2019.
99. Namba, K. *et al.*, Effects of Bifidobacterium longum BB536 administration on influenza infection, influenza vaccine antibody titer, and cell-mediated immunity in the elderly. *Biosci. Biotechnol. Biochem.*, 74, 5, 939–945, 2010.
100. Sharma, G. and Im, S.-H., Probiotics as a potential immunomodulating pharmabiotics in allergic diseases: current status and future prospects. *Allergy Asthma Immunol. Res.*, 10, 6, 575, 2018.
101. Hajavi, J. *et al.*, The immunomodulatory role of probiotics in allergy therapy. *J. Cell. Physiol.*, 234, 3, 2386–2398, 2019.
102. Lin, L. *et al.*, Gastrointestinal symptoms of 95 cases with SARS-CoV-2 infection. *Gut*, 69, 6, 997–1001, 2020.
103. Rosa, L. *et al.*, Lactoferrin: A natural glycoprotein involved in iron and inflammatory homeostasis. *Int. J. Mol. Sci.*, 18, 9, 1985, 2017.
104. Safaeian, L. *et al.*, Cytoprotective and antioxidant effects of human lactoferrin against H2O2-induced oxidative stress in human umbilical vein endothelial cells. *Adv. Biomed. Res.*, 4, 2015.
105. Li, S. *et al.*, Inhibition of HBV infection by bovine lactoferrin and iron-, zinc-saturated lactoferrin. *Med. Microbiol. Immunol.*, 198, 1, 19–25, 2009.
106. Ishikado, A. *et al.*, Liposomalization of lactoferrin enhanced it's anti-inflammatory effects via oral administration. *Biol. Pharm. Bull.*, 28, 9, 1717–1721, 2005.
107. Glinsky, G.V., Tripartite combination of candidate pandemic mitigation agents: Vitamin D, quercetin, and estradiol manifest properties of medicinal agents for targeted mitigation of the COVID-19 pandemic defined by genomics-guided tracing of SARS-CoV-2 targets in human cells. *Biomedicines*, 8, 5, 129, 2020.

9

Dietary Diversification and Its Impact on Human Health

Anindita Ray (Chakravarti)[1]* and Soumam Dutta[2]

[1]*Department of Food & Nutrition, Maharani Kasiswari College, Kolkata, India*
[2]*Food & Nutrition Division, University of Calcutta, Kolkata, India*

Abstract

Nutrition and diet play significant roles in the health and well-being of human beings. A nutritious diet is essential to ensure proper growth, development and maintenance at every stage of life. Not only the quantity of food but also the quality of diet has a significant impact on the overall health. Dietary diversity refers to the number of food groups consumed by an individual or a household over a reference time period. It is usually calculated by using a recall method and counting the number of unique food groups consumed. A higher score indicates better diversity. A diversified diet signifies the adequate consumption of essential nutrients. It is considered as an important indicator of dietary quality and access to a variety of food. Several studies show that a higher dietary diversity is associated with better health and reduced risk of various chronic diseases. Additionally, it is also associated with socioeconomic status and food security. Lower dietary diversity, on the other hand, impairs physical capacities, cognitive performance, and social health. It is thus essential to study the dietary diversity of individuals and households regularly so that necessary actions can be taken. Nanotechnology has revolutionized the food system and its applications can be noted from the farm to the fork. Thus, it may act as an important tool for improving dietary diversity and thereby improve nutritional security. In this article, we will highlight the importance of dietary diversification and the potential role of nanotechnology in this field.

***Keywords*:** Balanced diet, dietary diversity, nutritional adequacy, chronic diseases, nanotechnology

**Corresponding author*: anindita.ray.chakravarti@gmail.com

Tanima Bhattacharya and Shakeel Ahmed (eds.) Nanotechnology in Functional Foods, (223–242)

9.1 Introduction

Food and nutrition come under the basic needs of human beings. Out of the eight Millennium Development Goals (MDG, 2000) five were directly associated with optimal food and nutrition security namely, "eradication of extreme poverty and hunger," "achievement of universal primary education," "reduction of child mortality," "improvement of maternal health and combating HIV/AIDS," "Malaria and other diseases" [1]. Despite huge efforts, several MDG targets were not met by 2015, which gave birth to more ambitious Sustainable Development Goals (SDG, 2015) to be achieved by 2030 [2]. Again, of the seventeen SDGs, five are directly related to food and nutrition security namely, "no poverty," "zero hunger," "good health and well-being," "quality education," and "clean water and sanitation." Such intricate involvements of food security and nutrition with each and every aspect of development indicates its importance for ensuring individual, as well as community health. Reduced availability and accessibility of good quality food often leads to malnutrition [3]. While wasting, stunting and underweight appear to be the most alarming threat across the globe, the face of malnutrition is more complex and dramatically changing. Overnutrition, overweight, and obesity are now rising in many countries, including the developing countries [4]. Most importantly micronutrient deficiencies are further complicating these issues, giving rise to a "Triple Burden of Malnutrition" [5]. Hunger affects 8.9% people worldwide accounting for about 690 million people who go to bed on an empty stomach [6]. The number of undernourished people has increased globally by 10 million between 2018 and 2019. As compared to 2014, there are about 60 million more malnourished people at present [6]. The disappointing statistics indicate that "Zero Hunger" may not be achieved by 2030 along with many other important indicators [6]. The food and nutrition security of vulnerable populations may further deteriorate due to the COVID-19 outbreak [7]. India has witnessed significant improvements in life expectancy at birth as well as reduction in infant and maternal mortality since independence [8–10]. Despite such improvement, food insecurity is a major burden in India [11]. India ranked 94th of 107 countries in Global Hunger Index (2020) and 71st of 113 countries in Global Food Security Index (2019) [12, 13]. Similar to many other developing countries, India is also wounded by the triple edged sword of malnutrition [14]. The government has launched numerous health and nutrition-related programs since independence [15]. But till to date a significant portion of the Indian population is malnourished [16]. Especially, micronutrient deficiencies are creating huge troubles in the achievements of national goals. Deficiencies of vitamin A and iodine have improved over the years and are under control in recent times due to effective prophylaxis programs [16]. But the clinical and subclinical deficiencies of many other vitamins and minerals are alarming. Especially anemia is affecting around 58.6% under five children, 53.2% nonpregnant women, 50.4% pregnant women, and 22.7% men [16]. Along with India, the falling ranks of many other Asian, African, and Latin American countries in the Global Hunger Index is a major threat to global health [17]. To overcome all such national and global challenges, it is essential to ensure a healthy and balanced diet for the entire population. Lack of diversity and variety in the diet limits the availability of optimal nutrients thereby causing deficiency, excess, or imbalance of certain nutrients. Ensuring proper dietary diversity may serve as an effective way to promote the nutritional status of the population. The EAT-Lancet Commission recently published a call for the transformation food systems for a dietary shift toward "planetary health diets" to improve both human health and environmental sustainability. It focuses on the need to incorporate nutrition and sustainability into a global agenda for food transformation [18]. This is only possible if our diets become more diversified and is able to provide "enough for all."

Recent advances in nanoscience and nanotechnology have opened a new avenue which focuses on developing nanomaterials for improving human health [19]. Nanotechnology, in particular, has several applications in the food industry and has the power to transfigure the entire food system. It has numerous important roles to play in the food cycle, from the farm to the fork [20]. Development of novel nanofoods is one of the thrust areas of research in food science and nutrition. Nanofood can be defined as food that has been cultivated, manufactured, processed, modified, or packaged using nanotechnology [20]. With time, the nanofoods are rapidly reaching the supermarkets and kitchen tables from the laboratory. Although, it is still unclear whether nanotechnology is big enough to tackle problems like food insecurity and malnutrition.

The present article thus targets to highlight the importance of dietary diversification and its impact on human health. The possible applications of nanotechnology will be discussed.

9.2 Dietary Diversity—What Does It Indicate?

Defining a healthy diet is often challenging. A diet superior in quantity and quality are essential for human health and well-being. One of most important qualitative indicators of dietary adequacy is dietary diversity. Dietary diversity is a globally acknowledged component of a healthy diet. Dietary variety is another term which is often used interchangeably with dietary diversity. It is widely used for assessing individual, as well as household dietary quality. Additionally, it is a popular indicator used for determining effectiveness of ongoing nutrition, health and agriculture-related programs. In simple words, dietary diversity refers to the number of unique food groups consumed by an individual or a household over a reference time period [21]. Incorporation of a wide variety of food within and across different food groups is considered to be an important factor which determines dietary quality. The major food groups include "Cereals, Millets & Pulses," "Milk & Animal Products," "Fruits & Vegetables," and "|Oils, Fats & Nuts." [22]. All these food groups are rich in particular type of nutrients. A diverse diet with a correct balance of all these food groups ensures the availability of all essential nutrients necessary for promoting good health. Besides these major food groups, various other group of foods maybe considered like vitamin A-rich food, iron-rich food, folic acid-rich food, calcium-rich food, spices, and so on. Dietary diversity is a major concern for the poor population, especially in the developing countries, since availability and accessibility of a wide variety of food depends on the socioeconomic status and affordability. It is found that a major proportion of the population in developing countries predominantly consume starchy food products [23]. The intake of vegetables and fresh fruits are quite below the recommended levels in many countries [24]. There is also an upward shift in the consumption of processed, packaged, and convenience foods [23]. Such nondiverse diets are often deficient in many micronutrients. The bioavailability of such nutrients is another issue. Additionally, high consumption of fats, salt, and sugar is a serious problem, even in the developing countries, due to recent shifts in dietary and lifestyle pattern [25]. It is thus essential to promote the concept of dietary diversity to ensure an optimal quality of diet. Dietary diversification is the strategy to improve consumption of a variety of food from different food groups in order to promote micronutrient status and bioavailability of diets. Dietary diversification is a vital component of various health and nutrition-related programs, which aim to promote micronutrient consumptions. Traditional food systems may contribute a lot to dietary diversification, thereby improving various health-related parameters. Exploring traditional food systems is thus essential. For example, the role of Mediterranean dietary pattern and Japanese diet is well researched whereas the role of Indian traditional diets is not evaluated to its fullest potential [26]. Making common people aware of various traditional recipes and ingredients will not only diversify their diet but also improve palatability. Additionally, newer concepts of diversification

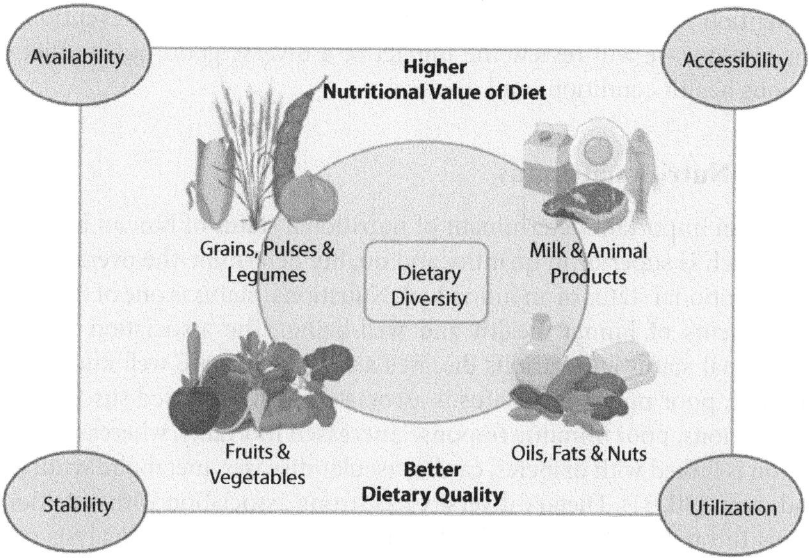

Figure 9.1 Dietary diversity. Inclusion of a wide variety of food increases nutritional value of the diet and dietary quality. These are in turn based on the four major pillars of food security- Availability, Accessibility, Utilization and Stability.

like substituting plant-based relishes with small, dried fish relishes; maize-based porridges prepared with maize flour soaked to reduce its Hexa (IP-6)- and Penta (IP-5)-inositol phosphate content instead of conventional porridges are being constantly used [27]. Figure 9.1 gives a conceptual overview of dietary diversity.

9.3 Impact of Dietary Diversity on Health and Well-Being

Dietary diversity signifies the quality of diet. Significant positive associations can be observed between dietary diversity, Nutrient Adequacy Ratio (NAR) and Mean Adequacy Ratio (MAR) of diet [28]. So, dietary diversity can be used as a proxy indicator of dietary quality. A healthy balanced diet is highly essential for general health and well-being as well as for preventing specific health issues. The disease process in the natural history of any disease involves two major phases namely, (1) prepathogenesis phase and (2) pathogenesis phase [29]. Different levels of preventions and modes of interventions are required at different phases of disease progression. A good quality diet can play important role not only in the primary level of prevention but also in the secondary and tertiary level of prevention. In this section, we will review the impact of a diverse good quality diet on various health conditions and diseases.

9.3.1 Nutritional Status

Diet is an important determinant of nutritional status of human beings. A diet which is superior in quantity and quality determine the overall health and nutritional status of an individual. Nutritional status is one of the major components of human health and well-being. The association between nutritional status and various diseases and disorders are well known. For instance, poor nutritional status is associated with increased susceptibility to infections, poor immune response, increased mortality, whereas overnutrition is linked with diabetes, cardiovascular diseases, metabolic syndrome and so on [30, 31]. Dietary diversity has strong association with nutritional status throughout the life cycle of a human being. An early analysis of 11 Demographic and Health Surveys by Arimond *et al.* (2004) indicated significant association between dietary diversity of children and their nutritional status (height for age), independent of socioeconomic status [32]. Since then, many other studies have indicated the impact of dietary diversification on infant and child nutritional status along with other age groups (Table 9.1). It is found that a higher dietary diversity is associated with better nutritional status and health outcome. But, in some studies, higher dietary diversity was found to be associated with overweight and obesity. Thus, it is essential to reach minimum dietary diversity with proper portion control and emphasis should be given on healthy food choices.

9.3.2 Cardiovascular Diseases

Cardiovascular diseases are the leading causes of mortality and morbidity worldwide. It is also a major contributor to disability. Especially, ischemic heart diseases and strokes are important public health concerns. The odds of developing cardiovascular diseases increase with unhealthy lifestyle patterns [43]. Especially, diet and nutrition play important role in the development of these conditions. A diet high in refined carbohydrate, saturated fat, trans fat, excessive sodium, added sugars, processed food and low in fruits, vegetables, nuts, whole grains, fibre eventually results in cardiovascular complications [44]. These complications maybe associated with hypertension, hyperlipidemia, dyslipidemia, hypertriglyceridemia, hypercholesterolemia, hyperglycemia, hyperuricemia, and so on. Cerebrovascular complications are also associated with these conditions [43]. Besides diet,

Table 9.1 Impact of dietary diversity on nutritional status.

Age group	Location	Study outcome	Reference
Infants & Young Children (6–23 months)	India	Limited diversity of complementary food was found to be a strong predictor of nutritional status.	[33]
	Tanzania	The risk of wasting, stunting, underweight decreased with an increase in number of food groups consumed.	[34]
Under Five Children (6–59 months)	Ghana	Positive association is found between the consumption of vitamin A rich food, other vegetables, egg and weight-for-age score.	[35]
	Burkina Faso	Dietary diversity was associated with stunting but not with wasting and mid-upper arm circumference (MUAC).	[36]
Schoolers (6-9 years)	India and Iran	Higher dietary diversity was observed among overweight and normal weight children. Lower diversity was found among underweight children.	[37]
Adolescent Girls (15–19 years)	Ethiopia	Inadequate dietary diversity was significantly associated with underweight.	[38]
Adolescents (13–17 years)	Ethiopia	Dietary diversity was found to be an independent predictor of stunting an underweight.	[39]
Adults (18 years and above)	Brazil	Dietary diversity score was negatively associated with underweight and positively related to excess body weight.	[40]
	China	Higher dietary diversity was associated with excess energy intake and obesity.	[41]
	India	Linear association was found between dietary diversity and nutritional outcome.	[42]

other lifestyle factors like lack of physical activity, smoking, binge eating, alcoholism, stress may lead to such complications [43]. It is, thus, crucial to ensure a good quality diet which can supply adequate nutrition coupled with a healthy lifestyle for normal functioning of the cardiovascular system. Dietary diversity is an important factor associated with cardiovascular health. In a population-based cross-sectional study with adult participants at Tehran, cardiovascular disease risk was found to be inversely related with dietary diversity score. There was a decreased risk of hypertension, high low-density lipoprotein (LDL), hypercholesterolemia and diabetes with increase in dietary diversity [45]. In another population-based study, lower serum triglyceride levels, systolic blood pressure, and higher serum adiponectin levels were observed in the top quartile of dietary diversity score. Additionally, the components of metabolic syndrome were significantly higher in the lower quartiles [46]. Similar results were obtained in a study on rural and urban Korean adults where lower dietary diversity was related with higher blood pressure, serum triglyceride and LDL levels [47].

9.3.3 Diabetes Mellitus

Diabetes mellitus is a metabolic condition characterized by raised blood glucose levels. It can be classified in several categories like type 1 diabetes mellitus (T1DM), type 2 diabetes mellitus (T2DM), gestational diabetes mellitus (GDM) and other types of diabetes like maturity onset diabetes of the youngs (MODY), Latent autoimmune diabetes of the adults (LADA), diabetes due to other conditions like endocrinopathies, use of certain drugs, and so on. The two major types are T1DM and T2DM. T1DM is characterized by destruction of pancreatic beta cells whereas T2DM is characterized by insulin resistance. Lifestyle factors highly important contributor to diabetes mellitus, especially T2DM [48, 49]. Apart from diabetes, pre-diabetes is also another public health issue. This condition is characterized by Impaired Glucose Tolerance (IGT) an Impaired Fasting Glycemia (IFG) [50]. Diabetes mellitus if not properly managed may lead to various complications like hypoglycemia, ketoacidosis, retinopathy, neuropathy, nephropathy, cardiomyopathy, ulceration, delayed wound healing, and so on [48]. A diet high in refined carbohydrate, added sugars, empty calories, processed food products coupled with a sedentary and unhealthy lifestyle may lead to T2DM development. Several studies have confirmed the association between dietary diversity and diabetes mellitus (DM), especially Type 2 DM (T2DM). In a cross-sectional study with T2DM patients it was found that only 34.5% participants achieved a desirable dietary diversity whereas only 1.8% participants consumed ideal servings from each food group [51]. In a prospective study with 23,238 participants it was observed that higher total dietary diversity reduced the risk of developing T2DM by 30% [52]. Another study with pre-diabetic individuals found that higher dietary diversity was associated with lower fasting blood glucose levels and other components of metabolic syndrome like lower serum triglyceride, waist circumference and higher HDL-cholesterol. This may prevent the development of diabetes from pre-diabetes [53].

9.3.4 Cancer

Cancer is a condition characterized by uncontrolled cell growth which may invade or spread to other parts of the body. There are more than hundred different kinds types of cancer, which vary significantly in terms of their nature and response to treatment. Tumor is an abnormal proliferation of cells. A tumor can be either malignant or benign. Benign tumors are not able to invade surrounding tissues and spread to distant body parts whereas, malignant tumors can both invade surrounding tissues and spread to distant areas via circulatory systems. The aetiology of cancer is multifactorial and includes a wide range of factors like genetics, heredity, certain viral infections, exposure to carcinogens, smoking tobacco, poor diet and lifestyle factors [54]. Dietary diversity is an important factor associated with cancer development [55]. Inverse relationship has been observed between dietary diversity and gastric cancer risk. The higher quartiles of dietary diversity, especially fruits and vegetables, were found to be related to a lower risk of gastric cancer [56]. Similarly, in a case-control study with bladder cancer patients, higher total dietary diversity and fruit diversity were found to reduce cancer risk [57]. A greater dietary diversity, especially vegetable diversity, decreased the risk of developing breast cancer. The effects are found to be moderate [58]. Other studies have confirmed similar association between dietary diversity and risk of colon cancer, [59] colorectal cancer, [60] oesophageal cancer, [61] laryngeal cancer, [62] oral and pharyngeal cancer [63]. However, in a recent study significant association was not found with prostate cancer [64].

9.3.5 Cognitive Functions

Cognitive impairment is one of the frequently encountered problems, especially among the older adults. It involves difficulty in remembering certain things, learning new concepts, spatial orientation, temporal orientation, language problems, difficulty in decision making, and so on. The symptoms may range from mild to severe. The underlying causes are very complex and involves genetics, emotional issues, social problems, malnutrition, dietary factors, chronic stress and other lifestyle factors [65]. Several studies have found significant association between dietary diversity and cognitive status of individuals. In a population-based study with 8751 elderly persons (age >= 65 years) in China, it was found that poor dietary diversity was associated with poor global cognitive functions [66]. Another prospective study with Chinese adults revealed that higher dietary diversity score was associated with better memory and inversely associated with self-reported decline in memory [67]. In a longtudinal study with Japanese older adults (age 60-81 years), it was observed that an increase in the diversity of diets was protective against decline in cognition [68]. Another longitudinal study with Japanese community dwellers (age 40-89 years) showed that increased dietary diversity was significantly related to decreased hippocampal atrophy which is a major centre for the processing of memory in the brain [69]. Such evidences show a possible role of dietary diversification in the promotion of cognitive functions and memory. Figure 9.2 represents the overall impact of dietary diversification on health and well-being.

Figure 9.2 Impact of dietary diversification on health and well-being.

9.4 How to Measure Dietary Diversity?

Measurement of dietary diversity maybe a challenging task as it is a qualitative factor. Different approaches can be followed to obtain data regarding dietary diversity. The usual measurement procedures involve simple counting of number of unique food or food groups consumed over a reference time period by an individual or a household (Dietary Diversity Score) [21]. It is the simplest method as well as cost effective. The measurement maybe complicated by the use of different classification systems of food and different reference periods. Such classifications maybe done by the researchers based on the research objectives or maybe based on the local food grouping system of the country. The reference period mentioned in the guidelines of Food and Agricultural Organization (FAO) for measuring household or individual dietary diversity is one day [70]. But this period may range between 1 to 15 days [21]. Some scales

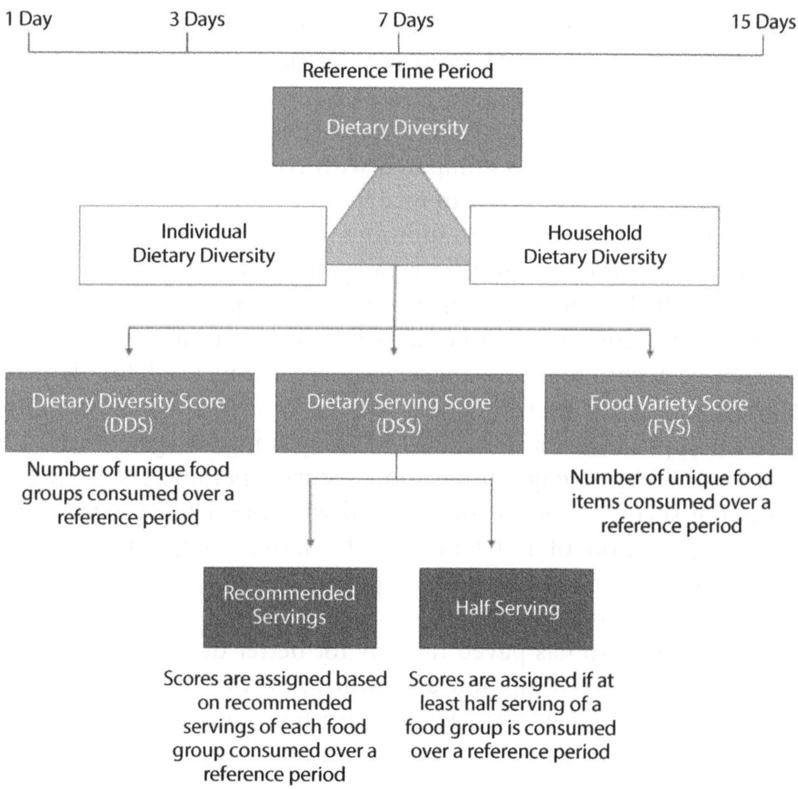

Figure 9.3 Measures of dietary diversity.

only consider the number of unique food items consumed over a reference period (Food Variety Score). Some scales consider the number of servings of each food group consumed to assign specific scores for each recommended serving (Dietary Serving Score) [21, 28]. Whereas some scales also count only those food groups from which at least half a serving is consumed during the reference period [21, 28]. Use of such different scales for measuring dietary diversity may create complication while comparing these studies. Figure 9.3 represents different measures of dietary diversity.

9.5 The Future: Using Nanotechnology as a Key in Improving Dietary Diversification

Nanotechnologies have taken the world of food by storm and perhaps it is one of the most effective sustainable method that can be used for the purpose of diversifying food. Nanotechnology has huge applications in the field of agriculture [71]. Nutrient deficiency in crops is a serious issue and affects human health negatively, especially in the rural areas. Nano-fortification of crops is an area of current research all over the world. Nano-fertilizers, as compared with conventional fertilizers, are better equipped to help crops to incorporate nutrients from soil which are not easily lost and these are additionally target-oriented [72]. These, if are in the form of bio-fertilizers or green nano-fertilizers, can also do away with the controversies of toxicity and pollution significantly as well. Other commonly used agrochemicals can also be developed using these technologies. Nanosensors are being used for detecting pathogens, monitoring soil conditions and crop growth. Several other potential applications are there which may improve agricultural productions thereby strengthening food security parameters [73]. Food processing sector is also revolutionized with the help of nanoscience. Nanoencapsulation of nutrients and bioactive compounds is major area of research and has numerous applications [74]. Nano-based food materials are not only making it possible to improve the nutrient bioavailability but also has paved the way for better delivery mechanisms that can enhance assimilation and utilization of nutrients and functional ingredients [74]. This may in turn help to combat the curse of malnutrition in developing countries. Bio-nano particulate nutrients are being produced throughout the world to enhance diversification in food, which is one of the primary concerns for the entire human race today [75]. Several nutritional supplements are developed using nanotechnology for special medical purposes, veterinary supplements and sports supplements [76]. The commonly used nanocarriers for dietary supplements are, nanogels, nanofibers, nanoemulsions, nanocapsules, nanoliposomes, solid lipid nanoparticles, core-shell nanoparticles, and so on [76]. Moreover, innovative food packaging and nano-biosensors are now ensuring the provision of safer food to the population at large and techniques are improving everyday [73]. For instance, nanofilms and nanoclays can be used to prevent food spoilage and oxygen absorption. Nanoparticles containing antimicrobial compounds are being used in packaging materials for improving shelf life [73].

Despite significant progresses in nanotechnology-based agriculture and food production numerous challenges are observed till to date. A major issue associated with nanofoods is the toxicity aspect [72]. Nanomaterials may cause toxicity issues depending on various factors out of which method of their synthesis is an important one. Chemically synthesized nanoparticles are found to have greater toxic effects due to the slow release of various chemicals used in the synthesis [72]. However, the biologically synthesized nanomaterials are relatively safer for the organisms and environment [72]. Application of green technologies may further improve the outcomes [77]. Consumer acceptance of nanofoods is another important factor. Consumers are more likely to accept nanofoods if the perceived benefits outweigh the perceived risks. It is observed that nanotechnology is more acceptable to the consumers when used in food packaging and processing as compared to direct incorporation into the food products [78]. Although, acceptance may increase if the application of nanotechnology can develop cheaper and safer food products. Development of specific policies and regulations for nanofood products is also essential. Moreover, proper information regarding the benefits and risks associated with these products should be disseminated to the consumers to develop trust and confidence.

Thus, it can be understood that nanotechnologies have a huge role to play in developing new methods and approaches in the field of dietary diversification, like all other fields and can bring about a betterment in the scenario. Figure 9.4 represents the application of nanotechnology in various sectors for dietary diversification.

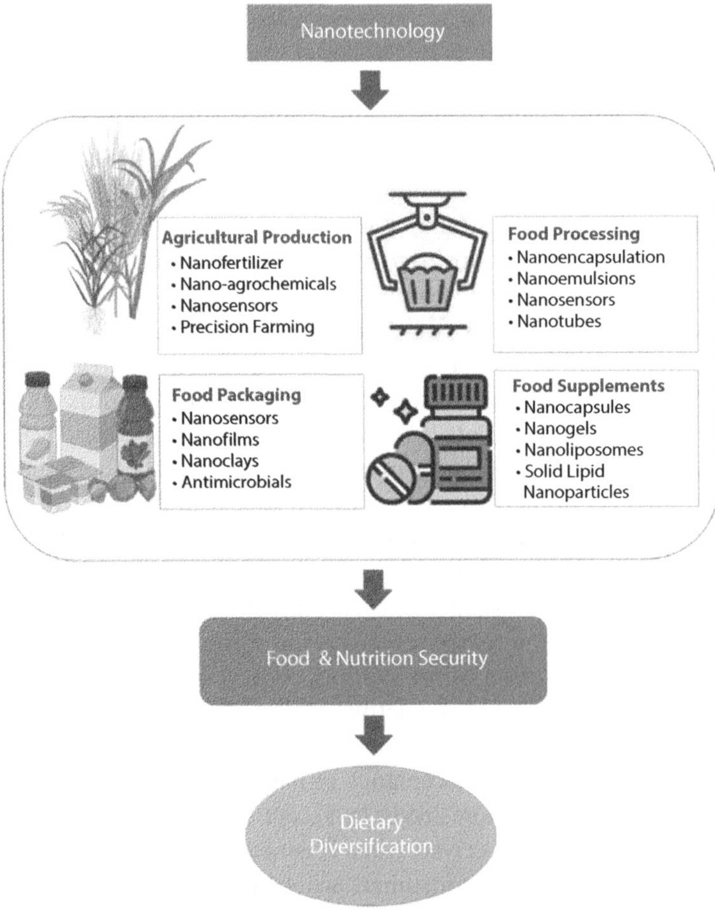

Figure 9.4 Application of nanotechnology for dietary diversification. Nanotechnology has revolutionized the food system. Application of nanotechnology may improve food production, availability, accessibility, utilization and stability. This in turn may lead to dietary diversification and thereby ensure nutritional security.

9.6 Conclusion

There has been a general shift in the pattern of dietary intake in the past two decades; consumption of energy-dense food, fatty food and a low intake of well-balanced food has led to the increase in the burden of disease load. The predominant dietary profile all over the world is proving to be inadequate to meet the physiological needs of the human life. To combat this situation, steps have to and are being taken in the lines of agriculture, aquaculture, animal husbandry, and improved household as well as commercial food processing techniques to increase the availability and absorption of nutrients. Several studies have shown the positive and direct relation between dietary diversification and improved nutritional status. Such has been proved in case of "modern day" diseases like cancer, diabetes, cardiovascular diseases, and so on. Several interventions are proven to be fruitful as well. Diversification of diets can be achieved by improving agricultural production, strengthening food security parameters, kitchen gardening or vegetable plot development, incorporation of a wide variety of food items and healthy cooking methods within the families. Additionally, improvements in health services, supplementation programs and food fortification may play an important role in fighting against malnutrition. Moreover, the advances in nanotechnology are found to have huge impact on the food system and these technologies should be used more efficiently to diversify human diets and ensure nutritional security of the population.

References

1. Lomazzi, M., Borisch, B., Laaser, U., The Millennium Development Goals: Experiences, achievements and what's next. *Glob. Health Action*, 7, 23695, 2014.
2. Kumar, S., Kumar, N., Vivekadhish, S., Millennium Development Goals (MDGs) to Sustainable Development Goals (SDGs): Addressing unfinished agenda and strengthening sustainable development and partnership. *Indian J. Community. Med.*, 41, 1, 2016.
3. Ramachandran, P., Food & nutrition security: Challenges in the new millennium. *Indian J. Med. Res.*, 138, 373, 2013.
4. Bhurosy, T. and Jeewon, R., Overweight and obesity epidemic in developing countries: A problem with diet, physical activity, or socioeconomic status? *ScientificWorldJournal*, 2014, 964236, 2014.
5. Blankenship, J.L., Rudert, C., Aguayo, V.M., Triple trouble: Understanding the burden of child undernutrition, micronutrient deficiencies, and overweight in East Asia and the Pacific. *Matern. Child Nutr.*, 16, Suppl 2, e12950, 2020.
6. United Nations, Goal 2: Zero Hunger, United Nations, New York City, USA, 2021. https://www.un.org/sustainabledevelopment/hunger/.
7. Ma, N.L., Peng, W., Soon, C.F. *et al.*, Covid-19 pandemic in the lens of food safety and security. *Environ. Res.*, 193, 110405, 2021.
8. Kumari, M. and Mohanty, S.K., Caste, religion and regional differentials in life expectancy at birth in India: Cross-sectional estimates from recent National Family Health Survey. *BMJ Open*, 10, e035392, 2020.
9. Bhatia, M., Dwivedi, L.K., Ranjan, M., Dixit, P., Putcha, V., Trends, patterns and predictive factors of infant and child mortality in well-performing and

underperforming states of India: A secondary analysis using National Family Health Surveys. *BMJ Open*, 9, e023875, 2019.
10. Goli, S. and Jaleel, A.C., What is the cause of the decline in maternal mortality in India? Evidence from time series and cross-sectional analyses. *J. Biosoc. Sci.*, 46, 351, 2014.
11. Upadhyay, R.P. and Palanivel, C., Challenges in achieving food security in India. *Iran. J. Public Health*, 40, 31, 2011.
12. Global Hunger Index, India, 2021.
13. Global Food Security Index, *Ranking and Trends*, Economist Impact, London, UK, 2020, https://foodsecurityindex.eiu.com/index.
14. Kumar, P., Chauhan, S., Patel, R., Srivastava, S., Bansod, D.W., Prevalence and factors associated with triple burden of malnutrition among mother-child pairs in India: A study based on National Family Health Survey 2015-16. *BMC Public Health*, 21, 391, 2021.
15. Grover, A. and Singh, R.B., Health policy, programmes and initiatives, in: *Urban Health and Wellbeing*, p. 251, 2019.
16. International Institute for Population Sciences, National family health Survey (NFHS-4), pp. 2015-2016, International Institute of Population Sciences, Mumbai, India, http://rchiips.org/nfhs/NFHS-4Reports/India.pdf.
17. Global Hunger Index, 2020 Global Hunger Index by Severity, 2020, https://www.globalhungerindex.org/ranking.html.
18. Custodio, M.C., Ynion, J., Samaddar, A. *et al.*, Unraveling heterogeneity of consumers' food choice: Implications for nutrition interventions in eastern India. *Glob. Food Sec.*, 28, 100497, 2021.
19. Zdrojewicz, Z., Waracki, M., Bugaj, B., Pypno, D., Cabała, K., Medical applications of nanotechnology. *Postepy Hig. Med. Dosw.*, 69, 1196, 2015.
20. Sekhon, B.S., Food nanotechnology - an overview. *Nanotechnol. Sci. Appl.*, 3, 1, 2010.
21. Ruel, M.T., Operationalizing dietary diversity: A review of measurement issues and research priorities. *J. Nutr.*, 113, 3911S, 2003.
22. National Institute of Nutrition, *Dietary guidelines for Indians*, National Institute of Nutrition, Hyderabad, India, 2011.
23. Kelly, M., The nutrition transition in developing Asia: Dietary change, drivers and health impacts, in: *Eating, Drinking: Surviving. SpringerBriefs in Global Understanding*, P. Jackson, W. Spiess, F. Sultana, (Eds.), Springer, Cham, 2016.
24. Frank, S.M., Webster, J., McKenzie, B. *et al.*, Consumption of fruits and vegetables among individuals 15 years and older in 28 low- and middle-income countries. *J. Nutr.*, 149, 1252, 2019.
25. Dasgupta, R., Pillai, R., Kumar, R., Arora, N.K., Sugar, salt, fat, and chronic disease epidemic in India: Is there need for policy interventions? *Indian J. Community Med.*, 40, 71, 2015.

26. Nair, M.K., Augustine, L.F., Konapur, A., Food-based interventions to modify diet quality and diversity to address multiple micronutrient deficiency. *Front. Public Health*, 3, 277, 2016.
27. Gibson, R.S. and Hotz, C., Dietary diversification/modification strategies to enhance micronutrient content and bioavailability of diets in developing countries. *Br. J. Nutr.*, 85, S159, 2001.
28. Rathnayake, K.M., Madushani, P., Silva, K., Use of dietary diversity score as a proxy indicator of nutrient adequacy of rural elderly people in Sri Lanka. *BMC Res. Notes*, 5, 469, 2012.
29. Park, K., *Park's textbook of preventive and social medicine*, pp. 39–40, Bhanot Publishers, Jabalpur, India, 2019.
30. Wells, J.C., Sawaya, A.L., Wibaek, R. *et al.*, The double burden of malnutrition: Aetiological pathways and consequences for health. *Lancet*, 395, 75, 2020.
31. Childs, C.E., Calder, P.C., Miles, E.A., Diet and immune function. *Nutrients*, 11, 1933, 2019.
32. Arimond, M. and Ruel, M.T., Dietary diversity is associated with child nutritional status: Evidence from 11 demographic and health surveys. *J. Nutr.*, 134, 2579, 2004.
33. Divya, N., Rajanish, K.V., Malavika, J., Sharma, A., The study of dietary diversity score in children between 6 months to 23 months: A hospital based study. *Int. J. Contemp. Pediatr.*, 5, 1053, 2018.
34. Khamis, A.G., Mwanri, A.W., Ntwenya, J.E., Kreppel, K., The influence of dietary diversity on the nutritional status of children between 6 and 23 months of age in Tanzania. *BMC Pediatr.*, 19, 518, 2019.
35. Frempong, R.B. and Annim, S.K., Dietary diversity and child malnutrition in Ghana. *Heliyon*, 3, e00298, 2017.
36. Sié, A., Tapsoba, C., Dah, C. *et al.*, Dietary diversity and nutritional status among children in rural Burkina Faso. *Int. Health*, 10, 157, 2018.
37. Hooshmand, S. and Udipi, S.A., Dietary diversity and nutritional status of Urban Primary School Children from Iran and India. *J. Nutr. Disord. Ther.*, S12, 001, 2013.
38. Gonete, K.A., Tariku, A., Wami, S.D., Akalu, T.Y., Dietary diversity practice and associated factors among adolescent girls in Dembia district, northwest Ethiopia, 2017. *Public Health Rev.*, 41, 23, 2020.
39. Jikamo, B. and Samuel, M., Does dietary diversity predict the nutritional status of adolescents in Jimma Zone, Southwest Ethiopia? *BMC Res. Notes*, 12, 402, 2019.
40. Bezerra, I.N. and Sichieri, R., Household food diversity and nutritional status among adults in Brazil. *Int. J. Behav. Nutr. Phys. Act.*, 8, 22, 2011.
41. Zhang, Q., Chen, X., Liu, Z., Varma, D.S., Wan, R., Zhao, S., Diet diversity and nutritional status among adults in southwest China. *PloS One*, 12, e0172406, 2017.

42. Nithya, D.J. and Bhavani, R.V., Dietary diversity and its relationship with nutritional status among adolescents and adults in rural India. *J. Biosoc. Sci.*, 50, 397, 2018.
43. Olvera Lopez, E., Ballard, B.D., Jan, A., *Cardiovascular Disease*, StatPearls Publishing, Treasure Island (FL, 2021.
44. Casas, R., Castro-Barquero, S., Estruch, R., Sacanella, E., Nutrition and Cardiovascular Health. *Int. J. Mol. Sci.*, 19, 3988, 2018.
45. Azadbakht, L., Mirmiran, P., Esmaillzadeh, A., Azizi, F., Dietary diversity score and cardiovascular risk factors in Tehranian adults. *Public Health Nutr.*, 9, 728, 2006.
46. Farhangi, M.A. and Jahangiry, L., Dietary diversity score is associated with cardiovascular risk factors and serum adiponectin concentrations in patients with metabolic syndrome. *BMC Cardiovasc. Disord.*, 18, 68, 2018.
47. Kim, J., Lee, S., Shin, Y., Kim, Y., Dietary diversity is associated with the risk of hypertension in rural and urban middle-aged Koreans (P18-074-19). *Curr. Dev. Nutr.*, 3, P18–074–19, 2019.
48. Sapra, A. and Bhandari, P., *Diabetes Mellitus*, StatPearls Publishing, Treasure Island (FL, 2021.
49. Feingold, K.R., Atypical Forms of Diabetes, in: *Endotext*, K.R. Feingold, B. Anawalt, A. Boyce, (Eds.), MDText.com, Inc., South Dartmouth (MA, 2019.
50. Alvarez, S., Coffey, R., Algotar, A.M., *Prediabetes*, StatPearls Publishing, Treasure Island (FL, 2021.
51. Tiew, K.F., Chan, Y.M., Lye, M.S., Loke, S.C., Factors associated with dietary diversity score among individuals with type 2 diabetes mellitus. *J. Health Popul. Nutr.*, 32, 665, 2014.
52. Conklin, A., II, Monsivais, P., Khaw, K.T., Wareham, N.J., Forouhi, N.G., Dietary diversity, diet cost, and incidence of Type 2 diabetes in the United Kingdom: A prospective cohort study. *PloS Med.*, 13, e1002085, 2016.
53. Gholizadeh, F., Moludi, J., Lotfi Yagin, N. *et al.*, The relation of dietary diversity score and food insecurity to metabolic syndrome features and glucose level among pre-diabetes subjects. *Prim. Care Diabetes*, 12, 338, 2018.
54. Cooper, G.M., *The Cell: A Molecular Approach*, Sinauer Associates, Sunderland (MA), 2000.
55. Key, T.J., Bradbury, K.E., Perez-Cornago, A., Sinha, R., Tsilidis, K.K., Tsugane, S., Diet, nutrition, and cancer risk: What do we know and what is the way forward? *BMJ*, 368, m511, 2020.
56. La Vecchia, C., Muñoz, S.E., Braga, C., Fernandez, E., Decarli, A., Diet diversity and gastric cancer. *Int. J. Cancer*, 72, 255, 1997.
57. Isa, F., Xie, L.P., Hu, Z. *et al.*, Dietary consumption and diet diversity and risk of developing bladder cancer: Results from the South and East China case-control study. *Cancer Cause. Control*, 24, 885, 2013.
58. Fernandez, E., Negri, E., La Vecchia, C., Franceschi, S., Diet diversity and colorectal cancer. *Prev. Med.*, 31, 11, 2000.

59. Slattery, M.L., Berry, T.D., Potter, J., Caan, B., Diet diversity, diet composition, and risk of colon cancer (United States). *Cancer Causes Control.*, 8, 872, 1997.
60. Fernandez, E., Negri, E., La Vecchia, C., Franceschi, S., Diet diversity and colorectal cancer. *Prev. Med.*, 31, 11, 2000.
61. Lucenteforte, E., Garavello, W., Bosetti, C. *et al.*, Diet diversity and the risk of squamous cell esophageal cancer. *Int. J. Cancer*, 123, 2397, 2008.
62. Garavello, W., Lucenteforte, E., Bosetti, C. *et al.*, Diet diversity and the risk of laryngeal cancer: A case-control study from Italy and Switzerland. *Oral. Oncol.*, 45, 85, 2009.
63. Garavello, W., Giordano, L., Bosetti, C. *et al.*, Diet diversity and the risk of oral and pharyngeal cancer. *Eur. J. Nutr.*, 47, 280, 2008.
64. Cano-Ibáñez, N., Barrios-Rodríguez, R., Lozano-Lorca, M. *et al.*, Dietary diversity and prostate cancer in a Spanish adult population: CAPLIFE study. *Nutrients*, 12, 1694, 2020.
65. Dhakal, A. and Bobrin, B.D., *Cognitive Deficits*, StatPearls Publishing, Treasure Island (FL), 2021.
66. Yin, Z., Fei, Z., Qiu, C. *et al.*, Dietary diversity and cognitive function among elderly people: A population-based study. *J. Nutr. Health Aging*, 21, 1089, 2017.
67. Zhang, J., Zhao, A., Wu, W. *et al.*, Dietary diversity is associated with memory status in Chinese adults: A prospective study. *Front. Aging Neurosci.*, 12, 580760, 2020.
68. Otsuka, R., Nishita, Y., Tange, C. *et al.*, Dietary diversity decreases the risk of cognitive decline among Japanese older adults. *Geriatr. Gerontol. Int.*, 17, 937, 2017.
69. Otsuka, R., Nishita, Y., Nakamura, A. *et al.*, Dietary diversity is associated with longitudinal changes in hippocampal volume among Japanese community dwellers. *Eur. J. Clin. Nutr.*, 75, 946, 2020.
70. Kennedy, G., Ballard, T., Dop, M.C., *Guidelines for Measuring Household and Individual Dietary Diversity*, Food and Agricultural Organization, Italy, 2010.
71. Prasad, R., Bhattacharyya, A., Nguyen, Q.D., Nanotechnology in sustainable agriculture: Recent developments, challenges, and perspectives. *Front. Microbiol.*, 8, 1014, 2017.
72. Elemike, E.E., Uzoh, I.M., Onwudiwe, D.C., Babalola, O.O., The role of nanotechnology in the fortification of plant nutrients and improvement of crop production. *Appl. Sci.*, 9, 499, 2019.
73. Sonkaria, S., Ahn, S.H., Khare, V., Nanotechnology and its impact on food and nutrition: A review. *Recent Pat. Food Nutr. Agric.*, 4, 8, 2012.
74. Pateiro, M., Gómez, B., Munekata, P.E.S., Barba, F.J., Putnik, P., Kovačević, D.B., Lorenzo, J.M., Nanoencapsulation of promising bioactive compounds to improve their absorption, stability, functionality and the appearance of the final food products. *Molecules*, 26, 1547, 2021.

75. Jafari, S.M. and McClements, D.J., Nanotechnology approaches for increasing nutrient bioavailability. *Adv. Food Nutr. Res.*, 81, 1, 2017.
76. Jampilek, J., Kos, J., Kralova, K., Potential of nanomaterial applications in dietary supplements and foods for special medical purposes. *Nanomaterials (Basel)*, 9, 296, 2019.
77. Verma, A., Gautam, S.P., Bansal, K.K., Prabhakar, N., Rosenholm, J.M., Green nanotechnology: Advancement in phytoformulation research. *Medicines (Basel)*, 6, 39, 2019.
78. Giles, E.L., Kuznesof, S., Clark, B. *et al.*, Consumer acceptance of and willingness to pay for food nanotechnology: A systematic review. *J. Nanopart. Res.*, 17, 467, 2015.

10

Effect of Nanofoods on Human Health

Luxita Sharma

Amity Medical School, Dietetics and Applied Nutrition, Amity University, Haryana, India

Abstract

The term "nanofood" is defined as the food that is being grown, produced, or processed and packed using the techniques of nanotechnology. Nanoparticles are being used extensively in the development of nanofoods and food materials. Due to the increase in inventions and use of patentable technology in the field of nanofoods is leading to the concerns about the health and safety of consumers. The unique quality of nanoparticles is to increase area per unit volume. The example that can explain is use of gold, which is inactive otherwise, but once it is converted to nanoparticles, it inhibits as highly reactive species and develops very different and innovative properties. And these properties make gold nanoparticles useful in medicines to medical imaging technologies. The nanoparticles react with the biological system also like the human cells, breakdown of lipids, proteins, etc. So, it is being shown on many studies that nanomaterial can cross cell membranes and can activate inflammatory and other immune responses. But still, we require more data to prove the adverse or toxic effects of nanoparticles. In this chapter, the various impacts of nanoparticles on food and their impact on health will be explained.

Keywords: Nanotoxicity, nanofoods, food packaging, human health, nanoscience

10.1 Introduction

Nanofood is a process followed since many centuries and food structures also are present at nanoscale. The Food and Drug Administration has given definition of nanoparticles of food with the measurements less than micrometer scale and these shows certain exclusive properties.

Email: lshrama@ggn.amity.edu

Tanima Bhattacharya and Shakeel Ahmed (eds.) Nanotechnology in Functional Foods, (243–260)

Many patents have been filed by food industry and food technologists for the use of invented nanomaterials for food which are in high demand across the globe. The food production and supply are always being associated with the human health and food safety parameters. The benefits of nanoscience applicability in foods are enormous in terms of production, packaging, storage, and transportation. As the name suggests, nanoparticles are the particles of very small size with very minute dimensions. Technically speaking, these are the materials with dimensional specifications in the nanoscale, i.e., under 100 nm [1]. The literal meaning of the word "nano" implies small, tiny, and atomic in nature [2]. The term "nano" belongs to Greek origin and means "dwarf," and one nanometer is equivalent to one billionth of a meter. The different nanoforms of nanomaterials and nanoparticles include in Table 10.1.

Table 10.1 Different nanomaterials and nanoparticles.

Name	Definition
Nanoparticles	Particle of matter with 1 to 100 nm
Nanotubes	Hollow Tubular nanostructure with dimensions of nanometers
Fullerenes	An allotrope of carbon
Nanofibers	Fibers with diameters in the nanometer range.
Nanowhiskers	A Type of crystal nanofiber with a diameter of less than 100 nm
Nanosheets	*A two-dimensional nanostructure with thickness in a scale ranging from 1 to 100 nm*

All these forms of nanomaterials differ from one another in some way or the other. To throw some light on this, the nanoparticles are a three-dimensional entity, which have the size of 100 nm or less. Nanotube and fullerenes differ in their structure, in which the former has a cylindrical arrangement, and the latter has its molecules arranged in a spherical form. The length to diameter ratio of nanofibers is 3:1, while nanowhiskers are finer and may be varying length in micrometers. Lastly, nanosheets have a single dimension in nanorange [3].

Among nanotubes (Figure 10.1), use of carbon nanotubes has been considered the most vital, although boron and molybdenum nanotubes are also used widely [28]. In fact, a nanoparticle is a small entity that can work as a whole unit for transportation activities and has its own specific properties. In current scenario, nanoparticle research is a rapidly growing field of research. Nanoparticles have reported utilization in various fields, in fact in all the fields that we can think of. It has been reported that these small particles have immense utility in the fields of biomedical, optical, and electronic areas of research.

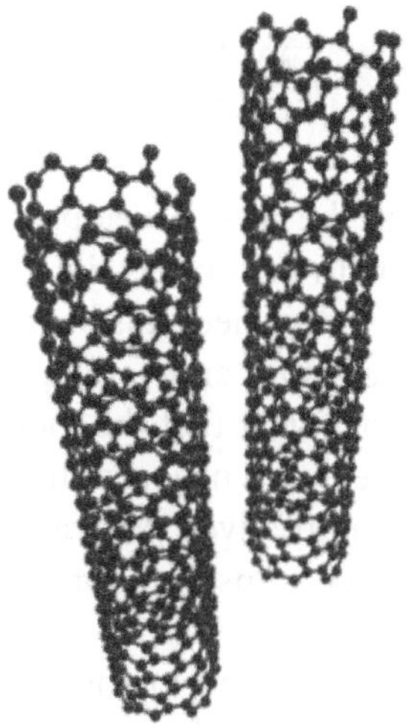

Figure 10.1 Nanocarbon tubes. Pixabay.com.

10.1.1 History

Invention of nanoparticles and their subsequent utilization is a long history. The identification of nanoparticles is as old as 9th century and were first discovered in Mesopotamia. During those days, they were utilized by artisans to give a sparkling and shiny effect on the pot surfaces. The shine and shimmer thus generated on pottery items is supposed to be due to presence of metallic film, which was applied to the clean and clear surface of a glazing. The potters during those times created these tiny nanoparticles by incorporating various metallic salts and oxides like copper and silver with various substances, like vinegar and clay, and used them on the surface of a pottery item that was previously glazed [4].

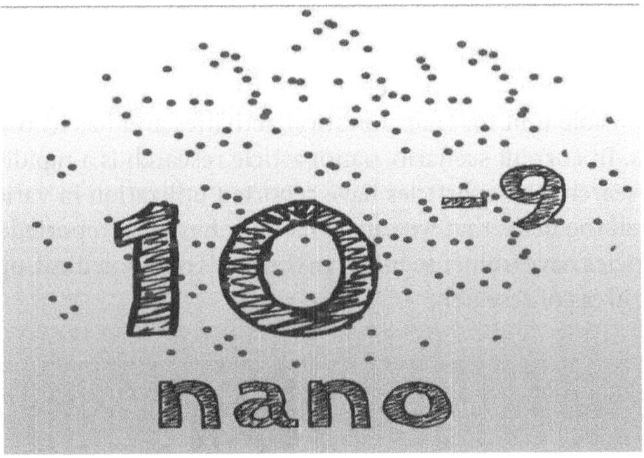

Food Industry: Owing to the latest developments that are rapidly taking in all the fields associated with day-to-day life, the domain of food science and technology is also not untouched. It is being utilized in arenas related to food preservation, processing, and food packaging. For many years now, it has been a well-known fact that the food industry over the ages is thriving on the idea of developing new products, creating a niche market for that product, investing in its advertising, and after that, reaping profits especially in terms of financial gains. Nanoprocessed food products have a big demand in the food market.

After this comes the idea of developing something that comes with health benefits as well. Thus, comes the concept of food quality in picture. This also involves keeping the health of population in mind, for whose utility the product is being manufactured. These small-sized microparticles can lead for both qualitative and mass production of high-quality, healthier, cost-effective, and safe food items. This new technology is far superior to the primitive food production and processing technologies with the great benefit of generally enhanced shelf lives [5].

Technically speaking, there are two varied approaches to nanotechnology [6], which are:

- "Top-down" approach: it includes as the name suggests starting by use of materials larger in size and making it to a small nanoscale structure with very minute and precise patterns. The top-down approach includes several methods like dispergating, fluidizing, homogenizing,and emulsifying technologies.

- "Bottom-up" approach: also known as molecular nanotechnology. It involves building both the organic and inorganic structures from the basic unit that is the atom. The various processes involved in bottom-up technology includes precipitation/condensation, evaporation techniques, controlled synthesis of sols gels [7, 8].

The term "self-assembly" implies the utilization of natural forces for nanostructure assembling.Thesecharacteristicscan thus be utilized to regulate the organization of material.

10.2 Types of Nanoparticles in Foods

10.2.1 Inorganic Nanoparticles

As the name is suggesting, the particles are inorganic in nature and exhibit the properties of being amorphous or crystalline basically. These particles have different tendencies to dissolve under different solvents [9].

10.2.1.1 Silver Nanoparticles

The use of silver nanoparticles is different and can be used as antimicrobial agents. These are extensively used on the food packaging [9].

10.2.1.2 Iron Oxide Nanoparticles

Nanoparticles made from iron oxide find use as colorants and bio available iron sources [9].

10.2.1.3 Zinc Oxide

These particles are being utilized in zinc supplements and functional foods apart from being source of zinc. Zinc is also a trace element essential in nature [9].

10.2.1.4 Titanium Dioxide

Titanium particles have specific optical properties like brightness and lightness. Thus, these particles have functional utilities [9].

10.2.1.5 Silicon Dioxide Nanoparticles

Nanoparticles having siliconfind utility as anticaking agents. Thus, they give increased flow properties.

10.2.2 Organic Nanoparticles

These particles can be in either state of matter, i.e., liquid, semi-solid (gel), or solid (crystalline/amorphous) according to their composition and atmospheric conditions. Structure wise, they can have spherical or non-spherical structures [2].

10.2.2.1 Lipid Nanoparticles

These nanoparticles consist of neutral lipids like mono/di/triacylglycerol and terpenes. The post digestibility results depend on hydrolysis by digestive enzymes, such as lipase and phospholipase.

10.2.2.2 Protein Nanoparticles

This kind of nanoparticles have a combination of protein nanoparticles joined together by physical interactions. They aid by creation of optimum delivery systems to encapsulate, protect, and deliver various commonly used bioactive agents like colors, flavors, preservatives, vitamins, minerals, and nutraceuticals [2, 3].

10.2.2.3 Carbohydrate Nanoparticles

These particles are created by breaking larger particles like starch, cellulose into smaller ones. These can be spherical or nonspherical as per the origin [2, 3].

Among both the types of nanomaterials, the organic ones have higher biodegradability as compared with the inorganic ones and hence the organic nanomaterials are used in bio nanocomposites [10].

10.2.3 Surface Functionalized Nanomaterials

Apart from organic and inorganic nanoparticles, it is another type, which find its application in the food industry. These nanomaterials add a specific functionality to the medium, which can range from imparting antimicrobial activity or enhancing the shelf life because of preservation. The functionalized ENMs (Engineered nanomaterials) find major applicability in food packaging materials, where they act as animpediment against moisture, gases, and volatile compounds along with providing mechanical strength [11].

10.3 Applications

Nanotechnology has wide applications in the food industry. Especially pertaining to products, it has its applicability in food formulations, food packaging, food, and nutrient delivery. The processing of food is being completed using the nanobiotechnology and many reactions that are done for the heat mass transfer and molecular synthesis [12]. Other applications include nanoparticles, nanoemulsions, nanocomposites, and nanostructured aerials. To elaborate, we can say that nanotechnology does offer some seemingly potential benefits associated with the quality and safety of the foods that we consume [12].

10.3.1 Contamination Sensors

The lightning in a proper manner and a concentrated flash of light is important to show the presence of bacteria like *E. coli*.

10.3.2 Disinfectant Packages

The food films which are used for packaging could be eaten and are made up of good flavoring agents such as herbs extracts and oils. The films are also extracted from the nanomaterials of minerals of zinc, calcium, or other materials that can be potentially killing pathogens.

10.3.3 Enhanced Food Storage Capacity

Barriers that are nanotechnology enhanced can help by keeping oxygen-sensitive foods fresher.

10.3.4 Enhanced Nutrient Delivery

Encapsulation with the help of nanoparticles improves the solubility of food nutrients/components several organic and inorganic compounds, including antioxidants, vitamins, essential fatty acids, etc.

10.3.5 Green Packages

Nanofibers produced from outer shells of lobsters. These possess antimicrobial properties and are biodegradable too.

10.3.6 Pesticide Reduction

Cloths have been developed which are saturated with nanofibers. They work by slowly releasing the pesticides, as a result the need for chemical spraying is diminished which results in reduced chemical leaching into the water source [3].

10.3.7 Tracking, Tracing, and Brand Protection

Nanoparticles can be utilized to create the bar codes that have the capability to label every product and may help in tracing outbreaks.

10.3.8 Texture

The food products are made with low fat consumption in the laboratories to enhance their spread ratio and stability. These properties improve the building of the nanosized gems [5].

10.3.9 Flavor

Now comes the important aspect that is flavor which can support the making of the foods that can have altered tasting and, they can act as enhancers of sweet/salt taste perception.

10.3.10 Recognition and Eradication of Pathogens

Nanocarbohydrate particles can bind with any available strain's bacteria. This helps in the identification, binding, capturing, and ultimately elimination of the deadly microorganisms [3, 5, 6].

10.4 Use of Nanotechnology in Food Science and Technology

As stated above the science involved in nanotechnology is associated with majority of food processing techniques. Various edible items that are meant for human consumption, like ice creams, aerated drinks, wafers, chocolate compounds, are being presented as healthy food items by doing variations in the macro nutrient compositions especially, proteins, facts, fibers, and carbohydrates. Nanotechnology is also helping in increasing the shelf life of food items based on principle of controlling the air and moisture content of the packaged food product. Usage of nano-coatings which are as thin as 5 nm also serves to increase the keeping qualities of many food products [29].

Nanoparticles help the food industry by [13]

- Improving flow properties of the developed product.
- Altering colors and making it more appealing.
- Imparting required stability.
- Modifying flavors and the associated nutritional properties of the food item.

Aluminosilicate compounds are known to have properties of anticaking agents hence addition of this compound in the powders/beverages will improve the solubility and moisture would not harm the product. Titanium dioxide is being used as whitening and brightening the colors additive.

10.4.1 Nanofood Safety

The utilization of nanotechnology is the discovery of any chemical pollutants or pathogens of any variety. This is the main property that can be used in food safety as nanoparticles can kill the pathogens that are present as food components. Apart from that, these sensors also have inbuilt infrared and magnetic materials that can identify any minor quantity of pathogenic organisms.

A single nanosensor is capable of sensing million microbes and, they can identify the micro spaces that holds the pathogens for spoiling the dish and can cause various disorders. This helps in their identification and further denaturation [14].

Nanosensors can be amalgamated and hybridized with other technologies for rapid detection of certain microorganism. For example, use of a hybrid nanosensor for detection of *E. coli* in milk. Here, the nanosensor has been incorporated with magnetic resonance and fluorescence [29].

Thus, with the nanotechnology application processes in the food analysis will improve the food safety and quality by leaps and bounds.

10.4.2 Smart Food Packaging System

The smart packaging and food safety is being improved day by day so that a better, innovative, and new product can be produced and distributed in the market. Microbes were always a serious concern for the food industries, but as mentioned in the above para nanosensors are coming up as a solution as the new food safety improvisation. The newly designed nanosensors are capable to do the tasks such as food analysis, detection of colors, clean water, etc. This can be carried out by their specialization in toxin detection. Now these also require a specific transportation requirements [14].

Now the important industry of food that catches the contamination is the meat industry. Meat requires a specific temperature and a particular taste/smell/freshness. So, as we are discussing sensor-based indicators that can be of use in meat industries and these sensors can identify the temperatures and the microbes hidden in the food materials. Therefore, when we add these sensors in the food then the results are amazing, and we can rely on the identification of specific attributes like antioxidant, antimicrobial, and scavenging properties as well. Then this leads to an increased keeping quality of packaged products [8].

Nanotechnology has also enabled the production of biodegradable packaging materials for food, which are environment friendly in nature (Figure 10.2). For instance, use of a silicate-based nanostructure for food packaging [30].

It has also been reported that the new and smart packaging techniques that have come up also aid in retaining the food quality. It also helps in maintaining the packaging integrity, thus protecting the packaged food item from any kind of external environmental stimuli and maintenance of a positive viable temperature. All these critical points are examined through the production process and ultimately help ensure distribution of quality food products. Now a day's specific bar codes have been developed with the help of nanoparticles. These barcodes are known as nanobarcodes and are used to create identification tags which have the composite information about the food product under study [15].

10.4.3 Antimicrobial Food Packages

Applications of antimicrobial packages in food sectors

- Food storage cans cured with silver.
- Slicing boards and kitchen utensils cured with silver.
- Silver treated plastic storage bags, Crockery items and electronics used in kitchen as well
- Zinc oxide treated stick food wrappings.

10.4.4 Applications of Nano-Particles

Numerous nanoparticles find their application in the food industry includes [16].

10.4.4.1 Low-Density Lipoproteins

Fish oil is used to produce nanoparticles. The food industry extensively uses the food additives so these nanoparticles can be used for hiding the odor or smell of the tuna. Microencapsulation is the process used in this type of application. Canola Oil is another example, which is used to make

Figure 10.2 Packaging materials. Pixabay.com.

Vitamin E nanoparticles though nano-emulsification. It is used due to its better oxidative stability.

10.4.4.2 Biodegradables Biopolymeric Nanoparticles

The encapsulated form of the polylactic acids form to produce the biodegradables biopolymeric nanoparticles. The drug delivery, vaccine storage and delivery is its important function.

10.4.4.3 Liposomes

Phospholipids are the major component of the liposomes. Their role is assimilating disinfectant tools for foods. Some of the nanostructured liposomes are also produced via nanoencapsulation and used as antioxidants in food. These can also be used for detection of toxins in food. Using techniques like flour immunoassay, fluorescent-labeled liposomes can be used to detect concentration of toxins as low as 1 Nm [6].

10.4.4.4 Fine Emulsion Droplets

Fine emulsion droplets are made by reducing the size of fat globules using homogenization/microfluidization techniques.

10.4.4.5 High Intensity Ultrasound Waves

Ultrasound emulsification is used in oil and water nanoemulsions to change the characteristic of food.

10.5 Disadvantages of Nanoparticles

The technology development has its own pros and cons, this area of nanotechnology is mounting very fast. The toxic impacts are studied but then also we cannot ignore them [17].

10.5.1 Harmful to Human Health

Nanoparticles as such are present on the on the exteriors of the many of food packaging but are not harmful in any way to human health. But the worry comes with the various combinations of these nanoparticles with the food inside or the composition of the food, which is edible also and that might exhibit health repercussions.

The mechanism of absorption and resultant circulation in the body can have cytotoxic effects.

Nanomeans micro so no wonder they can reach inside the skin or can be absorbed by the openings of the human body. Nanoparticles can be ingested also and then the reactions can occur as they reach bodily fluids and cells. Nanoparticles are highly reactive species and hence they can cross the capillaries and various membrane barriers. This implies that nanoparticles have various types of toxicokinetic and toxicodynamic properties. Many studies have revealed that nanoparticles can have strong toxic reactions on immune system and can impact the internal organs like heart, liver, and kidneys.

Along with that a lack of awareness among consumers, government policies and protocols along with specific diagnostic methods especially designed for nanotechnological hazard assessment are superior warranties for better understanding toxicity and its regulation.

There is an inverse relation between the toxicity and size of metal-based nanoparticles. Smaller nanoparticles will have a higher toxicity.

Nanoparticles have the potential to bind proteins and enzymes. This combination results in ROS stimulation & reduced oxidative stress.

Dearth of research-based evidence on the impact of nanoparticles on our body and several physiological processes leave a scope for various extensive studies.

The genotoxicity caused by the nanomaterials should be tested so that individuals are less impacted. Genotoxicity caused because of nanoparticles can be either primary or secondary in nature.

10.5.1.1 Primary Genotoxicity

The nanoparticle has direct link with the genomic DNA. It has antiinflammatory reactions, and the examples are cases of silver nanoparticles, asbestos etc.

10.5.1.2 Secondary Genotoxicity

The nanoparticles that results in the inflammatory reactions and cause the genetic damage. So, the response is generated in the body and the activation of the macrophages and neutrophils happens [18].

10.6 Guidelines for Safety Issues

The technology developed should have no side effects and be harmless for the human body, but the Nanoparticles may show some potential risks over human health. The size of the nanoparticles plays an important role as it is the property that allows them to interact with human cells. They may possess threats even of the studies show no risks involved, this is the most ambiguous part [18, 19]. When a study was done on the rats with insertion of carbon via nanotubes showed that fibrosis and oxidative stress was increased [20].

The serious effects of the nanoparticles on the genome and proteome of plants can be their growth hinderances [20]. The regulatory agencies which are responsible for the recognition of Nanoparticles such as Food and Drug Administration has amended some of the guidelines and has included the safest rules that abide by their nontoxicity of any kind [22].

The Institute of Food Science & Technology recommended the use of E numbering system whenever the additives were used in the foods. So E_N should be used as symbol where N is nanoparticles [23].

There is very positive interaction of nanoparticles with the food chain will result in less hazards for the human health. The products having nanoformulations are lethal to flora and fauna, and till now no dedicated specific rules and regulations have been introduced for food and agriculture sector.

The regulatory bodies have laid down the standards which include the safe use of the factors that contribute to the functions of the nanoparticles such as size, administering technique, variety, dimensions, physical and chemical properties [24].

Regulatory bodies of the European Union have highlighted that the nanotechnology-based food products at first undergo safety and hazard assessments before they are declared deemed fit for human/animal consumption.

The absence/lack of rules and regulations associated with food in several countries can be attributed to comparatively miniscule information available with regards to exposure, availability, and toxicity to human race.

For effective nanotechnological applications comprehensive law abiding and government regulated rules and framework is required which have been formulated after rigorous toxicological scrutiny. The need of the hour is a detailed and well planned/acceptable global regulatory system for planned and controlled utilization of nanoparticles in the rapidly growing food-based organizations [25, 26].

10.7 Conclusion and Future Perspective

As has been discussed in previous paragraphs nanotechnology has played a pivotal role in the food and edible products industry. Though there is relevant data that suggest that quality of food produced diminishes with advanced processing techniques applied from time to time. The long-term impacts of the storage of the food products or grains has given the boom to the food industry as the shelf life of the food products is being largely impacted. The quality of the food should not be affected when it is being stored for long duration of times. The attributes, such as color, texture and flavor, should be maintained and preserved [25].

The use of nanotechnology in the arena of food manufacturing benefits the consumers by providing them with the relevant information on the state of the food product/item packaging along with its nutritive value and ensuring enhanced food safety by ensuring detection of food borne pathogens. Certain compounds present in food, which are bioactive in nature and are protective against various diseases, but at the same time, they are water repelling/hydrophobic in nature, as a result, they have diminished their bioavailability and stability. For such cases, the nanotechnology-based delivery systems have an impact by increasing the availability to the body and ensure target-oriented delivery of food-based bioactive compounds [23].

- The food items with nanomaterials as base give a significant amount of challenge to both administration and industries. Ultimately keeping the end user/customer's confidence and acceptance in relation to nanofood items locally available.

- Nanocolloidal particles have been reported to be widely utilized in some way or the other in ensuring food quality, keeping the nutrition intact, safety of food as well as processing and packaging of food.
- These properties and associated behaviors of colloidal particles are vital especially with the context of designing foods which are safe for consumption and possess high standards of quality.
- The production of nanoparticles is being done on a global level but still, the standards and regulatory measure required for its intended utilization in food products is lacking.
- As per the available data the utilization of nanoparticles as a food ingredient is more harmful as compared to nanoparticles being utilized in food packaging.
- During the process of production of the food products, there are possibilities that nanomaterials, which are primarily the components of the food, can enter the ecological system. These nanomaterials then act as a threat. So, we want to convey those proper studies are required so that the possibilities of nanomaterials being a threat decreases, and they can be used effectively as part of the food chain.
- As of today, the number of in vivo studies done specifically to study effects of nanofoods on human and animal health is very less. Thus, specifically designed labels and the marketing regulations for nanofoods should be done. This will aid in enhancing the acceptance by potential consumers.

The nanobiotechnology has encouraged the use of smaller and very sensitive technology. The applicability of its use in the food industry in terms of food packaging and food safety is well known. There are encouraging results that are shown from various studies for example the nanomaterials can protect the food from moisture, gases, fats, bad odors or after tastes and flavors. The nanomaterials have given excellent solutions in terms of bioactive compounds and their use in the industries.

And yes, we also agree that the advance in nanotechnology is carving new paths but also carries many challenges with it. But our regulatory bodies are working to make the guidelines for the safety issues and consumer satisfaction is targeted. The priority is being taken to test and vet a product before it is offered to the consumers [5]. Thus, to conclude that the nanotechnologies should be utilized appropriately and in an intended manner so that the hazards are not posed. Consequently, the quality assurance will ensure the usefulness of the product for the human health and wellness.

References

1. Murthy, S.K., Nanoparticles in modern medicine: State of the art and future challenges. *Int. J. Nanomedicine*, 2, 2, 129–141, 2007.
2. García, M., Forbe, T., Gonzalez, E., Potential applications of nanotechnology in the agro-food sector. *Ciência Tecnol. Alime.*, 30, 3, 573–581, 2010.
3. Cushen, M., Carry, J., Morris, M., Cruz-Romero, M., Cummins, E., Nanotechnologies in the food industry–Recent developments, risks and regulation. *Trends Food Sci. Technol.*, 24, 24–36, 2012.
4. Mandal, A., What are Nanoparticles?, 2019. Retrieved from: https://www.news-medical.net/life-sciences/What-are-Nanoparticles.aspx.
5. Singh, T., Singh, T., Shukla, S., Kumar, P., Wahla, V., Bajpai, V.K., Rather, I.A., Application of nanotechnology in food science: Perception and overview. *Front. Microbiol.*, 8, 2017. DOI=10.3389/fmicb.2017.01501
6. Nile, H. *et al.*, Nanotechnologies in food science: Applications, recent trends, and future perspectives. *Nano Micro Lett.*, 12, 45, 2020.
7. Bhushan, B., Luo, D., Schricker, S.R., Sigmund, W., Zauscher, S., *Handbook of Nanomaterials Properties*, Springer, Berlin/Heidelberg, Germany, 2014.
8. Brayner, R., Fievet, F., Coradin, T., A danger or a promise? A chemical and biological perspective. Springer, London, UK, 2013, in: *Nanomaterials: A Danger or a Promise? A Chemical.*
9. Groves, K., Potential benefits of micro and nanotechnology for the food industry: Does size matter?, *New Food Mag.*, 4, 49–52, 2008.
10. Sahoo, M., Vishwakarma, S., Panigrahi, C., Kumar, J., Nanotechnology: Current applications and future scope in food. *Food Front.*, 2, 1, 1–20, 2020.
11. Puitheti, S., Application of nanotechnology in food, nutraceuticals and pharmaceuticals. *E-J. Sci. Technol.*, 2, 10, 2015.
12. David, J. and Hang, X., Is nano safe in foods? Establishing the factors impacting the gastrointestinal fate and toxicity of organic and inorganic food-grade nanoparticles. *NPJ Sci. Food*, 6, 2017.
13. Nastu, P., Top 10 uses of nanotechnology in food. 2009. Retrieved from: https://www.environmentalleader.com/2009/02/top-10-uses-of-nanotechnology-in-food/.
14. Renton A., Welcome to the world of nanofoods. The Observer Guardian Unlimited, UK, 2006 December 13, guardian.co.uk., Available from: http://observer.guardian.co.uk/foodmonthly/futureoffood/story/01971266,00.html; http://nanofood.khu.ac.kr/article1.html
15. Sulaiman, M.A., Use of nanotechnology in food processing, packaging and safety review. *Afr. J. Food Agric. Nutr. Dev.*, 10, 2719–2739, 2010.

16. Branton, D., Deamer, D.W., Marziali, A., Bayley, H., Benneret, S.A., The potential and challenges of nanopore sequencing. *Nat. Biotechnol.*, 26, 1146–1153, 2009.
17. Walia, N., Dasgupta, N., Ranjan, S., Ramalingam, C., Gandhi, M., Food—Rade nanoencapsulation of vitamins. *Environ. Chem. Lett.*, 17, 991–1002, 2019.
18. Gonzalez, N. and Johnston, L., Safety of engineered nanomaterials. *Chem. Int.*, 40, 28–29, 2018.
19. Yao, S., Swetha, P., Zhu, Y., Nanomaterial-enabled wearable sensors for healthcare. *Adv. Healthc. Mater.*, 7, 1–27, 2018.
20. Higashisaka, K., Rizwan, M., Ali, S., Qayyum, M.F., Ok, Y.S., Adrees, M., Ibrahim, M., Zia-ur-Rehman, M., Farid, M., Abbas, F., Effect of metal and metal oxide nanoparticles on growth and physiology of globally important food crops: A critical review. *J. Hazard. Mater.*, 322, 2–16, 2007.
21. Wen, K.P., Chen, Y.C., Chuang, C.H., Chang, H.Y., Lee, C.Y., Tai, N.H., Accumulation and toxicity of intravenously-injected functionalized graphene oxide in mice. *J. Appl. Toxicol.*, 35, 1211–1218, 2015.
22. Food and Drug Administration, FDA Regulation of Nanotechnology Products, 2004. http: //www.fda.gov/nanotechnology/regulation.html.
23. Maynard, A.D. and Kuempel, E.D., Airborne nanostructured particles and occupational health. *J. Nanopart. Res.*, 7, 6, 587–614, 2005.
24. Meetoo, D.D., Nanotechnology and the food sector: from farm to the tables. *Emir. J. Food Agric.*, 23, 5, 387–403, 2011.
25. Chen, J., Zheng, J., Decker, E.A., Mcclements, J., Improving nutraceutical bioavailability using mixed colloidal delivery systems: Lipid nanoparticles increase tangeret in bioaccessibility and absorption from tangeretin-loaded zein nanoparticles. *RSC Adv.*, 5, 73892–73900, 2015.
26. Buzea, C., Nano materials and nano particles: Sources and toxicity. *Biointerphases*, 2, 17–71, 2007.
27. Ameta, S.K., Rai, A.K., Hiran, D., Ameta, R., Ameta, S.C., Use of nanomaterials in food science, in: *Biogenic Nano-Particles and their Use in Agro-ecosystems*, M. Ghorbanpour, P. Bhargava, A. Varma, D. Choudhary, (Eds.), Springer, Singapore, 2020.
28. Anzar, N., Hasan, R., Tyagi, M., Yadav, N., Narang, J., Carbon nanotube - A review on synthesis, properties and plethora of applications in the field of biomedical science. *Sensors Int.*, 1, 2020.
29. Bratovcic, A., Nanomaterials in food processing and packaging, its toxicity and food labeling. *Act. Sci. Nutr. Health*, 4, 9, 07–13, 2020.
30. Shafiq, M., Anjum, S., Hano, C., Anjum, I., Abbasi, B.H., An overview of the applications of nanomaterials and nanodevices in the food industry. *Foods*, 9, 2, 148, 2020.

11

Nanofunctional Foods: Current Advances and Perspective

Hitesh Chopra

Chitkara College of Pharmacy, Chitkara University, Rajpura (Punjab) India

Abstract
During the past decade, there have been exponential increase in demand and consumption in the use of functional foods, leading to an improvement in average human health and performance. But they tend toward being low bioavailable and getting digested through the first-pass metabolism. So there is a need to develop dosage forms of such moieties to face these issues. Many researchers had developed nanoscale formulations through nanoencapsulation and entrapment in polymer matrices and helping to target the food to the target site. It has led to increasing in consumer acceptance due to comparative betterment of organoleptic parameters. The nanoformulations of food and their products has been developed, such as liposomes, multiple emulsions, nanoparticles, etc, and are quite promising in nature. This chapter represents various ways of delivering functional foods to target organs through the use of nanotechnology, followed by their fate and metabolism.

Keywords: Functional foods, nanotechnology, delivery systems, nanoencapsulation, bioactive compounds

11.1 Introduction

Functional foods are those foods that apart from nutritional value also help in fighting against diseases and promote proper growth and development. These foods are not essential to the diet and do not prevent and fight against the diseases. As such, there is no universal definition for functional foods,

Email: Chopraontheride@gmail.com

Tanima Bhattacharya and Shakeel Ahmed (eds.) Nanotechnology in Functional Foods, (261–292)

the first trend of functional food was developed by Japan in the 1980s. When the country was facing the problems of health care costing, the government devised a regulatory system for the people that approve certain foods matching its qualification criteria and helping in the improvement of the nation's aging population. Those foods were known as Food for Special Health Use (FOSHU).

Practically the functional foods are naturally derived from either plants or animal sources. They are capable to deliver benefits apart from basic nutrition, such as lycopene from tomatoes, the DHA and EPA from fish oils, etc. The concentration of functional foods can be enhanced via either variation in breeding conditions or genetic variations, eg, the beta carotene-rich rice [1–3], *Super Banana* [4], carotene-rich *Actinidiadeliciosa* [5], carotene-rich *Citrussinensis* [6], etc. For animal-based products, the diet of animals is generally altered for increasing the level of linolenic acid or omega 2 acids in milk and meat products. Some of the foods are added to each other to work synergistically for instance the addition of probiotics *Bifidobacterium* spp. and *Lactobacillus acidophilus* to yogurt tend to increase the activity of each other and thus beneficial to each other [7].

Generally, the natural products are delivered through the use of aqueous solutions as it increases the acceptability, bioavailability, and palatability. But solubility is a hurdle toward the use of functional foods administered orally as such and is also susceptible to metabolism before reaching the systemic circulation. Moreover, some of them get oxidized or rancid when coming in contact with high-humidity surroundings. Therefore, to deliver the bioactive compounds to target sites has remained as a major hurdle.

Nanotechnology has the potential to overcome this hurdle and can provide various possible resources for it. It can be defined as science dealing with the design, production, and application of structure at the nanoscale level. Many researchers are working on process of formation of high performance delivery bases for the encapsulation of food bioactive. Encapsulation not only helps in protecting against the metabolism but also enhances the solubility, stability, bioavailability somewhere. In comparison to the macro-based technique, the nanoscale provides more surface area with improved controlled release and solubility. Many other advantages include enhancement in bioavailability and greater precision for entrapment efficiency. It has been estimated that the global population is all set to hit the eight billion mark by the year 2024, so it is obvious that there shall be an enormous rise in the surge of food demand and shall heavily depend on the agriculture- food industry. Use of nanotechnology shall surely play the game changing role in agriculture and food sectors.

In the food industry, the use of nanotechnologies has become visible, such as the development of micelles, nanoemulsions, biopolymeric nanoparticles, ribosomes, etc. Many of the companies working in the market, such as Heinz, Nestle, Hershey, etc., are working on the development of nano-enabled foods and food packaging. Nestle and Unilever are working on the development of low-fat ice cream based on the nanoemulsions. The formulators have also developed iron- or zinc-based nanostructures aiming to improved bioavailability and minimizing color changes in the final product. Recently, the advances in a controlled and stable release of novel food delivery systems have been reported [8–22].

11.2 Nanotechnology Techniques for Functional Foods

There are various techniques to deliver the functional foods to the requisite target, some of them are described below (Figure 11.1).

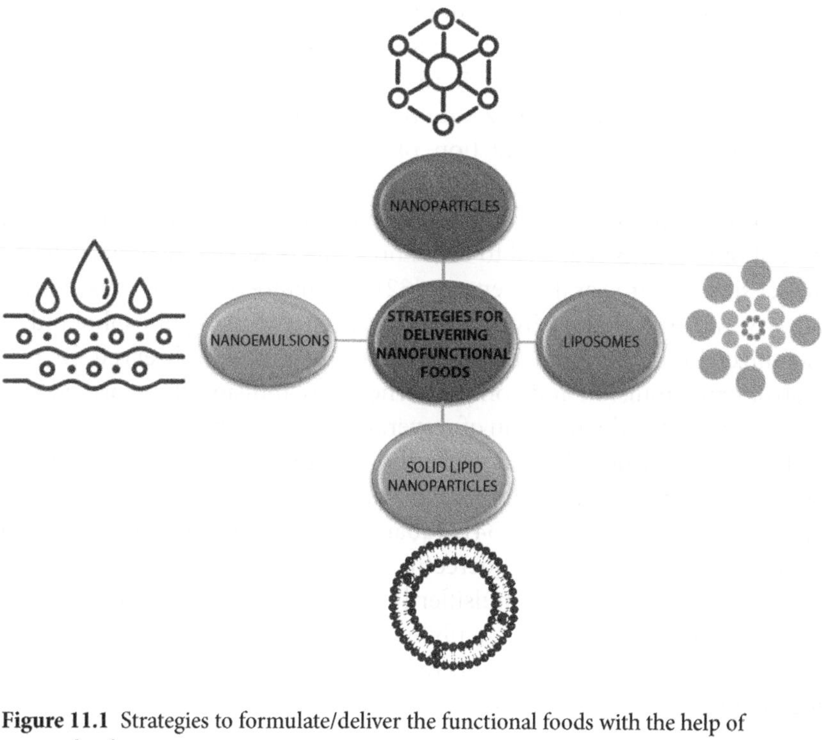

Figure 11.1 Strategies to formulate/deliver the functional foods with the help of nanotechnology.

11.2.1 Nanoemulsions

Nanoemulsion can be defined as the emulsions of nanosize which aim to improve the delivery of the active ingredient. They are thermodynamically stable isotopic systems in which two different phases are mixed with the help of an emulsifying agent or surfactant or cosurfactant.

Numerous researchers worked on encapsulating the carotenoids, such as beta carotene [8–12], lycopene [13–15], lutein [16–21], and astaxanthin [22–26]. In the food industry, most of the emulsions are produced by the high-pressure homogenization method [27]. With this method, the nanoemulsion of carvacrol, mandarin, bergamot, and lemon was prepared. The primary emulsion is exposed to high-pressure homogenization at 200 MPa for five cycles to obtain a nanoemulsion of 133 to 200 nm. The emulsion showed inhibitory action against *E. coli* and *S. typhimurium* and when incorporated with chitosan was able to affect the shelf life in positive manner [28]. Similarly, jackfruit which is rich in carotenoids was subjected to a high-pressure homogenization process and o/w nanoemulsion was obtained and exhibited antioxidant activity [29]. The researchers fabricated the nanodispersion with narrow PDI of Lycopene for their use in the beverage industry [30]. The homogenization procedure also affected the zeta potential and turbidity of nanodispersion. The nanoemulsions of lentils were also prepared using 50 to 300 MPa pressure. It was found that with the increase in homogenization the viscosity decreased [31].

Another method of production of nanoemulsion is ultrasonication method. The agitation of frequency greater than 20 kHz sound waves is used to break the coarse droplets into nanoemulsions. Some minimum amount of energy is required to cause breaking of large size of particles to smaller size, required as input energy [32]. The nanoemulsion of resveratrol and resveratrol-cyclodextrin complex was prepared using phospholipid as a stabilizer [33]. The average size of 20 and 24 nm was observed for resveratrol nanoemulsion and complex. The nanoemulsion thus formed was able to prevent the degradation of resveratrol on exposure to UV radiation.

The nanoemulsion loaded with the curcumin has better bioaccessibility as compared to commercial supplement formulation prepared using synthetic ingredients [34]. Iqbal *et al.*, worked out with method to enhance the solubility of quercetin and curcumin using the modified starch (HI-Cap-100) as an emulsifier [35]. It was found that the quercetin and curcumin showed higher solubility in the case of 5% modified starch-based nanoemulsion with enhancement by 2.07 and 39 times respectively when compared to oil and water.

Kharat *et al.* analyzed the effect of various constituents such as surfactant and gums on the nanoemulsion of curcumin [36]. It was found that Quillajasaponin-based nanoemulsion showed rapid degradation after 15 days. It was due to the presence of conjugated double bonds, that are generating the free radicals and promoting the peroxidation reaction.

Maurya *et al.*, fabricated nanoemulsion of Vitamin D3 based on the phase inversion method [37]. The nanoemulsion was based on the caprylic/caprylic triglyceride, leciva S70, Kolliphor HS15, Vitamin D3, and sodium chloride solution. A total of 41 formulations was plotted on nanoparticle formation zone and only 12 out of them were able to qualify. The formulation so optimized showed sustained release of Vitamin D3, evidencing the higher bioavailability. The formulation was also found suitable for milk-based beverages.

Another group of researchers prepared vitamin D3 nanoemulsion from Pea proteins as an emulsifier [38]. The nanoemulsion prepared showed a controlled size and high entrapment efficiency. The variation in the particle size resulted in cellular uptake efficiency. The transport efficiency across the Caco-2 cells was higher in the case of small-sized nanoemulsions with a particle size of 233 nm and comparatively to free vitamin D3 it was about 5.3 times greater.

Park *et al.*, worked out nanoemulsion of turmeric using ultrasonication and spray drying technique [39]. The milk fortified with the nanoemulsion showed stability for over 21 days at 4°C showed no significant change in particle size and zeta potential and no effect was observed on curcumin content. The *in vitro* analysis showed the release of less than 10% of curcumin in the stomach while the intestine showed almost 70% of release.

Researchers developed nanoemulsions based on sunflower oil as a carrier for ω-3 fatty acid as the base and tween 80 and isopropanol as a surfactant mixture [40]. The droplets of size 100-200 nm were prepared and PDI less than 0.3 for about 28 days, but fatty acid content decreased by approximately 15% upon storage. The bioaccessibility was about 53% as compared with 34% of release of plain oil blends. The formulation could act as supportive therapies for improving ω-3 fatty acid nutrition. The work done by various other researchers for encapsulation of bioactives using nanoemulsions has been tabulated in Table 11.1.

11.2.2 Nanoparticles

Nanoparticles (NPs) have been defined as the particle with size ranging from 1 to 100 nm in diameter. They have been used widely due to their novel features. They can be synthesized from various bulk materials and

Table 11.1 Methods used for entrapping bioactive through nanoemulsion technique.

Bioactive compound	**Aim of study**	**Particle size**	**Reference**
Vitamin D	To prepare the nanoemulsion of Vitamin D	<200 nm	[41]
	Fortification of cheese with Vitamin D3	NA	[42]
	To observe the effect of carrier oil on the bioavailability of Vitamin D3	NA	[43]
	To prepare nanoemulsion of vitamin D, using the edible polymers for human consumption	174 ± 7 nm; 26 ± 3 nm	[44]
	To study the antibacterial and genotoxic effect of nanoemulsions	166.2, 118.0, 170.8 and 40.52 nm	[45]
	Fabrication of plant-based nanoemulsion of Vitamin D3	0.34 μm	[46]
	To develop nanoemulsion for the fortification of beverages.	39.12 ± 0.33–64.11 ± 1.93 nm	[37]
PUFA	Formulation of nanoemuslions for the application in personal care and pharmaceutical products	>1000 nm at low surfactant levels, <200 nm at high surfactant levels	[47]
Salmon oil	To study the effect of antioxidants on insolubility in water and lipid phases	160–207 nm	[48]

(Continued)

Table 11.1 Methods used for entrapping bioactive through nanoemulsion technique. (*Continued*)

Bioactive compound	Aim of study	Particle size	Reference
DHA algae oil	To produce the nanoemulsion using ultrasound waves and to protect it from gastric degradation	Less than 1000 nm	[49]
DHA algae oil	To increase the stability and lipid digestibility of nanoemulsion	148 nm; 206 nm, 760 nm	[50]
High-oleic palm oil (1–20% *w/w*)	To study the effect of whey protein denaturation on stability of nanoemulsion	163.7–2268i.0 nm	[51]
Flaxseed/high DHA algae oil	To increase the bioavailability of flax seed and DHA	192 nm; 182 nm	[52]
Fish oil	To check the correlation of bioavailable fish oil and lipopolysaccharide induced production of nitric oxide in rat PBMCs.	89.7 ± 27.7 nm	[53]
Lycopene-enriched tomato extract	To study the effect of reduction of particle size of lycopene on oral bioavailability	96 ± 12 nm	[54]
Astaxanthin and lycopene	To produce the linseed nanoemulsions with enhanced carotenoid bioavailability	1, 4 and 10 cycles at 70 MPa (210, 168, 164 nm); 100 MPa: 242, 134, 126 nm	[55]
Lycopene	To study the interaction between carrier lipid phase and formulation of lipophilic bioactive compound.	2.6 μm; 5.4 μm	[56]

(*Continued*)

Table 11.1 Methods used for entrapping bioactive through nanoemulsion technique. (*Continued*)

Bioactive compound	Aim of study	Particle size	Reference
Lutein	To fabricate nanoemulsion of lutein using low energy method	190–270 nm with MCT, 192 nm no lutein, 205 nm with 0.2% lutein, 270 nm with 1.2% lutein	[57]
β-carotene	To retard the rate of degradation of nanoemulsion for long term storage	78 nm	[58]
β-carotene	To fabricate the nanoemulsion of beta carotene using layer by layer technology	9.24 ± 0.16–276.77 ± 17.70 nm	[59]
Spirulina oil	To evaluate the effect of biopeptides on the nanoemulsions of spirulina oil	222.9 ± 3.4–466.9 ± 5.3 nm	[60]
β-carotene	To prepare high release nanoemulsion for intestinal region from whey proteins	579.45–1829.50 nm	[61]
	To observe the effect of anthocyanidin derived from Lotus seedpod and nanoemulsions of beta carotene.	NA	[62]
	To study the outcome of amorphous OSA modified starch and unsaturated lipid carrier on the formulation of *Lavandula angustifolia* essential oil-based nanoparticle	123.9–185.6 nm	[63]

(*Continued*)

Table 11.1 Methods used for entrapping bioactive through nanoemulsion technique. (*Continued*)

Bioactive compound	Aim of study	Particle size	Reference
β-carotene	To evaluate the effect of environmental conditions and thermal stress on the shelf life and stability of nanoemulsion.	300 nm	[64]
All trans retinoic acid	To develop and evaluate the parenteral formulation of all trans retinoic acid	<236.8 ± 26.9 nm	[65]
Vitamin A	To formulate the o/w/o nanoemulsion of vitamin A	-	[66]
All trans retinoic acid	Formulate the PLGA microspheres for brain tumor targeting	-	[67]
Vitamin A	To formulate the nanoemulsion of Vitamin A and C	475.7 nm	[68]
Vitamin A	To compare the effect of micelles and nanoemulsion on the fate of administration	<300 nm	[69]
Vitamin A in palmitate/peanut oil	To improve the antioxidant properties of nanoemulsion form of Vitamin A	115 nm	[70]
Vitamin A	To formulate nanoformulation of Vitamin A that can withstand the gastric environment	5–20 μm	[71]
Vitamin K1	To formulate nanoemulsion using low energy method	<253.9 nm	[72]

(*Continued*)

Table 11.1 Methods used for entrapping bioactive through nanoemulsion technique. (*Continued*)

Bioactive compound	Aim of study	Particle size	Reference
Vitamin K	To formulate nanoemulsion with enhanced antitumor activity	<119.3 ± 1.3 nm	[73]
Vitamin E	To study the changes on nanoemulsion from thermal radiations during chilled stored stability study	NA	[74]
Vitamin E	To formulate more efficacious system for delivery of vitamin E	<100 μM	[75]
Vitamin E	To formulate nanoemulsion using spontaneous emulsification method	<50 nm	[76]
Vitamin E	To prepare vitamin E nanoemulsion enriched with low fat for dietary supplement	<1000 nm	[77]
Vitamin E	Formulation of nanoemulsion of vitamin E with biopolymers	<0.38 μM	[78]
Thyme oil (1% *w/v*)	To evaluate the antimicrobial effect of thyme oil nanoemulsion	82.5 nm; 125.5 nm	[79]
Thymol and eugenol	To evaluate the antimicrobial effect of thyme oil nanoemulsion	55 (eugenol) and 75 (thymol) nm	[80]

(*Continued*)

Table 11.1 Methods used for entrapping bioactive through nanoemulsion technique. (*Continued*)

Bioactive compound	Aim of study	Particle size	Reference
Clove and lemongrass oil	To evaluate the antifungal effect of volatile oil nanoemulsion	76.73 nm	[81]
Bergamot oil and sweet orange oil	To evaluate the sensory parameters of mixed oil emulsions	30 nm (5% oil phase); 190 nm (15% oil phase)	[82]
Peppermint oil	To evaluate the antibacterial effect of volatile oil nanoemulsion	<200 nm	[83]
Thyme oil	To evaluate antibacterial property of nanoemulsion	163 nm	[84]
Oregano, thyme, lemongrass and mandarin essential oils (1% *w/v*)	To study the long-term stability of nanoemulsions stabilized by high methoxyl pectin and a nonionic surfactant (Tween 80).	<50 nm	[85]
Lemongrass oil	To evaluate the effects of oil concentration, number of passes and calcium chloride concentration on the physical, mechanical and transport properties of oil-alginate composite films.	53 ± 5 nm, 46 ± 7 nm 23 ± 2 nm, 7.35 ± 1.67 nm	[86]

their action is dependent on the chemical composition and size/shape of particles. They can be magnetic, thermo sensitive, pH-sensitive, ion-sensitive, etc.

Biopolymers are the polymers derived from natural sources and have characteristic biodegradation and biocompatibility with foods. They generally have various sites available for the cross linking action. The various biopolymers used for nanoparticles preparation includes chitosan, Starch, Cellulose and Sodium hyaluronate etc. They also provide the enteric coating and protect the degradation of foods from various intestinal enzymes and mechanisms [88–90].

Using the ionic gelation technique, the chitosan-based nanoparticles were prepared [88]. The nanoparticles were able to deliver the heparin across intestinal epithelial cells via Para cellular pathway. Via pipette various concentration of heparin and chitosan were added to heparin solution. At pH 6.0, the chitosan's amino group and anionic group of heparin interact. At a certain ratio of both components, the NPs were formed with a diameter of 173.9 nm and surface zeta potential was 19.3 mV. The release pattern showed pH-sensitive behavior as follows: At pH 1.2, less than 2% of entrapped material released and showed the presence of strong bonds between ionic chitosan and heparin. While at pH 6.6 to 7.0, the rest of the heparin was released indicating the breakage of bonds.

Another researcher prepared pH-sensitive acetylated starch nanoparticles from the nanoprecipitation of acetylated banana starch [89]. From the reaction between native banana starch and acetic anhydride acetylated banana starch was obtained and later dispersed to curcumin acetone solution. The loaded NPs showed a particle size of 201 nm and a zeta potential of −3.7 mV. The *in vitro* results showed pH-dependent release at pH 2.0, about 40% of curcumin was released in 2 h while at pH 8.0, and the remaining curcumin was released in 2 h.

The folic acid-based ligand composed of folic acid/polyethylene glycol/polyethylene mine was bonded to the curcumin-loaded alginate NPs [90]. The ligand inhibited the diffusion of Ca2+ ions. The pH-dependent release was observed at pH 1.2(2% of curcumin released) while at pH 7.4 (10% of the curcumin released). Folate-based NPs for delivery of curcumin has also been fabricated for targeted delivery in breast cancer treatment [91]. The NPs of size ranging from 119 to 127 nm were prepared and showed a loading efficiency of 96.47%. The release pattern showed pH-responsive behavior. With a decrease in pH from 7.4 to 5.0, the release rate of NPs increased.

β-lactoglobulin (BLG), has shown prominent results in the field of bioactive compounds acting as a carrier, and was stable near to the acidic gastric environment and protease enzyme [92–94].

Yi *et al.* fabricated BLG dextran-based NPs for delivering β-carotene using the homogenization method [95]. The reaction between BLG and dextran was hindered by stearic interference and hence prevention the flocculation at neutral pH. The encapsulation efficiency was 98.4% with a loading capacity of 1.1%. At pH 2.0, only 5.4% of loaded moiety was released despite the presence of pepsin while in the case of trypsin and pH 7.0, 51.8% of drug released.

Madalena *et al.* developed BLG-based nanoparticles for delivering the riboflavin in a controlled manner in the GI system [96]. At pH 6.0, the NPs were obtained and with a spherical shape with the particle size of 172 ± 1 nm and low PDI. The NPs were stable throughout the stomach and in the duodenum, they got degraded. In the stomach, the Riboflavin release was 11% while in the duodenum, jejunum, and ileum a % release of 35%, 38%, and 5% were obtained respectively.

Spigelman *et al.*, developed (−)-Epigallocatechin-3-gallate (EGCG)-based NPs with BLG [97]. The EGCG is oxidation sensitive and water-soluble in nature, while BLG is nonresponsive to the degradation by oxidation. The author observed that with an increase in the ECG concentration the particle size increased. The NPs with loading efficiency of 60% to 70% were observed and were loading on BLG concentration. Characterization results showed that molecular level complexation occurred. The results of simulated digestion showed limited release of EGCG suggesting as vesicle for protection against stomach environment, while in intestine environment showed sustained release.

Zimet *et al.* formulated nanocomplexed form of DHA, low methoxy pectin and BLG near to pH 4.5 [98]. It was observed that with the addition of pectin, negatively charged species were formed and contained about 166 times higher DHA than surrounding serum. Then diluted nanoparticle dispersion was prepared with 0.05% BLG and DHA in the molar ratio of 1:2 and particle size ~100 nm. The NPs dispersion was stressed under accelerated conditions and 5% to 10% of DHA lost at 100 h at 40°C.

Chen *et al.* developed a system for encapsulating the poor water-soluble bioactive flavonoid tangeretin and BLG [99]. The Tangeretin and zein (Organic phase) along with BLG (Aqueous phase) were mixed to form NPs and followed by the freeze-drying process and converting into powder form. When the powder was dispersed in water, it forms colloidal dispersion, which can be further used for compatible food products. The work done by various other researchers for encapsulation of bioactives using nanoparticle approach has been tabulated in Table 11.2.

Table 11.2 Literature survey of various bioactives nanoencapsulated.

Name of drug	Method of preparation	Outcome of study	Reference
Curcumin	Using nanosponge aerosol as mould	Improve bioaccessibility	[100]
	Wet-milling technique	Improvement in antimicrobial activity and increase in solubility	[101]
	Sonication method	Treatment of bacterial disease	[102]
	Ionic gelation technique	272.9 nm particle size and 12.05 mV zeta potential	[103]
Resveratrol	Supercritical antisolvent technique	Enhanced oral absorption	[104]
	Oil/water (O/W) emulsion technique	Treatment of nonalcoholic fatty liver	[105]
	Seed growth method	Enhancement in anti-inflammatory and antitumor effects	[106]
	Rotary evaporation method	Treatment of prostate cancer	[107]
Beta carotene	Homogenization-evaporation method	Delivery of beta carotene to intestine	[108]
Lutein	Hot sonication method	Increased lutein accumulation in cornea	[109]
	Double emulsion solvent evaporation method	Enhancement in lutein uptake by body	[110]

(Continued)

Table 11.2 Literature survey of various bioactives nanoencapsulated. (*Continued*)

Name of drug	Method of preparation	Outcome of study	Reference
	O/W emulsion solvent evaporation	Treatment for hypoxic ischemic encephalopathy	[111]
Vitamin C	Via conjugation with phenylboronic acid	Concentration dependent action of Vitamin C	[112]

11.2.3 Liposomes

Liposomes are polar lipid-based spherical structures that have both hydrophilic and hydrophobic heads and chains [113]. The lower tail made of lipids interacts with the bilayer structures, having the aqueous interior layer layering the aqueous phase. Therefore, both hydrophobic and hydrophilic compounds can be encapsulated at the same time [114]. The release of entrapped compounds occurs through diffusion through the bilayers, or maybe due to vesicle disruption either through pH, osmotic pressure, temperature, and ionic strength. They have also shown applications in protecting sensitive material from degradation, encapsulation of incompatible materials together, odor and taste masking, etc. Due to the encapsulation process, there has been a significant reduction in the number of additives required.

Cheng *et al.* evaluated the effect of the method of preparation of curcumin liposomes on particle size and release kinetics [115]. The liposomes were prepared using the pH-driven, ethanol injection and thin film method and TEM images as shown in Figure 11.2. During *in-vitro* digestion, the pH driven formed liposomes showed quite similar results as observed for thin film method, and it was quite better when compared with ethanol method in terms of bioaccessibility.

Odeh *et al.* formulated the curcumin-based liposomes using remote loading technique using the secondary and tertiary amino acid modified β-cyclodextrins (βCDs), using them as carrier and solubility enhancers [116] as shown in Figure 11.3. The chemical modification of βCDs was done to form the ionizable weak base and guest host complex later. Using the pH gradient action, these were loaded to preformed liposomes of DPPC/cholesterol. The encapsulation efficiencies of CRM-βCD showed about 5 times

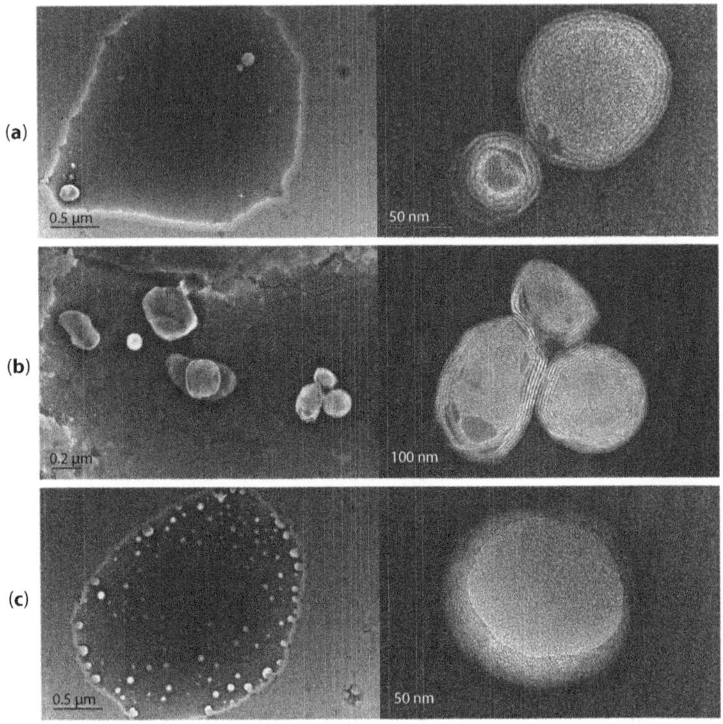

Figure 11.2 TEM images of liposomes entrapping the curcumin produced via (a) pH-driven method, (b) thin film method, and (c) ethanol injection method. (Reproduced with permission from Ref [115]).

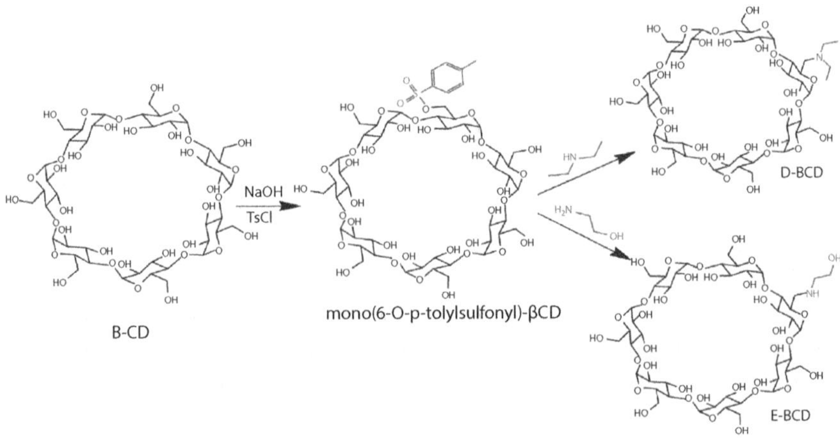

Figure 11.3 Method of synthesis of modified cyclodextrin. (Reproduced with permission from Ref [116]).

increase compared to the normal liposomes and ultimately increased the bioavailability.

Huang *et al.* developed curcumin and resveratrol liposome coencapsulation technique [117]. The co-encapsulated curcumin and resveratrol liposomes showed the lowest particle size, low PDI, and high encapsulation efficiency. The results were compared with liposomes containing separate entities. The characterization results showed that curcumin was located in the hydrophobic acyl chain region of liposomes while resveratrol oriented itself in the polar head region this orientation helped in the stabilization of liposomes.

11.2.4 Solid Lipid Nanoparticles (SLNs)

The solid lipid nanoparticles (SLNs) can be defined as the particles, of submicron size lipid emulsions consisting of solid lipid layer. They have various properties such as small size, large surface area, high drug loading percentage, etc. Whereas nanostructured lipid carriers (NLCs) has come out as second-generation molecule of lipid nanoparticles to overpower the pits and falls of first-generation SLNs. The use of biodegradable and compatible liquids makes them suitable for their use in the delivery of bioactive.

Various bioactives have been delivered through the use of solid lipid nanoparticles. Fathi *et al.* developed SLNs based on Hesperetin [135]. The SLNs so developed had encapsulation efficiency of 39.90% to 63.08%. The release kinetics of Hesperetin was evaluated in the GI tract and showed a mixture of Fickian and dissolution behavior. The samples were then fortified with milk, and it was found that it could mask the bitter taste and remove the poor solubility of Hesperetin.

Ramalingam *et al.* developed SLNs for the increasing the stability and oral bioavailability of curcumin [136]. The SLNs showed an increase in bioavailability of curcumin when compared with oral suspensions. Researchers also developed curcumin-loaded SLNs (CLEN) and contained 15 mg curcumin per ml, which is one of the highest increases in the solubility of curcumin in an aqueous medium [137]. They found that the oral bioavailability of curcumin in CLEN was 69.78 times higher than free curcumin and 9.00 times greater than the marketed formulation. And formulation also showed 104, 13.3, and 10 times enhanced stability at pH 6.8, 1.2, and 7.4, respectively, as shown in Figure 11.4. Work done by various other researchers for encapsulation of bioactives using liposomal approach has been tabulated in Table 11.3.

Table 11.3 Reported bioactivity of liposomes used for functional foods.

Name of bioactive	Polymer matrix	Outcome of study	Reference
Quercitin	Polyethyleneglycol-2000-distearoyl phosphatidyl ethanolamine (PEG2000-DSPE)	Inhibited growth of glioma cells	[118]
	Egg phosphatidylcholine, cholesterol	Intranasal delivery of Quercitin to CNS	[119]
	polyethylene glycol 4000	Anti-tumor activity	[120]
	Refined soy lecithin Lipoid S100, cholesterol	Treatment of acute liver injury in rats	[121]
	DSPC and cholesterol	Treatment of cancer through parenteral route	[122]
Curcumin	1-palmitoyl-2-oleoyl-phosphatidylcholine	Anti-inflammatory activity	[123]
	L-β,γ-dimyristoyl-α-lecithin and DMPG	Anti-inflammatory activity	[124]
	DPPC and MPEG-2000-DPPE Na	As adjuvant therapy in type 1 DM	[125]
	Soybean lecithin (SPC>90%) and cholesterol	Treatment of lung cancer via inhalation	[126]
	Bovine serum albumin, soybean lecithin, cholesterol	Long circulatory curcumin liposomes	[127]
	Lipoid S75	High bioaccessibility and stable storage	[128]

(Continued)

Table 11.3 Reported bioactivity of liposomes used for functional foods. (*Continued*)

Name of bioactive	Polymer matrix	Outcome of study	Reference
Vitamin C	Soy phosphatidylcholine	Increased stability of orange juice	[129]
	Mixed natural phospholipids	Increased oral biaccessibility	[130]
	Cholesterol Chol, PC, DOTAP, and DSPG	Increased skin penetration	[131]
Lutein	Egg yolk phosphatidylcholine, polyoxyethylene sorbitan monooleate	Enhanced stability of lutein	[132]
	L-α-phosphatidylethanolamine, stearylamine	Highest drug encapsulation and slowed release of drug	[133]
	Hydrogenated soy phosphatidylcholine	Enhanced aqueous stability	[134]

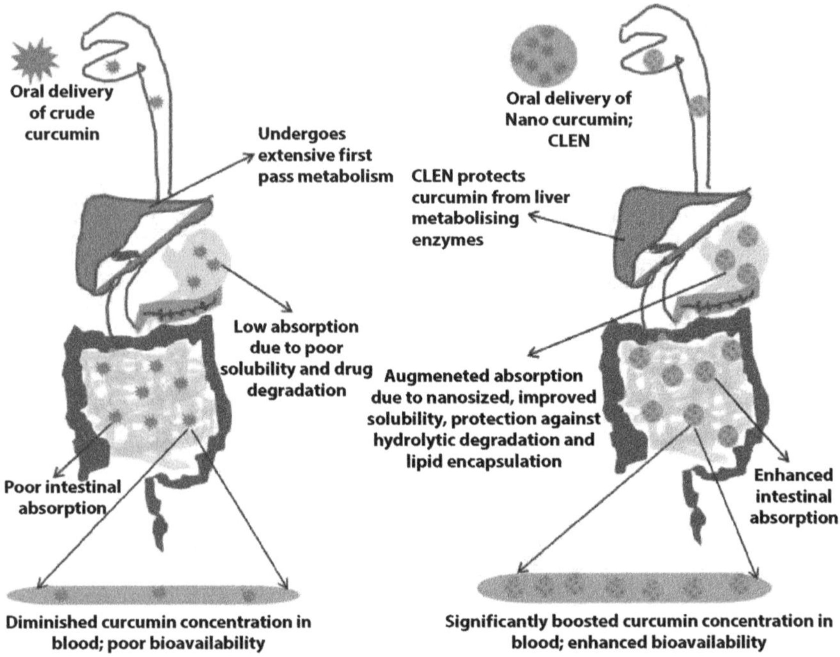

Figure 11.4 Image representing the benefits of encapsulated formulation of curcumin as CLEN. Reproduced with permission from Ref. [137].

11.3 Conclusion and Future Directions

The delivery of bioactive has always been an issue for formulators. Their poor physical properties and chemical degradation nature cause a lot of concern. Though the use of nanotechnology some issues have been irradiated still many promises are yet to be delivered. Various nanosystems, such as nanoemulsions, nanoparticles, liposomes, SLNs, have shown benefits for the various functional foods, but their long-term stability and scale-up to the production scale is still under scanner. Some of the issues, such as precise medicine, can be resolved by the use of nanotechnology and three-dimensional (3D) printing. 3D printing has shown its benefits for the pharmaceutical drug delivery system but a field of foods, vitamins, and nutrients is still untouched. Surely, with the use of nanotechnology and 3D printing, many of the problems shall be resolved.

References

1. Owens, B., Golden Rice is safe to eat, says FDA. *Nat. Biotechnol.*, 36, 559–560, 2018.
2. Vuong, L.T., Dueker, S.R., Murphy, S.P., Plasma β-carotene and retinol concentrations of children increase after a 30-d supplementation with the fruit Momordicacochinchinensis (gac). *Am. J. Clin. Nutr.*, 75, 5, 872–879, 2002.
3. Tang, G., Qin, J., Dolnikowski, G.G., Russell, R.M., Grusak, M.A., Golden rice is an effective source of vitamin A. *Am. J. Clin. Nutr.*, 89, 6, 1776–1783, 2009.
4. Waltz, E., Vitamin A super banana in human trials. *Nat. Biotechnol.*, 32, 857, 2014.
5. Kim, M., Kim, S.C., Song, K.J., Kim, H.B., Kim, I.J., Song, E.Y., Chun, S.J., Transformation of carotenoid biosynthetic genes using a micro-cross section method in kiwifruit (Actinidiadeliciosa cv. Hayward). *Plant Cell Rep.*, 29, 12, 1339–1349, 2010.
6. Pons, E., Alquézar, B., Rodríguez, A., Martorell, P., Genovés, S., Ramón, D., Rodrigo, M.J., Zacarías, L., Peña, L., Metabolic engineering of β-carotene in orange fruit increases its *in vivo* antioxidant properties. *Plant Biotechnol. J.*, 12, 1, 17–27, 2014.
7. Lourens-Hattingh, A. and Viljoen, B.C., Yogurt as probiotic carrier food. *Int. Dairy J.*, 11, 1-2, 1–17, 2001.
8. Yuan, Y., Gao, Y., Zhao, J., Mao, L., Characterization and stability evaluation of β-carotene nanoemulsions prepared by high pressure homogenization under various emulsifying conditions. *Int. Food Res. J.*, 41, 1, 61–68, 2008.
9. Moreira, J.B., Goularte, P.G., Morais, M.G.D., Costa, J.A.V., Preparation of beta-carotene nanoemulsion and evaluation of stability at a long storage period. *Food Sci. Technol.*, 39, 3, 599–604, 2019.
10. Luo, X., Zhou, Y., Bai, L., Liu, F., Deng, Y., McClements, D.J., Fabrication of β-carotene nanoemulsion-based delivery systems using dual-channel microfluidization: Physical and chemical stability. *J. Colloid Interface Sci.*, 490, 328–335, 2017.
11. Fan, Y., Gao, L., Yi, J., Zhang, Y., Yokoyama, W., Development of β-carotene-loaded organogel-based nanoemulsion with improved *in vitro* and *in vivo* bioaccessibility. *J. Agric. Food Chem.*, 65, 30, 6188–6194, 2017.
12. Teixé-Roig, J., Oms-Oliu, G., Ballesté-Muñoz, S., Odriozola-Serrano, I., Martín-Belloso, O., Improving the *In Vitro* bioaccessibility of β-Carotene using pectin added nanoemulsions. *Foods*, 9, 4, 447, 2020.
13. Zhao, C., Wei, L., Yin, B., Liu, F., Li, J., Liu, X., Wang, J., Wang, Y., Encapsulation of lycopene within oil-in-water nanoemulsions using lactoferrin: Impact of carrier oils on physicochemical stability and bioaccessibility. *Int. J. Biol. Macromol.*, 153, 912–920, 2020.

14. Kim, S.O., Ha, T.V.A., Choi, Y.J., Ko, S., Optimization of homogenization–evaporation process for lycopene nanoemulsion production and its beverage applications. *J. Food Sci.*, 79, 8, N1604–N1610, 2014.
15. Butnariu, M.V. and Giuchici, C.V., The use of some nanoemulsions based on aqueous propolis and lycopene extract in the skin's protective mechanisms against UVA radiation. *J. Nanobiotechnology*, 9, 1, 1–9, 2011.
16. Vishwanathan, R., Wilson, T.A., Nicolosi, R.J., Bioavailability of a nanoemulsion of lutein is greater than a lutein supplement. *Nano Biomed. Eng.*, 1, 1, 38–49, 2009.
17. Murillo, A.G., Aguilar, D., Norris, G.H., DiMarco, D.M., Missimer, A., Hu, S., Smyth, J.A., Gannon, S., Blesso, C.N., Luo, Y., Fernandez, M.L., Compared with powdered lutein, a lutein nanoemulsion increases plasma and liver lutein, protects against hepatic steatosis, and affects lipoprotein metabolism in guinea pigs. *Nutr. J.*, 146, 10, 1961–1969, 2016.
18. Steiner, B.M., Shukla, V., McClements, D.J., Li, Y.O., Sancho-Madriz, M., Davidov-Pardo, G., Encapsulation of Lutein in Nanoemulsions Stabilized by Resveratrol and Maillard Conjugates. *J. Food Sci.*, 84, 9, 2421–2431, 2019.
19. Zhao, C., Shen, X., Guo, M., Stability of lutein encapsulated whey protein nano-emulsion during storage. *PloS One*, 13, 2, e0192511, 2018.
20. Li, J., Guo, R., Hu, H., Wu, X., Ai, L., Wu, Y., Preparation optimisation and storage stability of nanoemulsion-based lutein delivery systems. *J. Microencapsul.*, 35, 6, 570–583, 2018.
21. Lim, C., Kim, D.W., Sim, T., Hoang, N.H., Lee, J.W., Lee, E.S., Youn, Y.S., Oh, K.T., Preparation and characterization of a lutein loading nanoemulsion system for ophthalmic eye drops. *J. Drug Deliv. Sci. Technol.*, 36, 168–174, 2016.
22. Khalid, N., Shu, G., Holland, B.J., Kobayashi, I., Nakajima, M., Barrow, C.J., Formulation and characterization of O/W nanoemulsions encapsulating high concentration of astaxanthin. *Int. Food Res. J.*, 102, 364–371, 2017.
23. Shanmugapriya, K., Kim, H., Saravana, P.S., Chun, B.S., Kang, H.W., Astaxanthin-alpha tocopherol nanoemulsion formulation by emulsification methods: investigation on anticancer, wound healing, and antibacterial effects. *Colloids Surf. B.*, 172, 170–179, 2018.
24. Haung, H.Y., Wang, Y.C., Cheng, Y.C., Kang, W., Hu, S.H., Liu, D., Xiao, C., Wang, H.M.D., A novel oral astaxanthin nanoemulsion from haematococcuspluvialis induces apoptosis in lung metastatic melanoma. *Oxid. Med. Cell. Longev.*, 2020, 1–13, 2020.
25. Kim, D.M., Hyun, S.S., Yun, P., Lee, C.H., Byun, S.Y., Identification of an emulsifier and conditions for preparing stable nanoemulsions containing the antioxidant astaxanthin. *Int. J. Cosmet. Sci.*, 34, 1, 64–73, 2012.
26. Sun, R., Xia, N., Xia, Q., Non-aqueous nanoemulsions as a new strategy for topical application of astaxanthin. *J. Dispers. Sci. Technol.*, 41, 1–12, 2020.
27. Liu, J., Bi, J., Liu, X., Zhang, B., Wu, X., Wellala, C.K.D., Zhang, B., Effects of high pressure homogenization and addition of oil on the carotenoid bioaccessibility of carrot juice. *Food Funct.*, 10, 1, 458–468, 2019.

28. Severino, R., Ferrari, G., Vu, K.D., Donsì, F., Salmieri, S., Lacroix, M., Antimicrobial effects of modified chitosan based coating containing nanoemulsion of essential oils, modified atmosphere packaging and gamma irradiation against Escherichia coli O157: H7 and Salmonella Typhimurium on green beans. *Food Control*, 50, 215–222, 2015.
29. Ruiz-Montañez, G., Ragazzo-Sanchez, J.A., Picart-Palmade, L., Calderón-Santoyo, M., Chevalier-Lucia, D., Optimization of nanoemulsions processed by high-pressure homogenization to protect a bioactive extract of jackfruit (Artocarpusheterophyllus Lam). *Innov. Food Sci. Emerg. Technol.*, 40, 35–41, 2017.
30. Shariffa, Y.N., Tan, T.B., Uthumporn, U., Abas, F., Mirhosseini, H., Nehdi, I.A., Wang, Y.H., Tan, C.P., Producing a lycopene nanodispersion: Formulation development and the effects of high pressure homogenization. *Int. Food Res. J.*, 101, 165–172, 2017.
31. Tabilo-Munizaga, G., Villalobos-Carvajal, R., Herrera-Lavados, C., Moreno-Osorio, L., Jarpa-Parra, M., Pérez-Won, M., Physicochemical properties of high-pressure treated lentil protein-based nanoemulsions. *LWT*, 101, 590–598, 2019.
32. Jafari, S.M., He, Y., Bhandari, B., Optimization of nano-emulsions production by microfluidization. *Eur. Food Res. Technol.*, 225, 5-6, 733–741, 2007.
33. Kumar, R., Kaur, K., Uppal, S., Mehta, S.K., Ultrasound processed nanoemulsion: A comparative approach between resveratrol and resveratrol cyclodextrin inclusion complex to study its binding interactions, antioxidant activity and UV light stability. *Ultrason Sonochem.*, 37, 478–489, 2017.
34. Zheng, B., Zhang, X., Peng, S., McClements, D.J., Impact of curcumin delivery system format on bioaccessibility: nanocrystals, nanoemulsion droplets, and natural oil bodies. *Food Funct.*, 10, 7, 4339–4349, 2019.
35. Iqbal, R., Mehmood, Z., Baig, A., Khalid, N., Formulation and characterization of food grade O/W nanoemulsions encapsulating quercetin and curcumin: Insights on enhancing solubility characteristics. *Food Bioprod. Process.*, 123, 304–311, 2020.
36. Kharat, M., Zhang, G., McClements, D.J., Stability of curcumin in oil-in-water emulsions: Impact of emulsifier type and concentration on chemical degradation. *Int. Food Res. J.*, 111, 178–186, 2018.
37. Maurya, V.K. and Aggarwal, M., A phase inversion based nanoemulsion fabrication process to encapsulate vitamin D3 for food applications. *J. Steroid Biochem. Mol. Biol.*, 190, 88–98, 2019.
38. Walia, N. and Chen, L., Pea protein based vitamin D nanoemulsions: Fabrication, stability and *in vitro* study using Caco-2 cells. *Food Chem.*, 305, 125475, 2020.
39. Park, S.J., Hong, S.J., Garcia, C.V., Lee, S.B., Shin, G.H., Kim, J.T., Stability evaluation of turmeric extract nanoemulsion powder after application in milk as a food model. *J. Food Eng.*, 259, 12–20, 2019.

40. Inapurapu, S.P., Ibrahim, A., Kona, S.R., Pawar, S.C., Bodiga, S., Bodiga, V.L., Development and characterization of ω-3 fatty acid nanoemulsions with improved physicochemical stability and bioaccessibility. *Colloids Surf. A Physicochem. Eng.*, 606, 125515, 2020.
41. Guttoff, M., Saberi, A.H., McClements, D.J., Formation of vitamin D nanoemulsion-based delivery systems by spontaneous emulsification: Factors affecting particle size and stability. *Food Chem.*, 171, 117–122, 2015.
42. Stratulat, I., Britten, M., Salmieri, S., Fustier, P., St-Gelais, D., Champagne, C.P., Lacroix, M., Enrichment of cheese with vitamin D3 and vegetable omega-3. *J. Funct. Foods*, 13, 300–307, 2015.
43. Ozturk, B., Argin, S., Ozilgen, M., McClements, D.J., Nanoemulsion delivery systems for oil-soluble vitamins: Influence of carrier oil type on lipid digestion and vitamin D3 bioaccessibility. *Food Chem.*, 187, 499–506, 2015.
44. Golfomitsou, I., Mitsou, E., Xenakis, A., Papadimitriou, V., Development of food grade O/W nanoemulsions as carriers of vitamin D for the fortification of emulsion based food matrices: a structural and activity study. *J. Mol. Liq.*, 268, 734–742, 2018.
45. Meghani, N., Patel, P., Kansara, K., Ranjan, S., Dasgupta, N., Ramalingam, C., Kumar, A., Formulation of vitamin D encapsulated cinnamon oil nanoemulsion: Its potential anti-cancerous activity in human alveolar carcinoma cells. *Colloids Surf. B.*, 166, 349–357, 2018.
46. Schoener, A.L., Zhang, R., Lv, S., Weiss, J., McClements, D.J., Fabrication of plant-based vitamin D 3-fortified nanoemulsions: Influence of carrier oil type on vitamin bioaccessibility. *Food Funct.*, 10, 4, 1826–1835, 2019.
47. Gulotta, A., Saberi, A.H., Nicoli, M.C., McClements, D.J., Nanoemulsion-based delivery systems for polyunsaturated (ω-3) oils: Formation using a spontaneous emulsification method. *J. Agric. Food Chem.*, 62, 7, 1720–1725, 2014.
48. Belhaj, N., Arab-Tehrany, E., Linder, M., Oxidative kinetics of salmon oil in bulk and in nanoemulsion stabilized by marine lecithin. *Process Biochem.*, 45, 2, 187–195, 2010.
49. Abbasi, F., Samadi, F., Jafari, S.M., Ramezanpour, S., Shargh, M.S., Ultrasound-assisted preparation of flaxseed oil nanoemulsions coated with alginate-whey protein for targeted delivery of omega-3 fatty acids into the lower sections of gastrointestinal tract to enrich broiler meat. *Ultrason. Sonochem.*, 50, 208–217, 2019.
50. Karthik, P. and Anandharamakrishnan, C., Enhancing omega-3 fatty acids nanoemulsion stability and *in-vitro* digestibility through emulsifiers. *J. Food Eng.*, 187, 92–105, 2016.
51. Ricaurte, L., de Jesús Perea-Flores, M., Martinez, A., Quintanilla-Carvajal, M.X., Production of high-oleic palm oil nanoemulsions by high-shear homogenization (microfluidization). *Innov. Food Sci. Emerg. Technol.*, 35, 75–85, 2016.

52. Lane, K.E., Li, W., Smith, C.J., Derbyshire, E.J., The development of vegetarian omega-3 oil in water nanoemulsions suitable for integration into functional food products. *J. Funct. Foods*, 23, 306–314, 2016.
53. Dey, T.K., Koley, H., Ghosh, M., Dey, S., Dhar, P., Effects of nano-sizing on lipid bioaccessibility and ex vivo bioavailability from EPA-DHA rich oil in water nanoemulsion. *Food Chem.*, 275, 135–142, 2019.
54. Ha, T.V.A., Kim, S., Choi, Y., Kwak, H.S., Lee, S.J., Wen, J., Oey, I., Ko, S., Antioxidant activity and bioaccessibility of size-different nanoemulsions for lycopene-enriched tomato extract. *Food Chem.*, 178, 115–121, 2015.
55. Sotomayor-Gerding, D., Oomah, B.D., Acevedo, F., Morales, E., Bustamante, M., Shene, C., Rubilar, M., High carotenoid bioaccessibility through linseed oil nanoemulsions with enhanced physical and oxidative stability. *Food Chem.*, 199, 463–470, 2016.
56. Raikos, V., Hayward, N., Hayes, H., Meroni, E., Ranawana, V., Optimising the ratio of long-to short-chain triglycerides of the lipid phase to enhance physical stability and bioaccessibility of lycopene-loaded beverage emulsions. *Int. J. Food Sci. Technol.*, 54, 4, 1355–1362, 2019.
57. Surh, J., Decker, E.A., McClements, D.J., Utilisation of spontaneous emulsification to fabricate lutein-loaded nanoemulsion-based delivery systems: factors influencing particle size and colour. *Int. J. Food Sci. Technol.*, 52, 6, 1408–1416, 2017.
58. Qian, C., Decker, E.A., Xiao, H., McClements, D.J., Inhibition of β-carotene degradation in oil-in-water nanoemulsions: influence of oil-soluble and water-soluble antioxidants. *Food Chem.*, 135, 3, 1036–1043, 2012.
59. Silva, H.D., Cerqueira, M. Â. P. R., Cunha-da-Silva, M., Vicente, A.A., Layer-by-layer deposition on-carotene oil-in-water nanoemulsions: An edible multi-layer system, in: *Proceedings of the XXII EncontroNacional SPQ: 100 anos de Químicaem Portugal*, Braga, Portugal, 3–6 July 2011.
60. Costa, A.M., Bueno, K.T.L., da Rosa, A.P.C., Costa, J.A.V., The antioxidant activity of nanoemulsions based on lipids and peptides from Spirulina sp. LEB18. *LWT*, 99, 173–178, 2019.
61. Liu, Y., Hou, Z., Lei, F., Chang, Y., Gao, Y., Investigation into the bioaccessibility and microstructure changes of β-carotene emulsions during *in vitro* digestion. *Innov. Food Sci. Emerg. Technol.*, 15, 86–95, 2012.
62. Chen, Y., Zhang, R., Xie, B., Sun, Z., McClements, D.J., Lotus seedpod proanthocyanidin-whey protein complexes: Impact on physical and chemical stability of β-carotene-nanoemulsions. *Int. Food Res. J.*, 127, 108738, 2020.
63. Sharif, H.R., Goff, H.D., Majeed, H., Shamoon, M., Liu, F., Nsor-Atindana, J., Haider, J., Liang, R., Zhong, F., Physicochemical properties of β-carotene and eugenol co-encapsulated flax seed oil powders using OSA starches as wall material. *Food Hydrocol.*, 73, 274–283, 2017.
64. Borba, C.M., Tavares, M.N., Macedo, L.P., Araújo, G.S., Furlong, E.B., Dora, C.L., Burkert, J.F., Physical and chemical stability of β-carotene

nanoemulsions during storage and thermal process. *Int. Food Res. J.*, 121, 229–237, 2019.
65. Hwang, S.R., Lim, S.-J., Park, J.-S., Kim, C.-K., Phospholipid-based microemulsion formulation of all-trans-retinoic acid for parenteral administration. *Int. J. Pharm.*, 276, 175–183, 2004.
66. Yoshida, K., Sekine, T., Matsuzaki, F., Yanaki, T., Yamaguchi, M., Stability of vitamin A in oil-in-water-in oil-type multiple emulsions. *J. Am. Oil Chem. Soc*, 76, 1–6, 1999.
67. Jeong, Y., Preparation of poly(?-lactide-co-glycolide) microspheres encapsulating all-trans retinoic acid. *Int. J. Pharm.*, 259, 79–91, 2003.
68. Kim, T.-I., Lim, D.-H., Kim, S.-B., Park, S.-M., Hur, T.-Y., Ki, K.-S., Kwon, E.-G., Vijayakumar, M., Kim, Y.-J., Kim, T.G., Preparation of nanoemulsions of vitamin A and C by microfluidization: efficacy on the expression pattern of milk-specific proteins in MAC-T Cells. *Molecules*, 24, 2566, 2019.
69. Benzaria, A., Chevalier, D., Picart, L., Hue, P., López-Pedemonte, T., Dumay, E., Intracellular fate of retinyl acetate-loaded submicron delivery systems by *in vitro* intestinal epithelial cells: A comparison between whey protein-stabilised submicron droplets and micelles stabilized with polysorbate 80. *Int. Food Res. J.*, 51, 679–692, 2013.
70. Choudhry, Q.N., Kim, M.J., Kim, T.G., Pan, J.H., Kim, J.H., Park, S.J., Park, J.-W., Kim, Y.-J., Saponin-based nanoemulsification improves the antioxidant properties of vitamin A and E in AML-12 cells. *Int. J. Mol. Sci.*, 17, 1406, 2016.
71. Tanglao, E.J., Kumar, A.B.N., Noriega, R.R., Punzalan, M.E., Marcelo, P., Development and physico-chemical characterization of virgin coconut oil-in-water emulsion using polymerized whey protein as emulsifier for Vitamin A delivery. *MATEC Web Conf*, vol. 268, p. 01002, 2019.
72. Campani, V., Biondi, M., Mayol, L., Cilurzo, F., Pitaro, M., De Rosa, G., Development of nanoemulsions for topical delivery of vitamin K1. *Int. J. Pharm.*, 511, 1, 170–177, 2016.
73. Shi, J., Zhou, S., Kang, L., Ling, H., Chen, J., Duan, L., Song, Y., Deng, Y., Evaluation of the antitumor effects of vitamin K 2 (menaquinone-7) nanoemulsions modified with sialic acid-cholesterol conjugate. *Drug Deliv. Transl. Res.*, 8, 1, 1–11, 2017.
74. Raikos, V., Encapsulation of vitamin E in edible orange oil-in-water emulsion beverages: Influence of eating temperature on physicochemical stability during chilled storage. *Food Hydrocoll.*, 72, 155–162, 2017.
75. Lv, S., Zhang, Y., Tan, H., Zhang, R., McClements, D.J., Vitamin E encapsulation within oil-in-water emulsions: Impact of emulsifier type on physicochemical stability and bioaccessibility. *J. Agric. Food Chem.*, 67, 5, 1521–1529, 2019.
76. Saberi, A.H., Fang, Y., McClements, D.J., Fabrication of vitamin E-enriched nanoemulsions: Factors affecting particle size using spontaneous emulsification. *J. Colloid Interface Sci.*, 391, 95–102, 2013.

77. El Kinawy, O.S., Petersen, S., Ulrich, J., Technological aspects of nanoemulsion formation of low-fat foods enriched with vitamin E by high-pressure homogenization. *Chem. Eng. Technol.*, 35, 5, 937–940, 2012.
78. Ozturk, B., Argin, S., Ozilgen, M., McClements, D.J., Formation and stabilization of nanoemulsion-based vitamin E delivery systems using natural biopolymers: Whey protein isolate and gum arabic. *Food Chem.*, 188, 256–263, 2015.
79. Xue, J., Davidson, P.M., Zhong, Q., Antimicrobial activity of thyme oil co-nanoemulsified with sodium caseinate and lecithin. *Int. J. Food Microbiol.*, 210, 1–8, 2015.
80. Ma, Q., Davidson, P.M., Zhong, Q., Nanoemulsions of thymol and eugenol co-emulsified by lauricarginate and lecithin. *Food Chem.*, 206, 167–173, 2016.
81. Sharma, A., Sharma, N.K., Srivastava, A., Kataria, A., Dubey, S., Sharma, S., Kundu, B., Clove and lemongrass oil based non-ionic nanoemulsion for suppressing the growth of plant pathogenic Fusariumoxysporum f. sp. lycopersici. *Ind. Crops Prod.*, 123, 353–362, 2018.
82. Yang, Y., Zhao, C., Tian, G., Lu, C., Li, C., Bao, Y., Tang, Z., Zheng, J., Characterization of physical properties and electronic sensory analyses of citrus oil-based nanoemulsions. *Int. Food Res. J.*, 109, 149–158, 2018.
83. Liang, R., Xu, S., Shoemaker, C.F., Li, Y., Zhong, F., Huang, Q., Physical and antimicrobial properties of peppermint oil nanoemulsions. *J. Agric. Food Chem.*, 60, 30, 7548–7555, 2012.
84. Ozogul, Y., Boğa, E.K., Akyol, I., Durmus, M., Ucar, Y., Regenstein, J.M., Köşker, A.R., Antimicrobial activity of thyme essential oil nanoemulsions on spoilage bacteria of fish and food-borne pathogens. *Food Biosci.*, 36, 100635, 2020.
85. Guerra-Rosas, M., II, Morales-Castro, J., Ochoa-Martínez, L.A., Salvia-Trujillo, L., Martín-Belloso, O., Long-term stability of food-grade nanoemulsions from high methoxyl pectin containing essential oils. *Food Hydrocoll.*, 52, 438–446, 2016.
86. Salvia-Trujillo, L., Rojas-Graü, M.A., Soliva-Fortuny, R., Martín-Belloso, O., Effect of processing parameters on physicochemical characteristics of microfluidized lemongrass essential oil-alginate nanoemulsions. *Food Hydrocoll.*, 30, 1, 401–407, 2013.
87. Banasaz, S., Morozova, K., Ferrentino, G., Scampicchio, M., Encapsulation of lipid-soluble bioactives by nanoemulsions. *Molecules*, 25, 17, 3966, 2020.
88. Das, R.K., Kasoju, N., Bora, U., Encapsulation of curcumin in alginate-chitosan-pluronic composite nanoparticles for delivery to cancer cells. *Nanomed.: Nanotechnol. Biol. Med.*, 6, 1, 153–160, 2010.
89. Acevedo-Guevara, L., Nieto-Suaza, L., Sanchez, L.T., Pinzon, M., II, Villa, C.C., Development of native and modified banana starch nanoparticles as vehicles for curcumin. *Int. J. Biol. Macromol.*, 111, 498–504, 2018.

90. Anirudhan, T.S., Anila, M.M., Franklin, S., Synthesis characterization and biological evaluation of alginate nanoparticle for the targeted delivery of curcumin. *Mater. Sci. Eng. C.*, 78, 1125–1134, 2017.
91. Esfandiarpour-Boroujeni, S., Bagheri-Khoulenjani, S., Mirzadeh, H., Amanpour, S., Fabrication and study of curcumin loaded nanoparticles based on folate-chitosan for breast cancer therapy application. *Carbohydr. Polym.*, 168, 14–21, 2017.
92. Świątek, S., Komorek, P., Turner, G., Jachimska, B., β-Lactoglobulin as a potential carrier for bioactive molecules. *Bioelectrochemistry*, 126, 137–145, 2019.
93. Bossios, A., Theodoropoulou, M., Mondoulet, L., Rigby, N.M., Papadopoulos, N.G., Bernard, H., Adel-Patient, K., Wal, J.M., Mills, C.E.N., Papageorgiou, P., Effect of simulated gastro-duodenal digestion on the allergenic reactivity of beta-lactoglobulin. *Clin. Transl. Allergy*, 1, 1, 6, 2011.
94. Broersen, K., Milk processing affects structure, bioavailability and immunogenicity of β-lactoglobulin. *Foods*, 9, 7, 874, 2020.
95. Yi, J., Lam, T., II, Yokoyama, W., Cheng, L.W., Zhong, F., Controlled release of β-carotene in β-lactoglobulin–dextran-conjugated nanoparticles' *in vitro* digestion and transport with Caco-2 monolayers. *J. Agric. Food Chem.*, 62, 35, 8900–8907, 2014.
96. Madalena, D.A., Ramos, Ó. L., Pereira, R.N., Bourbon, A., II, Pinheiro, A.C., Malcata, F.X., Teixeira, J.A., Vicente, A.A., *In vitro* digestion and stability assessment of β-lactoglobulin/riboflavin nanostructures. *Food Hydrocoll.*, 58, 89–97, 2016.
97. Shpigelman, A., Cohen, Y., Livney, Y.D., Thermally-induced β-lactoglobulin–EGCG nanovehicles: Loading, stability, sensory and digestive-release study. *Food Hydrocoll.*, 29, 1, 57–67, 2012.
98. Zimet, P. and Livney, Y.D., Beta-lactoglobulin and its nanocomplexes with pectin as vehicles for ω-3 polyunsaturated fatty acids. *Food Hydrocoll.*, 23, 4, 1120–1126, 2009.
99. Chen, J., Zheng, J., McClements, D.J., Xiao, H., Tangeretin-loaded protein nanoparticles fabricated from zein/β-lactoglobulin: Preparation, characterization, and functional performance. *Food Chem.*, 158, 466–472, 2014.
100. Ubeyitogullari, A. and Ciftci, O.N., A novel and green nanoparticle formation approach to forming low-crystallinity curcumin nanoparticles to improve curcumin's bioaccessibility. *Sci. Rep.*, 9, 1, 1–11, 2019.
101. Basniwal, R.K., Buttar, H.S., Jain, V.K., Jain, N., Curcumin nanoparticles: Preparation, characterization, and antimicrobial study. *J. Agric. Food Chem.*, 59, 5, 2056–2061, 2011.
102. Pandit, R.S., Gaikwad, S.C., Agarkar, G.A., Gade, A.K., Rai, M., Curcumin nanoparticles: physico-chemical fabrication and its *in vitro* efficacy against human pathogens. *3 Biotech.*, 5, 6, 991–997, 2015.
103. Suryani, Halid, N.H.A., Akib, N., II, Rahmanpiu, Mutmainnah, N., Preparation of curcumin nanoparticle by using reinforcement ionic gelation

technique, in: *AIP Conference Proceedings*, vol. 1838, AIP Publishing LLC, p. 020013, 2017.

104. Ha, E.S., Sim, W.Y., Lee, S.K., Jeong, J.S., Kim, J.S., Baek, I.H., Du, H.C., Heejun, P., Hwang, S.J., Kim, M.S., Preparation and evaluation of resveratrol-loaded composite nanoparticles using a supercritical fluid technology for enhanced oral and skin delivery. *Antioxidants*, 8, 11, 554, 2019.
105. Wan, S., Zhang, L., Quan, Y., Wei, K., Resveratrol-loaded PLGA nanoparticles: enhanced stability, solubility and bioactivity of resveratrol for non-alcoholic fatty liver disease therapy. *R. Soc Open Sci.*, 5, 11, 181457, 2018.
106. Lee, D.G., Go, E.B., Lee, M., Pak, P.J., Kim, J.S., Chung, N., Gold nanoparticles conjugated with resveratrol induce cell cycle arrest in MCF-7 cell lines. *Appl. Biol. Chem.*, 62, 1, 1–6, 2019.
107. Chaudhary, Z., Subramaniam, S., Khan, G.M., Abeer, M.M., Qu, Z., Janjua, T., Kumeria, T., Batra, J., Popat, A., Encapsulation and controlled release of resveratrol within functionalized mesoporous silica nanoparticles for prostate cancer therapy. *Front. Bioeng. Biotechnol.*, 7, 225, 2019.
108. Yi, J., Lam, T., II, Yokoyama, W., Cheng, L.W., Zhong, F., Beta-carotene encapsulated in food protein nanoparticles reduces peroxyl radical oxidation in Caco-2 cells. *Food Hydrocoll.*, 43, 31–40, 2015.
109. Liu, C.H., Chiu, H.C., Wu, W.C., Sahoo, S.L., Hsu, C.Y., Novel lutein loaded lipid nanoparticles on porcine corneal distribution. *J. Ophthalmol.*, 2014, 2014.
110. Renukuntla, J., Bolla, P.K., Kalhapure, R., Arnipalli, A., Franco, J., Meraz, C., Folate decorated polymeric nanoparticles for enhanced lutein uptake for treatment of age-related macular degeneration. *Investig. Ophthalmol. Vis. Sci.*, 60, 9, 2998–2998, 2019.
111. Bolla, P.K., Gote, V., Singh, M., Yellepeddi, V.K., Patel, M., Pal, D., Gong, X., Sambalingam, D., Renukuntla, J., Preparation and characterization of lutein loaded folate conjugated polymeric nanoparticles. *J. Microencapsul.*, 37, 7, 502–516, 2020.
112. Chakraborty, A. and Jana, N.R., Vitamin C-conjugated nanoparticle protects cells from oxidative stress at low doses but induces oxidative stress and cell death at high doses. *ACS Appl. Mater. Interfaces*, 9, 48, 41807–41817, 2017.
113. Singh, H., Thompson, A., Liu, W., Corredig, M., Liposomes as food ingredients and nutraceutical delivery systems, in: *Encapsulation Technologies and Delivery Systems for Food Ingredients and Nutraceuticals*, N. Garti, and D.J. McClements, (Eds.), pp. 287–318, Woodhead Publishing, Elsevier BV, Amsterdam, The Netherlands, 2012.
114. Tsai, W.C. and Rizvi, S.S., Simultaneous microencapsulation of hydrophilic and lipophilic bioactives in liposomes produced by an ecofriendly supercritical fluid process. *Int. Food Res. J.*, 99, 256–262, 2017.
115. Cheng, C., Peng, S., Li, Z., Zou, L., Liu, W., Liu, C., Improved bioavailability of curcumin in liposomes prepared using a pH-driven, organic solvent-free, easily scalable process. *RSC Adv.*, 7, 42, 25978–25986, 2017.

116. Odeh, F., Nsairat, H., Alshaer, W., Alsotari, S., Buqaien, R., Ismail, S., Awidi, A., Al Bawab, A., Remote loading of curcumin-in-modified β-cyclodextrins into liposomes using a transmembrane pH gradient. *RSC Adv.*, 9, 64, 37148–37161, 2019.
117. Huang, M., Liang, C., Tan, C., Huang, S., Ying, R., Wang, Y., Zhang, Y., Liposome co-encapsulation as a strategy for the delivery of curcumin and resveratrol. *Food Funct.*, 10, 10, 6447–6458, 2019.
118. Gang, W., Jie, W.J., Ping, Z.L., Ming, D.S., Ying, L.J., Lei, W., Fang, Y., Liposomal quercetin: Evaluating drug delivery *in vitro* and biodistribution *in vivo*. *Expert Opin. Drug Deliv.*, 9, 6, 599–613, 2012.
119. Priprem, A., Watanatorn, J., Sutthiparinyanont, S., Phachonpai, W., Muchimapura, S., Anxiety and cognitive effects of quercetin liposomes in rats. *Nanomed.: Nanotechnol. Biol. Med.*, 4, 1, 70–78, 2008.
120. Yuan, Z.P., Chen, L.J., Fan, L.Y., Tang, M.H., Yang, G.L., Yang, H.S., Du, X.B., Wang, G.Q., Yao, W.X., Wei, Y.Q., Liposomal quercetin efficiently suppresses growth of solid tumors in murine models. *Clin. Cancer Res.*, 12, 10, 3193–3199, 2006.
121. Liu, X., Zhang, Y., Liu, L., Pan, Y., Hu, Y., Yang, P., Liao, M., Protective and therapeutic effects of nanoliposomalquercetin on acute liver injury in rats. *BMC Pharmacol. Toxicol.*, 21, 1, 1–7, 2020.
122. Chen, K.T., Anantha, M., Leung, A.W., Kulkarni, J.A., Militao, G.G., Wehbe, M., Sutherland, B., Cullis, P.R., Bally, M.B., Characterization of a liposomal copper (II)-quercetin formulation suitable for parenteral use. *Drug Deliv. Transl. Res.*, 10, 1, 202–215, 2020.
123. Sinjari, B., Pizzicannella, J., D'Aurora, M., Zappacosta, R., Gatta, V., Fontana, A., Trubiani, O., Diomede, F., Curcumin/liposome nanotechnology as delivery platform for anti-inflammatory activities via NFkB/ERK/pERK pathway in human dental pulp treated with 2-hydroxyethyl methacrylate (HEMA). *Front. Physiol.*, 10, 633, 2019.
124. Kloesch, B., Gober, L., Loebsch, S., Vcelar, B., Helson, L., Steiner, G., *In vitro* study of a liposomal curcumin formulation (Lipocurc™): Toxicity and biological activity in synovial fibroblasts and macrophages. *In Vivo*, 30, 4, 413–419, 2016.
125. Bulboacă, A.E., Boarescu, P.M., Bolboacă, S.D., Blidaru, M., Feştilă, D., Dogaru, G., Nicula, C.A., Comparative effect of curcumin versus liposomal curcumin on systemic pro-inflammatory cytokines profile, mcp-1 and rantes in experimental diabetes mellitus. *Int. J. Nanomedicine*, 14, 8961, 2019.
126. Zhang, T., Chen, Y., Ge, Y., Hu, Y., Li, M., Jin, Y., Inhalation treatment of primary lung cancer using liposomal curcumin dry powder inhalers. *Acta Pharm. Sin. B.*, 8, 3, 440–448, 2018.
127. Wei, X.Q. and Ba, K., Construction a long-circulating delivery system of liposomal curcumin by coating albumin. *ACS Omega*, 5, 27, 16502–16509, 2020.

128. Cheng, C., Peng, S., Li, Z., Zou, L., Liu, W., Liu, C., Improved bioavailability of curcumin in liposomes prepared using a pH-driven, organic solvent-free, easily scalable process. *RSC Adv.*, 7, 42, 25978–25986, 2017.
129. Marsanasco, M., Márquez, A.L., Wagner, J.R., Alonso, S.D.V., Chiaramoni, N.S., Liposomes as vehicles for vitamins E and C: An alternative to fortify orange juice and offer vitamin C protection after heat treatment. *Int. Food Res. J.*, 44, 9, 3039–3046, 2011.
130. Davis, J.L., Paris, H.L., Beals, J.W., Binns, S.E., Giordano, G.R., Scalzo, R.L., Schweder, M.M., Blair, E., Bell, C., Liposomal-encapsulated ascorbic acid: Influence on vitamin C bioavailability and capacity to protect against ischemia-reperfusion injury. *Nutr. Metab. Insights*, 9, 25–30, 2016.
131. Maione-Silva, L., de Castro, E.G., Nascimento, T.L., Cintra, E.R., Moreira, L.C., Cintra, B.A.S., Valadares, M.C., Lima, E.M., Ascorbic acid encapsulated into negatively charged liposomes exhibits increased skin permeation, retention and enhances collagen synthesis by fibroblasts. *Sci. Rep.*, 9, 1, 1–14, 2019.
132. Tan, C., Xia, S., Xue, J., Xie, J., Feng, B., Zhang, X., Liposomes as vehicles for lutein: preparation, stability, liposomal membrane dynamics, and structure. *J. Agric. Food Chem.*, 61, 34, 8175–8184, 2013.
133. Elkholy, N.S., Shafaa, M.W., Mohammed, H.S., Biophysical characterization of lutein or beta carotene-loaded cationic liposomes. *RSC Adv.*, 10, 54, 32409–32422, 2020.
134. Tanaka, Y., Uemori, C., Kon, T., Honda, M., Machmudah, S., Kanda, H., Goto, M., Preparation of liposomes encapsulating β–carotene using supercritical carbon dioxide with ultrasonication. *J. Supercrit. Fluids*, 161, 104848, 2020.
135. Fathi, M., Varshosaz, J., Mohebbi, M., Shahidi, F., Hesperetin-loaded solid lipid nanoparticles and nanostructure lipid carriers for food fortification: Preparation, characterization, and modeling. *Food Bioprocess Technol.*, 6, 6, 1464–1475, 2013.
136. Ramalingam, P., Yoo, S.W., Ko, Y.T., Nanodelivery systems based on mucoadhesive polymer coated solid lipid nanoparticles to improve the oral intake of food curcumin. *Food Res. Int.*, 84, 113–119, 2016.
137. Gupta, T., Singh, J., Kaur, S., Sandhu, S., Singh, G., Kaur, I.P., Enhancing bioavailability and stability of curcumin using solid lipid nanoparticles (clen): A covenant for its effectiveness. *Front. Bioeng. Biotechnol.*, 8, 879, 2020.

12

Recent Advancement of Nanotechnology in Functional Foods: Applications Perspectives

Abhishek Kanugo[1], Rajat Goyal[2], Sanjay Sharma[3] and Rupesh Kumar Gautam[2*]

[1]SVKM NMIMS School of Pharmacy and Technology Management, Shirpur, Dhule, India
[2]MM School of Pharmacy, Maharishi Markandeshwar University, Sadopur-Ambala, Haryana, India
[3]SVKM NMIMS Shobhaben Pratapbhai Patel School of Pharmacy & Technology Management, Vile Parle West, Mumbai, Maharashtra, India

Abstract

Functional foods (FFs) are the prerequisite of any food or agro-based industry and play a vital role by providing nutrients, as well as in the prevention and treatment of several diseases, like cancer, diabetes mellitus, cardiovascular, respiratory, arthritis, and protection against COVID-19 viral infections. Conventional techniques have limitations in retaining nutritional benefits, preservation from environmental factors, and lack of stability. The advancement of nanotechnology offers several benefits to the food industry in terms of their characterization, fabrication, the preservation of foods (physical and chemical degradation), smart packaging, enhancing nutritional benefits, safety, etc. Moreover, novel nanomaterials and nanoparticles ensure improvement in solubility and bioavailability with achievable sustainable action at the targeted site, as well as a great contribution by smart and intelligent packaging. This book chapter specially focused on nanotechnology-based approaches used for applications, such as food processing, packaging, preserving, smart tracking, and extending the shelf-life of FFs. Nanoparticles containing organic and inorganic materials contribute to the potential applications in delivering functional foods. Outcomes of nanotechnology cover highly stable and more active antioxidant, antimicrobial, anticaking agent, etc. contributed extensively to the preservation of foodstuff. Following the market research report, the food market boosts globally by a CAGR of more than 24% during 2019 to 2023.

**Corresponding author*: drrupeshgautam@gmail.com

Tanima Bhattacharya and Shakeel Ahmed (eds.) Nanotechnology in Functional Foods, (293–328)

Keywords: Functional foods, nanotechnology, nanomaterials, nanosensor, smart packaging

12.1 Introduction

Nutritional benefits are achieved through food which plays a significant role in life and medicine. Ingestion of food has marked effects on the severity of diseases (acute and chronic), and metabolism, which results in enlargement in the catabolic activity leads to a higher incidence of morbidity and death. These problems can be easily corrected with the administration of functional foods (FFs). Functional foods are guarded with supplementary elements or nutrients envisioned to produce precise favorable health outcomes elaborated by the European Society of Clinical Nutrition and Metabolism (ESPEN) [1]. "Functional foods defined by US Department of agriculture as natural or processed foods that contain known or unknown biologically active compounds, effective and provide a clinically proven and documented health benefit for the prevention, management or treatment of chronic diseases." Functional foods comprise probiotic yogurt and fortified grain foodstuffs [2–4].

Foods play an immense role in maintaining bodily activities and good health by providing essential supplements like carbohydrates, vitamins, minerals, proteins, amino acids, etc. FFs having the potential for carrying functional ingredients or nutraceuticals that showed capability toward the prevention, management, and treatment of major noncommunicable diseases, such as cancer, diabetes, cardiovascular, dyslipidemia, obesity, etc. [5–8]. Moreover, FFs are equally effective against communicable diseases, including viral infections. The rich sources of FFs are fruits, vegetables, oils, nuts, legumes, etc. comprised of polyphenols, terpenoids, flavonoids, glycosides, alkaloids, unsaturated fatty acids, etc. These components are responsible for producing hypoglycemic, anti-inflammatory, antioxidant, antihyperlipidemic activities [9, 10].

FFs are also showed protection against the COVID-19 virus. Low immune power and lacking nutritional components, such as vitamins A, B6, B12, C, D, E, and elements like Zn, Fe, Se, Mg, and Cu are prone to susceptibility to viral infections. FFs boost the natural defense via antioxidants and possess protective action, which helps to prevent severe respiratory tract infections [11]. Antioxidants are capable of inactivating the attack of free radicals (superoxide, peroxide, hydroxyl radicals) on healthy tissues and thereby maintaining cellular protection, well-being and being recognized for the first-line defense mechanism of the body [12–17]. The FFs along with their potential benefits are depicted in Figure 12.1.

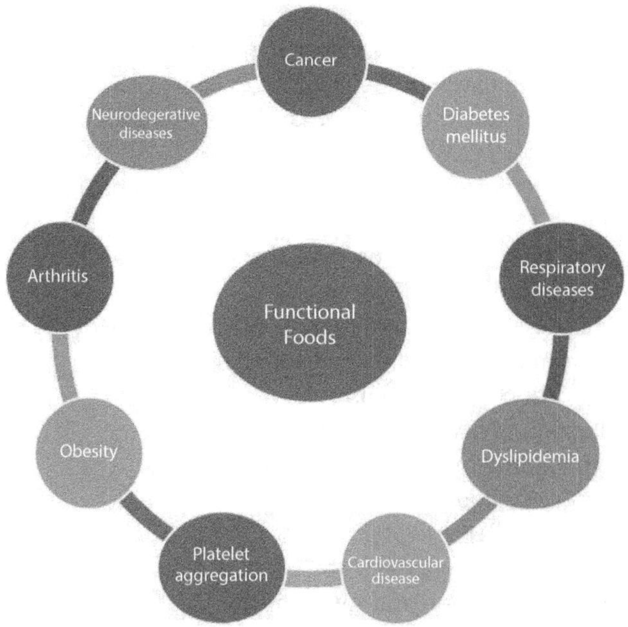

Figure 12.1 Role of functional foods in several diseases.

12.1.1 Role of Nanotechnology

The spoilage and wastage of foods are at prime concern for any food manufacturers. The spoilage of foods is mainly attributed due to physical, chemical, environmental factors and the presence of microorganisms, higher moisture content, oxygen, and carbon dioxide levels [18]. During storage and processing, oxidative degradation of foods occurs and ultimately affects polyunsaturated fats, lipids, carbohydrates, and proteins [19]. The preservation of foods at low temperatures can avoid degradation but affects the properties of foods, such as color, aspects, taste, smell, and rancid [20, 21]. The diseases caused by food infections or poisoning is a great concern, which imparts the health of an individual. Hence, prediagnosis and accurate recognition is prerequisite in the food sector. The detection of food-borne diseases by traditional methods, such as polymerase chain reaction (PCR), enzyme-linked immunosorbent assay (ELISA), lateral flow immunoassay, and various chromatographic techniques are effective but does not meet the requirement fully currently. To overcome several problems encountered during food manufacturing, storage, processing, and up to the hands of customers, there is a need for advanced technology that fulfills these requirements, and the best solution is nanotechnology

[22]. Nanotechnologies have gained the attention of manufacturers in the improvement of FFs. Nanotechnology is the science in which manipulation of matter, particularly in the size ranges from 1 to 100 nm, empowers the construction of constituents with distinct physical, chemical, and biological characteristics [23].

Advantages of nanotechnology in the field of functional foods [22, 24–27]

1. Rapid, sensitive, and cost-effective techniques
2. Unique food processing
3. Enhances solubility and thereby bioavailability of the foods
4. They can detect foodborne pathogens
5. Eradication of pathogens
6. Detection of heavy metals and pesticides
7. High loading capacity on the nanoparticles
8. Improved stability
9. Enhanced shelf-life of product
10. Improved controlled release
11. Reduced excretion and toxic effects
12. Improved antimicrobial and antioxidant properties.

12.1.2 Market Analysis of FFs

The potential of providing high nutritious value, a significant role in growth and development as well as minimizing the chances of several diseases by FFs is gaining the attention of the food industry worldwide. The global FF market size was 161.49 billion (2018), 174.6 billion (2019), and huge demand for FFs omega 3-fatty acids and probiotic, etc. later on due to the COVID-19 pandemic declared by WHO in 2020, the market of FFs declined sharply about 169.5 billion with a CAGR of −2.9% due to economic lockdown across the world for approximately 4 to 6 months. According to a market analysis report, the FFs market sharply rises by 276.68 billion between the periods of 2021 to 2024 by considering the sale from the top 25 players in the food market. The FF market of Asia Pacific predicted growth from 51 billion (2019) to 71 billion (2025) due to increasing demands [28–30].

12.2 Application of Nanotechnology in FFs

They have created a strong impact and need of an era in the field of FFs. Nanotechnology is benefited in several ways in the aspects of manufacturing, packaging, improving characteristics, preservation, and degradation of the products. The immense role of nanotechnology in FFs is described in individual sections.

12.2.1 Processing of FFs

Nanotechnology plays a significant contribution in the development of the processing of foods followed by production with safe packaging. Nanotechnology enables the manufacturers to provide quality foods in their nanoscale range and hence improved efficiency of digestion, absorption, metabolism, and bioavailability can be easily achieved. Improved taste, texture, consistency, masking unpleasant odor are avail by using nanotechnology [31, 32].

Nanotechnology has shown an enormous contribution to the food supply chain by minimizing wastage of food via postharvesting food processing, sensing ability, effective packaging, and tracking system. Manufacturing of FFs at the nanoscale, nanopatterning, nanofluid mechanics, effective monitoring of thermal and material movement, easily achievable by nanotechnology over conventional methods of manufacturing. Nanotechnology is found to be beneficial by providing quality and healthy foods as it maintains the quantity of fats, sugars, and salt at their minimal level thereby preventing several diseases associated with them in higher amounts [32, 33]. Uncontrolled heat transfer is one of the main hurdles in the manufacturing of FFs, which can be solved with nanofluidics (NFs). This technique enabled the manufacturers for maintaining the nutrients in the tomato juice after thermal processing [34].

Another advancement of nanotechnology is the incorporation of nanofilters, which are extremely useful in the clarification of beverages, elimination of tainted flavors, bitterness, and harmful pathogenic microorganisms from the FFs. Eradication of lactose from milk samples for intolerant people and color from the beetroot juices with preserving flavor easily achievable with nanofilters to maximize the safety of FFs. The use of nanocoating on the surface of foods facilitates cleaning, thereby improving safety in the processing of FFs [35, 36]. The incorporation of edible thin nanocoating up to 5 nm in several FFs serves as a barrier for entry of moisture and microorganisms, thereby prevents from food loss. The development of FFs requires a promising delivery system, which is capable of delivering functional materials at their targeted sites with a higher bioavailability and extended period. Nanotechnology complies with all the aspects of such delivery via nanoencapsulation, nanoemulsion, biopolymer matrices, and association colloids [37, 38].

The engineered nanomaterials showed significant contributions in the manufacturing, processing, and packaging of FFs. Nanomaterials containing inorganic oxide derivatives, such as SiO2, MgO, and TiO2, were approved by USFDA for minimizing lump generation, flavoring, and coloring materials [39]. Titanium dioxide is the most widely used in gum, sauces, cake icing, and candy as additive material and in many healthcare and personal use products. Possessing extraordinary properties, like whiteness, opacity, dispersibility, high refractive index, and colorant (E171-U.S. FDA) makes titanium dioxide unique. Several market brands contain titanium dioxide for several reasons at variable concentrations (Dickinson coconut curd, Mentos freshness gum, M&Ms, chocolate candy, Nestle coffee creamer, etc.) [40].

The current era of viral infection (covid-19 pandemic) increases the huge demands of FFs and those of natural origin like phytoconstituents belong to polyphenols. The first step in the processing of such bioactive compounds is their effective and complete extraction from the plants. Hence, in fulfilling huge demands of the world, manufacturers must have to use advanced technologies for faster extraction purposes like supercritical fluid extraction (SFE), pressurized liquid extraction (PLE), microwave-assisted extraction (MAE), enzyme-assisted extraction (EAE), ultrasound-assisted extraction (UAE), high-voltage electric discharge assisted extraction (HVED), etc. [41]. Recently, the use of nonthermal FF processing technology has attracted the manufacture for retaining superiority and prolonging the life of products by inactivation of microbes like High-pressure processing (HPP), cool plasma processing (CPP), ozone, etc. [42]. Emerging thermal technologies are also implemented by several food processing industries, such as ohmic heating, microwave heating, radiofrequency heating, infrared heating, refractance window, etc. [43].

Most of the FFs or nutraceuticals showed poor solubility, absorption, bioavailability, sensitivity to light, moisture, and temperature, degradation in the gastrointestinal tract, and extensive metabolism and instability. These problems can be easily overcome using nanotechnology-based approaches using nanocarriers. The incorporation of nanocarriers resulted in an enormous rise in the solubility of mostly hydrophobic compounds, thereby accelerating its absorption and bioavailability, improved stability for sensitivity and oxidation without affecting nutritional values [44, 45]. These nanocarriers are classified as, lipid-based (solid lipid nanoparticles [SLN], nanostructured lipid carriers [NLC], microemulsion), vesicular (nanoliposome, niosomes, cubosomes), polymer-based (polymeric micelles, polymeric conjugates, nanocapsules), inorganic nanoparticles (gold, silver, iron, titanium nanoparticles, etc.), carbon-based (carbon nanotubes, fullerenes) [37, 46]. Several nanocarriers incorporated in the processing, manufacturing of FFs are depicted in Table 12.1.

Table 12.1 Role of nanocarriers in functional foods [56, 57].

Active ingredient	Nanocarriers	Purpose	Applications	References
Flavonoids	β,γ-cyclodextrin	Processing of food	Marked antioxidant action, controlled release and stable	Zhang *et al.* (2017) [58]
Thymolcarvacrol	Soy lecithin as Nanoliposome	Processing of food	Higher encapsulation efficiency with a significant increase in antioxidant and antimicrobial action for Salmonella and S. aureus.	Engel *et al.* (2017) [59]
Quercetin	Chitosan-soy lecithin	Processing of food	Nanocapsules showed high entrapment efficiency with an enormous rise in antioxidant activity and stability	Hao *et al.* (2017) [60]
Gallic acid	Chitosan, bovine, gelatin, Maltodextrin	Processing of food	Attained 82% encapsulation with nanocarriers combination	Jeyakumari *et al.* (2018) [61]
Anthocyanin	Whey-protein-acacia gum	Processing of food	Possess high encapsulation efficiency, minimizes degradation, higher solubility, stability with antioxidant activity	Oancea *et al.* (2018) [62]
Anthocynidine	Whey-protein concentrate-Arabic gum	Processing of food	Gain higher stability with improved release	Pimental-Moral *et al.* (2018) [63]

(*Continued*)

Table 12.1 Role of nanocarriers in functional foods [56, 57]. (*Continued*)

Active ingredient	Nanocarriers	Purpose	Applications	References
Phenolic acid	Tween 20& 80-lecithin	Processing of food	Emulsion attained the highest encapsulation efficiency with stability	Velderrain-Rodriguez *et al.* (2019) [64]
7,8-dihydroxyflavone	Soy lecithin-bovine lactoferrin	Processing of food	Coated liposomes exhibited higher entrapment efficiency, antioxidant activity, stability, and trans-epithelial transport	Zhang *et al.* (2019) [65]
Curcumin	Soy lecithin-cationic guar gum	Processing of food	Coated liposomes showed an enormous rise in stability and protection against degradation	Aleman *et al.* (2019) [66]
Phenolic acids	Whey-protein	Processing of food	Greatest aqueous solubility and stability with higher antioxidant activity	Soleimanifer *et al.* (2020) [67]
Quercetin	Poly-D, L-lactide nanoparticles	Processing of food	Nanoparticles improved the solubility and stability of quercetin	Esfanjani *et al.* (2016) [68]
Epigallocatechin gallate (EGCG)	Chitosan-β-lactoblobulin	Processing of food	Nanoparticles facilitate the absorption of EGCG inside the intestine	Liang *et al.* (2016) [69]
Fruit extract of ribesnigrum	Silver nanoparticles	Processing of food	The incorporation of silver nanoparticles showed higher encapsulation with bactericidal, fungicidal, and anticancer activity	Dobruca *et al.* (2018) [70]

Nanoencapsulation is one of the promising techniques for the incorporation of a wide variety of bioactive ingredients by offering improved solubility, absorption, bioavailability, stability, and rapid digestion, safety towards oxidation, delivered in a controlled manner, retaining freshness and nutritional value [47]. Nanoencapsulation is the process containing nano-vesicular system in which the active ingredient is attached to the reservoir enclosed by the polymeric membrane or coating. A wide variety of vitamins, antioxidants, proteins, lipid compounds, carbohydrates was enclosed inside the nanoencapsulation. The active material was enclosed inside the hollow space in the form of solid, liquid, or dispersion [36]. Lee *et al.* (2017) developed nanoparticles containing silymarin and reported a marked increase in the solubility and antimicrobial action of silymarin [48, 49].

Many nutraceuticals products containing bioactive shelf-life extended after encapsulation. Another classic example is probiotic which was delivered in the gut without destruction [50]. Nanoemulsion has gained the attention of researchers being a thermodynamically stable system composed of oil and water stabilized with surfactant. This system was capable of enhancing the oral absorption and bioavailability of most of the bioactive compounds and easily scale-up [51–53]. Recently, many researchers have incorporated bioactive compounds into polymeric nanoparticles as they offer several advantages like protection from degradation, improved absorption, higher bioavailability, and sustained action, safe, biodegradable, biocompatible, and high drug loading capacity [54, 55].

12.2.2 Food Packaging

One of the most important aspects of ensuring food safety is food packaging. Packaging is extremely important in the food supply chain. The primary purpose of packaging is to serve as a container for food, allowing for effective transportation across the food supply chain, while also avoiding physical harm and protecting in contrast to manipulation and theft [71]. Food packaging is commonly used to maintain the quality of food and prolong its shelf life. A well-functioning food packaging provides the disinfected and nutritious food to users and serves as a vital connection among producers and customers, whereas food packaging having poor-quality causes a loss of food quality and consumer trust, in addition to the wastage of food and foodborne illnesses [21]. In industries, packaging of food is divided into three types including, (a) primary packaging, in which products comes into direct contact with content; (b) secondary packaging, which refers to the primary packaging's outer packaging and (c) tertiary packaging, which states to the delivery of warehouse storage/transport [72].

Recently, the packaging industry has concentrated on specific nanostructures and nanomaterials, such as nano-clay particulates, carbon nanotubes, and graphene nanosheets due to their ease of accessibility and processability, low cost and high performance. It is now possible to boost food packaging efficiency by the addition of more stable antimicrobial or antioxidant ingredients, enhancing the mechanical properties, and incorporating intelligent functions by using nanotechnology techniques [20]. Nanodelivery systems for bioactive have several advantages over conventional delivery systems including, enhanced active compound protection during processing and storage, as well as mitigation of hydrolysis and oxidation. They can be used to improve preservation, taste, flavor, and color delivery, as well as to create new sensory properties, extend shelf-life, improved handling, and lower operating costs [73]. Emulsification, solvent evaporation, coacervation, precipitation, electrospraying, and spray drying are only a few of the well-known techniques that can be used to create nano-delivery systems for food applications such as nanocapsules, nanospheres, and nanoemulsions [74]. Packaging may be labelled as "Improved", "Active" and "Intelligent" packaging, depending upon the form of packaging [20]. The relationship between smart, active, and intelligent packaging is described in Figure 12.2.

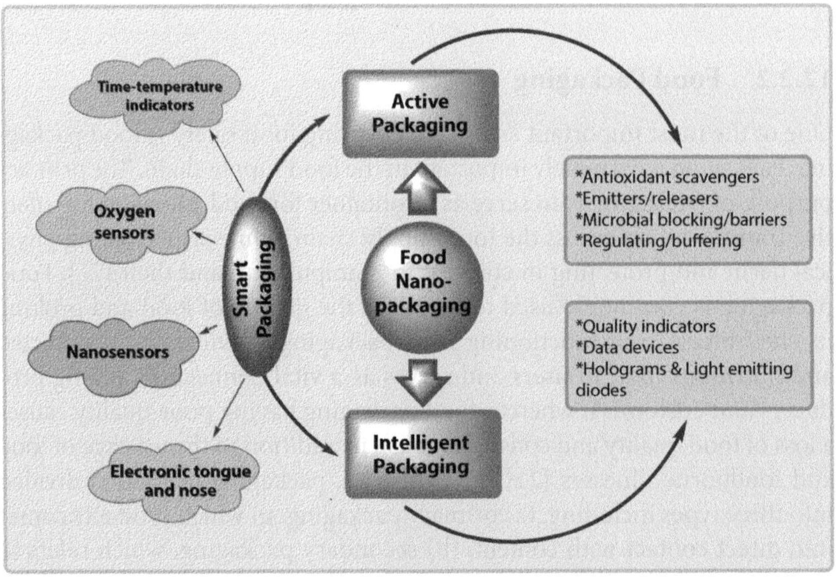

Figure 12.2 Relationship between smart, active and intelligent food nanopackaging.

12.2.2.1 Active Nanopackaging

Active packaging is a novel way to preserve or increase the shelf-life of food products, even though maintaining their consistency, protection, and dignity. According to the European regulation, (EC) No 450/2009, 'active packaging' is defined as a packaging system that integrates with food in such a way that "they intentionally insert the elements that would liberate or consume the substances into or from the processed food or the surrounding environment" [75].

Active packaging is a revolutionary term that is described as a form of packaging that alters the conditions of packaging to prolong the shelf-life and upgrade its physical or protective properties although preserving the product quality to promote food processing and consumption [76]. There are two types of active packaging methods: (1) active scavenging methods (i.e., absorbers) and, (2) active releasing methods (i.e., emitters). One of the most popular active packaging technologies is the use of oxygen scavengers, which have the goal of eliminating any residual oxygen from the package of food or enhancing the barrier properties by acting as an active barrier. When antimicrobial agents, O_2, preservatives, ethylene removers, water vapor absorbers, and other additives are combined with a polymer, it becomes more efficient in shelf-life extension and improving the consistency of food items [71].

12.2.2.2 Smart and Intelligent Nanopackaging

Smart packaging is a new food packaging system that combines active and efficient technologies [77]. It mentions the intelligent functions offered by nanomaterials like nanodevices and nanosensors in detecting the freshness and monitoring the fluctuations in packaging integrity [78].

They are well defined as smaller, low-cost labels and tags which are applied to the primary packaging (like bottles, pouches, and trays) and secondary packaging (like shipping containers), to promote the contact through the food supply chain and allow suitable steps to be engaged to accomplish the desired assistances in food quality and safety. Smart packaging devices are categorized into two types: (1) data carriers (i.e., radio frequency identification [RFID] tags and barcode labels) for storing and transmitting the data and, (2) package indicators (i.e., biosensors, time-temperature, and gas indicators) for monitoring the external environment and issuing alerts, when necessary [79].

Intelligent packaging is intended to track the conditions of packaged food or the atmosphere in which it is processed. This packaging technique is accomplished with sensing, detecting, tracing, recording, or collaborating the details regarding the quality of food and condition of the product, to help in the execution of shelf-life extension, to improve protection, provide information, and inform about the potential hazards during the whole food supply chain [80].

Applications of intelligent packaging:
Enhancing Safety and Biosecurity of Food: Intelligent food packaging is beneficial for tracing the goods and checking their circumstances, encouraging instantaneous data access, sharing and empowering the quick reply, and making suitable conclusions, particularly when combined with science-based principles. These assets are crucial for the safety of food packaging or the biosecurity approach.

Enhancing Quality and Convenience of Food: The latest implementation of intelligent food packaging has been largely concentrated on TTI labels usage to control temperature, even though many other applications are possible. It may be used in cooking appliances, such as an intelligent microwave oven device [79]. Several nanoparticles used in packaging of functional foods are depicted in Table 12.2.

12.2.2.3 Nano Barcodes for Product Authenticity

Around the world, barcodes are commonly used as product identification marks. Two-dimensional barcodes are commonly used because they are easy to locate and are more vulnerable to modification, injury, and falsification. Nanoparticle-based transparent barcodes, on the other hand, are difficult to manipulate. Nanodisk codes are advanced nanoparticle-based encoding method has recently been published. The nanodisk code is made up of two pairs of Raman scattering disks with surface enhancements that can be scanned with a Raman microscope [93]. For the finding of pathogens in food and biological samples and product labeling, invisible nanobarcodes, fluorescent DNA dendrimer nanobarcodes, and fluorescent-based barcode nanorods have been reported. Henceforth, nanotechnology may be used to build effective nanobarcode systems to certify the quality and safety of food in the future [31].

Over the years many principles have been established into widely agreed languages, known as "symbologies" to allow the barcodes to communicate with printers and scanners. UPC (Universal Product Code) barcode is an example of linear symbology that was first adopted in the 1970s and has since been widely used in supermarkets to assist with inventory management, stock reordering, and checkout [79].

Table 12.2 Nanoparticles used in packaging of FFs.

S. no.	Nanomaterials	Type	Applications	References
1.	TiO_2	Starch	Enhanced UV-protection, hydrophobicity, thermal properties, and drop in water vapor permeability	Goudarzi *et al.* (2017) [81]
2.	ZnO NPs	Chitosan	Enhanced mechanical properties (i.e., film thickness, tensile strength, barrier properties, and transparency)	Sanuja *et al.* (2015) [82]
3.	AgNPs and CNC	Cellulose	Amended mechanical properties, antioxidant and antibacterial	Wu *et al.* (2019) [83]
4.	ZnO NPs	Soybean protein	Intensification in microbial activity by 16% and increase tensile strength by 234%	Tang *et al.* (2019) [84]
5.	Nano ZnO	Chitosan coated on PE film	Improved antimicrobial activity and greater solubility	Al-Naamani *et al.* (2016) [85]
6.	Nanoclay	Polylactide (PLA)	Significant improvement of the oxygen barrier	Svagan *et al.* (2012) [86]
7.	CNC	Chitosan	Enhanced mechanical properties (like tensile strength and barrier properties)	Khan *et al.* (2012) [87]

(Continued)

Table 12.2 Nanoparticles used in packaging of FFs. (*Continued*)

S. no.	Nanomaterials	Type	Applications	References
8.	Bacterial based CNC	Gelatin	Improved mechanical properties, reduction in water affinity, edible and biodegradable	George *et al.* (2012) [88]
9.	TiO_2	Amylose starch/PVA	Enhanced mechanical properties and water resistance, antibacterial activity	Liu *et al.* (2015) [89]
10.	MgO	Chitosan	Improved UV protection, mechanical properties, flame retardant properties, thermal stability, and moisture barrier properties	Silva *et al.* (2017) [90]
11.	Palladium nanoparticles (PdNPs)	Poly(3-hydroxybutyrate) (PHB)	Oxygen scavenging	Cherpinski *et al.* (2018) [91]
12.	TiO_2 NPs	Polyacrylonitrile (PAN)	Ethylene scavenging, photocatalytic ethylene scavenging	Zhu *et al.* (2019) [92]

Reduced Space Symbology (RSS) is a new barcode symbology that was recently announced to meet the increasing demand for encrypting more data in a small space. The RSS-14 Stacked Omni-directional barcode encodes the entire 14-digit Global Trade Item Number (GTIN) and can be used for products like apples and oranges that need a narrow symbol due to space constraints. The RSS Expanded Barcode encrypts up to 74 alphanumeric characters and can be employed for mutable measure items (such as seafood and meat) where more data capacity is mandatory to encrypt the additional information including, batch/lot number, packed date, and package weight [79].

Individual packet labeling has been developed using fluorescent poly(p-phenylenevinylene)-based barcode nanorods [94]. Fluorescent DNA dendrimer nanobarcodes have been identified for the identification of anthrax, Ebola, *E. coli,* and extreme acute respiratory syndrome pathogens in food and biological samples [95]. In hospitals, barcodes may be used to ensure that the correct prescription is delivered to the correct patient at the correct time and dose. To ensure accuracy, the drug barcode and patient identification tags are scanned right before administration, and the information is sent over a wireless local area network (LAN) [79].

12.2.3 Preservation of Food

As most known pathogens are controlled and newer ones are created, wide-spectrum of food-borne infections fluctuates continuously over time. For *in situ* investigation of pollutants, detection and identification of pathogenic bacteria and monitoring of postprocessing food quality parameters, precise, rapid, and online sensing is critical. A biosensor is a small analytical device that helps in recording, detecting, and transmitting the data affecting biochemical responses. Some vital features of biosensors are their sensitivity, specificity, simplicity, reliability, and portability [79]. Food contaminating toxins, pathogenic bacteria, adulterants, pesticides, fertilizers, dyes, vitamins, taste, and smell may all be identified by nanosensors. Temperature-time indicators and oxygen indicators may be employed for observing the food freshness [93].

12.2.3.1 Nanosensor as Emerging Devices in the Food Industry

Nanosensors are critical in the food industry because they can detect and measure lower concentrations of mycotoxins, bacteria, organic compounds, and other chemicals. Controlling these microbes is difficult, but nanosensors can quickly detect toxins and pathogens during food processing and storage. [37]. Another possible application of nanotechnology is the development of nanosensors for the detection of food pollutants and pathogens in food systems. Nanosensors embedded in food packaging materials can aid in the detection of any physical, chemical, or biological changes that occur during the processing of food [96].

A bioreceptor that distinguishes a transducer and target analyte that transforms the biochemical indications into computable electrical responses are the two major components of this smart system [79]. Nanosensors can be developed using a diversity of methods, comprising molecular self-assembly, top-down lithography, and bottom-up assembly. Based on their applications in food analysis, nanosensors are categorized into nanoparticle-based nanosensors, optical nanosensors, and electrochemical nanosensors [97].

Researchers from all over the world have built several nanosensors to analyze the inclusive quality of food. The following are the examples of nanosensors: 1) By using conducting polymer nanoparticles Canadian Wheat Board Centre for Grain Storage Research, University of Manitoba, Canada detects the source and form of spoilage caused by bacteria, viruses, insects, rodents, and fungi. 2) The electronic tongue was developed to detect the freshness of food by detecting the gases emitted by spoiled food [98]. Nanosensor packaging is useful for tracing the internal and external conditions of food products, containers, and pellets in the food supply chain. Nanosensors in plastic packaging can detect gases emitted by food when it spoils and packaging changes its color to notify consumers. Additionally, the film comprising silicate nanoparticles can reduce the flow of oxygen into the package to prevent the inflow of moisture to maintain the freshness of food and to prevent the growth of molds within the refrigerator [99].

12.2.3.1.1 Detection of Toxins

Microbes that consume nutrients from food for their development produce food toxins as defense tools. Food products having high nutritional value are most probably susceptible to microorganism contamination. Microbes contaminate the food and produce the toxins in self-defense, causing a variety of human diseases. Food adulteration has a detrimental influence on human health. The toxins and adulterants can persuade to diarrhea, vomiting, paralysis, and even death. Being a prime and essential prerequisite of human beings, food protection is a major universal concern. The health issue has been exacerbated when certain chemicals or contaminants are used during battles. Therefore, there is a crucial need to produce rapid and responsive methods for the detection of microbes, harmful chemicals, and toxins. As a result, nanoparticles have been investigated as a means of detecting toxic chemicals and microbes with greater sensitivity [93].

Bacterial toxins can persuade cytotoxicity via alteration of biological activity and plasma membrane integrity. E.g., *E. coli* O157:H7 produces the Shiga toxin which inhibits the synthesis of protein and causes cell death and destructiveness, while *Listeria monocytogenes* produce listeriolysin O which create pores that lead to the phospholipid bilayer disruption and subsequent cell lysis. Furthermore, the toxins will persist in the food, atmosphere, and clinical samples even after the pathogens die. Toxicology screening should be performed as soon as possible to prevent intoxication [100].

Toxin detection methods currently available including Western blots, ELISA, antibody microarrays, antibody-coated polystyrene microbeads, and SPR (Surface Plasmon Resonance) biosensors, all are sensitive and multiplexable. These processes, however, are stagnant and labor-intensive, and they often necessitate homogeneous or filtered samples. Several nanomaterials like gold nanoparticles (GNPs), magnetic nanoparticles (MNPs), carbon nanotubes, gold nanorods, silver nanoparticles (SNPs), and quantum dots (QDs) are most widely used for detecting contaminants to increase the shelf life of foods products and boost the strategies of food packaging [100].

The coupling of quantum dots (QDs) and immunomagnetic separation (IMS) is used in the multiplex recognition of *Salmonella enteritidis* and *E. coli*. A TiO_2 nanowire package has been established for *Listeria monocytogenes* detection. Monoclonal antibodies are immobilized on the surface of TiO_2 nanowire to specifically capture *L. monocytogenes* [22]. For aflatoxin B1 identification, gold nanoparticles functionalized with the anti-aflatoxin antibodies are used. Likewise, aflatoxin M1 has been observed in milk samples using superparamagnetic beads encompassing anti-aflatoxin M1 antibodies and gold nanoprobes. Electrochemiluminescent sensors based on carbon nanotubes are developed for the oversensitive recognition of palytoxin in mussel meat [31, 101].

12.2.3.1.2 Detection of Food Pathogens

The detection of pathogenic bacteria in food items is mainly accomplished by the identification of bacterial genetic material or whole bacterial cells. Deoxyribonucleic acid (DNA) isolation and microorganism detection using nanoparticles takes less time and is more sensitive than traditional methods. Magnetic iron oxide nanoparticles are employed for the isolation of DNA of *Listeria monocytogenes*, a milk pathogenic bacterium, which is enumerated by using the polymerase chain reaction [93]. Polymerase chain reaction (PCR) is a useful method for detecting pathogens. Microarrays can be used in conjugation with PCR to act as a series of parallel dot-blots to aid in product detection and identification. For PCR-based microbial detection, 16s ribosomal ribonucleic acid is widely used as the discriminating marker. The PCR-based assay is expensive and the microarray method based on 16s ribosomal ribonucleic acid lacks the sensitivity. Therefore, the detection of 16s ribosomal ribonucleic acid using nanoparticles is simple, fast, and extremely sensitive [102].

One of the most effective nano biosensors, the nano bioluminescent spray provides a visual luminosity for easy identification of the pathogenic strains in various food items. This spray contains a variety of magnetic nanoparticles that counteract pathogenic bacteria in food to create a visible color that can be straightforwardly recognized. An array-based immunosorbent assay combined with liposomal nanovesicles has received significant attention in recent years due to its efficacy and specificity in detecting *L. monocytogenes*, *E. coli*, and *Salmonella* spp. Furthermore, pathogen detection in various mixed liquid food systems has also been accomplished by combining silicon-based nanosensors with proteins that vibrate at different frequencies, depending on their biomass [31].

The main advantage of microfluidic sensors is their miniature format and ability to detect compounds of interest quickly in microliters of necessary sample volume according to a nanosensor based on microfluidics and liposomes. Food spoilage, cereal grain samples, and the presence of *Ganoderma boninense* on palm oil tree trunks all are detected by using electronic noses [103].

12.2.3.1.3 Sensing Chemicals and Pesticides in Food

Nanoparticles have been also employed for detecting pesticides, fertilizers, and other harmful chemicals. Pesticides are commonly used in the agriculture field for the safeguard of crops and improvement in food production, but their noxiousness poses a threat to human health. Pesticides comprise a wide range of chemicals, such as insecticides, herbicides, fungicides, and plant growth regulators. Upon application, these residual pesticides persist in plants, soil, and water, resulting in long-term acquaintance and significant human health hazards [104].

Organophosphates (OPs) are the most commonly used group of pesticides, due to their high effectiveness for insect eradication and shorter lifetime in the environment under natural humidity and temperature conditions. As a result, there is a need for methods capable of directly detecting OPs at lower concentrations, such as those found in foods and drinking water. Pesticide detection is typically achieved by using high-accuracy analytical methods like LC, GC, MS, and others. These techniques are costly, time-consuming, and necessitate highly skilled technicians, and they are often unsuitable for field measurements [105].

Nanotechnology has great promises for solving many of the issues that currently occur with pesticides. Encapsulating pesticides inside nanoparticles, whose dimensions, structures, morphology, and surface properties can be tailored for specific applications, can increase their specificity and potency. For example, copper-based nanoparticles have been exposed to be more effective than standard copper-based pesticides for the protection of watermelon from root fungal infections [106]. The detection of organophosphorus pesticides in fruits and vegetables has been recorded by using an optical sensor that combines CdTe semiconductor quantum dots (QDs) with AChE via layer-by-layer assembly. Carbon nanotubes have been used to determine pesticides with ultrahigh sensitivity because of their higher electrocatalytic activity. Organophosphorus and carbamate pesticides have been detected by using gold nanoparticles as colorimetric and fluorometric sensors [107].

Organophosphate pesticides (e.g., chlorpyrifos, methyl parathion, and fenthion) are detected by the usage of CuO nanoparticles grown *in situ* over indium tin oxide (ITO) in a nonenzymatic process. The principle of detection is based upon the pralidoxime chloride (PAM) inhibition, coated over the pimelic acid-functionalized CuO nanostructures on the ITO surface [104]. Voltammetric biosensors are developed using the selective binding of the phosphate group of pesticide parathion to a zirconium dioxide/gold nanocomposite film electrode [108]. The detection of toxic cadmium ions has been confirmed by using MWCNT-silica nanocomposite-based potentiometer sensors [109]. The pesticide paraoxon has been detected by using cadmium selenide-zinc sulfide core-shell quantum dots [110].

12.2.3.1.4 Sensing the Quality of Key Food Ingredients

Vitamins and other important nutritional ingredients, such as antioxidant compounds, are needed for the normal functioning of various metabolic pathways in our bodies and their absence can lead to serious health problems such as cardiovascular disease, cancer, and anemia. In various processed foods, these main food ingredients are easily degraded. Numerous types of nanosensors have been stated to be conjugated with various nanomaterials to avoid degradation. MWCNT and SWCNT-ionic liquid nanocomposites are used for folic acid detection in fruit juices, wheat flour, and milk samples [93]. Similarly, MWCNT modified with N-(3,4-dihydroxyphenethyl)-3,5-dinitrobenzamide has been used for detecting the acetaminophen, ascorbic acid, and tryptophan, an essential amino acid [111].

Ascorbic acid (AA) is naturally found in most fresh fruits and their juices, and it is often combined with juices or soft drinks during production. It is important to the human diet because it aids in the creation of teeth, bone, connective tissue, and blood vessel walls, as well as the absorption of iron and amino acids. Ionic liquid nanocomposites based upon nickel oxide nanoparticles and carbon nanotubes have been employed for the recognition of vitamins and ascorbic acid [112]. Phenols and substituted phenols are the chief impurities in food, medical, and environmental matrices. Because of their higher toxicity, reliable analytical methods for evaluating sensitivity at lower levels in various matrices are needed. By immobilizing tyrosinase on colloidal gold modified with carbon paste electrode, an advanced renewable tyrosinase-based biosensor is used to detect phenol [113].

Food quality is determined by the levels of succinic acid, citric acid, L-malic acid, L-glutamic acid, D-sorbitol, sucrose, fructose, glucose, and hydrogen peroxide in stored food items. To monitor the quality of the abovementioned food constituents in various food products, several types of nanosensors in combination with iron, gold, silver, zirconium dioxide, tin dioxide, chitosan, nickel-platinum, prussian blue-gold and cuprous oxide nanoparticles have been used [31]. Thus, biosensors can be used as worthwhile tools for quality control, process regulator, and food safety determination [114]. Several nanoparticles used in preservation of functional foods are depicted in Table 12.3.

12.2.4 Nanomaterials in Improving the Texture, Appearance, Taste and Nutritional Values of Food

The aim of the research, in this case, is to boost the mechanical and physical properties of conventional resources used in the food packaging industry. This objective has been accomplished by formulating the improved nanocomposites via adding the reinforcing NPs (nanofillers) to polymers. The main function of nanomaterials is to provide the traditional polymers further sturdiness, abrasion resistance, and act as an efficient barrier in contradiction of light, water, moisture, and environmental gas such as oxygen [21].

Table 12.3 Nanoparticles used in preservation of FFs.

S. no.	Nanomaterial-based devices/ nanomaterials	Investigated food items/ matrix	Applications	References
1.	AgNPs/GNRs-based electrochemical sensors	Fruits and vegetables	Detection of methyl parathion	Govindasamy *et al.* (2017) [115]
2.	$CaCO_3$ NPs based Xanthine amperometric sensors	Fish	To detect the freshness of food	Shan *et al.* (2009) [116]
3.	Carbon nanotubes-based sensors	Meat	Detection of pathogenic bacteria in food	Yang *et al.* (2008) [117]
4.	AuNPs-based glassy electrode	Canned tuna	Detection of the food freshness	Cubukcu *et al.* (2007) [118]
5.	Cystamine-PAMAM modified gold electrochemical sensor	Peanuts extract peanuts-corn snacks	To detect and quantify ochratoxin A, aflatoxin B1, and aflatoxin B2	Castillo *et al.* (2015) [119]
6.	PAMAM-modified gold voltammetric immunosensor	Corn-containing the poultry bovine feeds corn flakes	Detection of the atrazine	Giannetto *et al.* (2014) [120]
7.	TiO_2/Pd nanostructure	Potato, onion, and cabbage	Detection of residual pesticides	Naser-Sadrabadi *et al.* (2020) [121]

(*Continued*)

Table 12.3 Nanoparticles used in preservation of FFs. (*Continued*)

S. no.	Nanomaterial-based devices/ nanomaterials	Investigated food items/ matrix	Applications	References
8.	Pd/Au nanocrystals	Chili sauce	Detection of adulterants	Zou *et al.* (2020) [122]
9.	Cu nanoparticles-based inks	Soft drinks	Detection of glucose, sucrose, and toxins	Pradela-Filho *et al.* (2020) [123]
10.	Au nanoparticles electro-deposited on graphene ribbons	Edible oils in the bakery industry	Detection of tert-butylhydroquinine	Delfino *et al.* (2020) [124]
11.	Au Nanoparticles	Skimmed milk	Detection of pathogens (Salmonella species)	Echegoyen *et al.* (2016) [125], Nguyen *et al.* (2020) [126]
12.	CeO_2 nanoparticles	Milk	Detection of mycotoxins	Goud *et al.* (2020) [127]
13.	Fe_3O_4 NPs	Milk	Detection of *S. enterica*	Wang *et al.* (2015) [128]
14.	Silica-coated Fe_3O_4 NPs	Milk powder, cheese	Detection of *C. sakazakii*	Zhao *et al.* (2013) [129]
15.	TiO_2 nanowire	Pure culture	Detection of *monocytogenes*	Wang *et al.* (2008) [130]

During food processing, nanoparticles have been used to enhance the nutritional qualities, improving color, taste and flavor, consistency, stability, flow properties, or extension of shelf-life. Without the use of surfactants or emulsifiers, nano-sized carrier structures or nanocapsules such as liposomes, protein-based carriers, and micelles have been used as nano-food additives, dietary supplements, masking the unpleasant taste, improving bioavailability, and allowing for better dispersion of insoluble additives. They exhibit excellent absorption of bioactive materials. These nanostructures are also used as anticaking agents, nano additives, and nutraceuticals delivery systems [131].

Electric tongue: A nano bioelectronic tongue sensor for bitter taste detection has been created by combining a carboxylated polypyrrole nanotube field-effect transistor with the human taste receptor protein, hTAS2R38. Remarkably, the nanotongue was able to detect the bitterness compounds such as propylthiouracil and phenylthiocarbamide with human-like efficiency [132]. This nanovesicle-based bioelectronic tongue act as an influential tool in detecting the sweeteners as an alternate to worthwhile and labor-intensive cell-based assays and sensory evaluation panels employed in the field of food and beverage industries [133]. A bioelectronic tongue based on floating electrodes has been developed for the detection of umami substances. In this analysis, nanovesicles comprising honeybee umami taste receptor, gustatory receptor 10 of Apismellifera, are hybridized with carbon nanotube field-effect transistors with floating electrodes (AmGr10). It may be used to detect the L-monosodium glutamate (MSG) in liquid foods as well [134].

12.2.5 Nanomaterials Improving Functional Property: Antioxidant, Antimicrobial Properties

Microbial contamination leads to pathogenic infections and deprived nutrition associated with the deterring foods. As a result, dealing with bacterial degradation is one of the most important topics in food production, manufacturing, shipping, and storage. Innovative nanoantimicrobials have been shown auspicious effects in preventing food degradation and thereby result in the extension of shelf-life. Antimicrobial activity has been proposed for a variety of metal and metal oxide nanomaterials. They produce an excessive amount of reactive oxygen species (ROS) as a result of their intrinsic physicochemical properties, resulting in oxidative stress and cell death [135].

12.2.5.1 Antimicrobial Packaging

Antimicrobial packaging is a recent concept in which an antimicrobial agent is incorporated into a polymer film to prevent the activities of specific microorganisms. Antimicrobial packaging results in the extension of the lag phase and reduction in the microorganism's growth. The primary goal of antimicrobial agents in food packaging is to ensure food protection, extend shelf-life, and maintain consistency [136]. Antimicrobial food packaging comprises of several forms such as the incorporation of sachets/pads containing volatile antimicrobials into the packages, the addition of volatile and nonvolatile antimicrobials directly into the polymers, adsorption or coating of antimicrobial agents on polymer surfaces, immobilization of antimicrobials to polymers via ionic or covalent linkages and usage of polymers that are inherently antimicrobial [137]. Antimicrobial agents, such as carbon dioxide, ethanol, chlorine dioxide, silver ions, bacteriocins, antibiotics, essential oils, and organic acids, have been established for the persistence of constraining the microorganism's growth in food packaging. Among them, peptides, plant extracts (such as rosemary extracts), and nisin have been widely used as antimicrobial agents [138].

12.2.5.2 Antioxidant Packaging: (also termed Oxygen-Scavenging Packaging)

The existence of oxygen triggers or accelerates the oxidative reactions, resulting in food degradation. The growth of aerobic microbes and molds is aided by oxygen. Oxidative reactions cause undesirable characteristics including, color changes, off-flavors, off-odors, and poor nutritious reliability. The usage of oxygen scavengers retards the oxidative reactions by eliminating the oxygen (residual and/or entering) from food packaging. The most popular scavenger is ferrous oxide. Certain nylons, sulfites, catechol, unsaturated hydrocarbons, photosensitive dyes, ligands, and enzymes like glucose oxidase are among the others [139].

Antioxidant active packaging can be used by controlling the amount of oxygen to which the product is exposed to improve the quality of food products and extend shelf-life. Antioxidant packaging schemes are divided into two categories: 1) independent antioxidant devices (including, pads, sachets, or labels containing the oxygen scavengers), and 2) antioxidant packaging materials. The most commonly used natural antioxidants are essential oils, tocopherols, and extracts of a plant from several herbs like oregano, tea, and rosemary [138].

12.6 Nanomaterials Accelerating Shelf-Life

Over recent decades, consumer's concern about the quality and safety of food items and additive contents has received a lot of attention. Fresh, high-quality foods that are unprocessed or minimally processed, do not contain preservatives, and have a long shelf-life are becoming increasingly popular. A food's shelf life is inextricably related to its packaging; both product conditions and the package should be considered. Active packaging is a new term that describes a packaging mode in which the package, product, and environment interact to extend the shelf-life, improve protection, or increase sensory properties while maintaining the product quality [76].

Edible nanomaterial coatings have also shown an increasing potential for food, fruits, and vegetables storage. These coatings are beneficial during transportation from the manufacturer to the store, as they keep the nutritional qualities intact without causing physical harm. Chitosan and chitosan-based additives provide a promising option for the extension of the shelf-life of the food products without negotiating their sensory qualities. Nanoclays and nanolaminates have also shown promising results in terms of improving gas barrier properties for efficient food packaging. Biofilms with enhanced tensile strength, thermal, mechanical, and antimicrobial properties have been discovered. As a result, when it comes to nanotechnology in food packaging, nanomaterials have become an important component [140].

12.7 Conclusion

Presently, the FFs are the fastest growing and demanding sector worldwide for delivering the nutrients and vital supplements needed for maintaining bodily activities. Currently, the life of people fetching busiest because of key tasks, work traditions, hard competition, and achieving the targets of life and all resulted in ignorance of dietary nutrients for the body which is prone to suffering from illness and diseases. FFs do not only provide nutritional supplements but also prevents from the attack of several diseases and infectious agents via antioxidant, prebiotics, vitamins, and minerals, etc. Increasing use of heavy pesticides and other agents are also responsible for diminishing the nutritive value of many foods which can be easily maintained with FFs. Nanotechnology plays a huge contribution in the field of development of FFs like processing, packaging, preservation, detection, improving the quality, texture, and retaining the taste and nutrients. Nanoparticles showed significant contributions in the manufacturing, packaging, processing, sensing, and extending of the shelf-life of the FFs. It was concluded that nanotechnology via nanomaterials or nanoparticles benefited several food industries in maintaining quality, protection, and safety through its sensing ability. Several nanoparticles (gold, silver, silica, magnetic, etc.) are recognized for functioning the role of biosensors; detects the availability of pathogens because of food contents, also in the detection of excessive concentration of heavy metals in the food. Moreover, these NPs also worked as antimicrobial, antioxidant, and extensively in regulating the quality of food and its preservation. Hence, it was concluded that nanotechnology has enormously contributed in beginning up to the consumptions of the FFs.

References

1. Cederholm, T., Barazzoni, R., Austin, P., Ballmer, P., Biolo, G., Bischoff, S.C., Compher, C., Correia, I., Higashiguchi, T., Holst, M., Jensen, G.L., Malone, A., Muscaritoli, M., Nyulasi, I., Pirlich, M., Rothenberg, E., Schindler, K., Schneider, S.M., de van der Schueren, M.A.E., Sieber, C., Valentini, L., Yu, J.C., Van Gossum, A., Singer, P., ESPEN guidelines on definitions and terminology of clinical nutrition. *Clin. Nutr.*, 36, 49–64, 2017.
2. Martirosyan, D.M. and Singh, J.A., New definition of functional food by FFC: What makes a new definition unique. *Funct. Foods Health Dis.*, 5, 209–223, 2015.
3. US Food & Drug Administration, *What is a dietary supplement*, 2015, (Website Link USFDA). http://www.fda.gov/AboutFDA/Transparency/Basics/ucm195635.htm.
4. Hunter, P.M. and Hegele, R.A., Functional foods and dietary supplements for the management of dyslipidaemia. *Nat. Rev. Endocrinol.*, 13, 278–288, 2017.
5. Alkhatib, A., Tsang, C., Tiss, A., Bahorun, T., Arefanian, H., Barake, R., Khadir, A., Tuomilehto, J., Functional foods and lifestyle approaches for diabetes prevention and management. *Nutrients*, 9, 1310, 2017.
6. Alkhatib, A., Tsang, C., Tuomilehto, J., Olive oil nutraceuticals in the prevention and management of diabetes: From molecules to lifestyle. *Int. J. Mol. Sci.*, 19, 2024, 2018.
7. Alkhatib, A., Antiviral functional foods and exercise lifestyle prevention of coronavirus. *Nutrients*, 12, 2633, 2020.
8. Granato, D., Barba, F.J., Bursać Kovačević, D., Lorenzo, J.M., Cruz, A.G., Putnik, P., Functional foods: Product development, technological trends, efficacy testing, and safety. *Annu. Rev. Food Sci. Technol.*, 11, 93–118, 2020.

9. Alkhatib, A. and Tuomilehto, J., Lifestyle diabetes prevention, in: *Encyclopedia of Endocrine Diseases*, pp. 148–159, Elsevier, Netherland, 2018.
10. Acquah, C., Agyei, D., Obeng, E.M., Pan, S., Tan, K.X., Danquah, M.K., Aptamers: An emerging class of bioaffinity ligands in bioactive peptide applications. *Crit. Rev. Food Sci. Nutr.*, 60, 1195–1206, 2020.
11. Calder, P., Carr, A., Gombart, A., Eggersdorfer, M., Optimal nutritional status for a well-functioning immune system is an important factor to protect against viral infections. *Nutrients*, 12, 1181, 2020.
12. Kumar, S., Narwal, S., Kumar, V., Prakash, O., α-glucosidase inhibitors from plants: A natural approach to treat diabetes. *Pharmacogn. Rev.*, 5, 19–29, 2011.
13. Adefegha, A.S., Phytochemistry and mode of action of some tropical spices in the management of type-2 diabetes and hypertension diabetes management view project reproductive toxicology view project. Artic. *Afr. J. Pharm. Pharmacol.*, 7, 7, 332–346, 2013.
14. Sisein, E.A., Biochemistry of free radicals and antioxidants. *Scholars Acad. J. Biosci.*, 2, 2, 110–118, 2013.
15. Nagarajappa, A.K., Pandya, D., Ravi, K.S., Role of free radicals and common antioxidants in oral health, an update. *Br. J. Med. Med. Res.*, 9, 1–12, 2015.
16. Shukla, G., Yaparthy, N., Rao, D.S., Kumar, C.J.S., Astashine capsules: World's most powerful antioxidant on Earth. *J. Pharm. Care Health Syst.*, 4, 10, 1077–1088, 2016.
17. Adefegha, S.A., Functional foods and nutraceuticals as dietary intervention in chronic diseases; Novel perspectives for health promotion and disease prevention. *J. Diet. Suppl.*, 15, 977–1009, 2018.
18. Rather, I.A., Koh, W.Y., Paek, W.K., Lim, J., The sources of chemical contaminants in food and their health implications. *Front. Pharmacol.*, 8, 830, 2017.
19. Domínguez, R., Pateiro, M., Gagaoua, M., Barba, F.J., Zhang, W., Lorenzo, J.M., A comprehensive review on lipid oxidation in meat and meat products. *Antioxidants (Basel)*, 8, 10, 429, 2019.
20. Bajpai, V.K., Kamle, M., Shukla, S., Mahato, D.K., Chandra, P., Hwang, S.K., Kumar, P., Huh, Y.S., Han, Y.K., Prospects of using nanotechnology for food preservation, safety, and security. *J. Food Drug Anal.*, 26, 1201–1214, 2018.
21. Alfei, S., Marengo, B., Zuccari, G., Nanotechnology application in food packaging: A plethora of opportunities versus pending risks assessment and public concerns. *Food Res. Int.*, 137, 109664, 2020.
22. Krishna, V.D., Wu, K., Su, D., Cheeran, M.C.J., Wang, J.P., Perez, A., Nanotechnology: Review of concepts and potential application of sensing platforms in food safety. *Food Microbiol.*, 75, 47–54, 2018.
23. National Nanotechnology Initiative, *Nanotechnology 101*, 2021, National Nanotechnology Initiative. https://www.nano.gov/nanotech-101.
24. Singh, H., Nanotechnology applications in functional foods; Opportunities and challenges. *Prev. Nutr. Food Sci.*, 21, 1–8, 2016.

25. Zhu, J., Sadu, R., Wei, S., Chen, D.H., Haldolaarachchige, N., Luo, Z., Gomes, J.A., Young, D.P., Guo, Z., Magnetic graphene nanoplatelet composites toward arsenic removal. *ECS J. Solid State SC.*, 1, M1–M5, 2012.
26. Kim, J.S., Kuk, E., Yu, K.N., Kim, J.H., Park, S.J., Lee, H.J., Kim, S.H., Park, Y.K., Park, Y.H., Hwang, C.Y., Kim, Y.K., Lee, Y.S., Jeong, D.H., Cho, M.H., Antimicrobial effects of silver nanoparticles. *Nanomed. Nanotechnol. Biol. Med.*, 3, 95–101, 2007.
27. Deligiannakis, Y., Sotiriou, G.A., Pratsinis, S.E., Antioxidant and antiradical SiO2 nanoparticles covalently functionalized with gallic acid. *ACS Appl. Mater. Interfaces*, 4, 12, 6609–6617, 2012.
28. *Functional Food Global Market Report 2020-30, COVID-19 Growth and Change*, 2021, Dublin, Ireland. https://www.researchandmarkets.com/reports/5027872/functional-food-global-market-report-2020-30.
29. *Global Functional Foods and Beverages Market 2020-2024*, 2021, https://www.researchandmarkets.com/reports/5011334/global-functional-foods-and-beverages-market-2020. (Website link, Dublin, Ireland).
30. *Functional Food Market Size and Share with Industry Overview by 2027*, 2021, (Website link, US Canada). https://www.alliedmarketresearch.com/functional-food-market.
31. Shafiq, M., Anjum, S., Hano, C., Anjum, I., Abbasi, B.H., An overview of the applications of nanomaterials and nanodevices in the food industry. *Foods*, 9, 148, 2020.
32. Nile, S.H., Baskar, V., Selvaraj, D., Nile, A., Xiao, J., Kai, G., Nanotechnologies in food science: Applications, recent trends, and future perspectives. *Nano-Micro Lett.*, 12, 3, 2020.
33. Anandharamakrishnan, C., Trends and impact of nanotechnology in agro-food sector, in: *Innovative Food Processing Technologies*, pp. 523–531, Elsevier, Netherland, 2021.
34. Jafari, S.M., Jabari, S.S., Dehnad, D., Shahidi, S.A., Effects of thermal processing by nanofluids on vitamin C, total phenolics and total soluble solids of tomato juice. *J. Food Sci. Technol.*, 54, 3, 679–686, 2017.
35. Leena, M.M., Bharathi, S.K.V., Moses, J.A., Anandharamakrishnan, C., Potential applications of nanofibers in beverage industry, in: *Nanoengineering in Beverage Industry*, pp. 333–368, Elsevier, Netherland, 2020.
36. Sekhon, B.S., Food nanotechnology-an overview. *Nanotechnol. Sci. Appl.*, 3, 1–15, 2010.
37. Yu, H., Park, J.Y., Kwon, C.W., Hong, S.C., Park, K.M., Chang, P.S., An overview of nanotechnology in food science: Preparative methods, practical applications, and safety. *J. Chem.*, 2018, 5427978, 2018.
38. Bratovčić, A., Odobašić, A., Ćatić, S., Šestan, I., Application of polymer nanocomposite materials in food packaging. *Croat. J. Food Sci. Technol.*, 7, 86–94, 2015.
39. He, X., Deng, H., Hwang, H., The current application of nanotechnology in food and agriculture. *J. Food Drug Anal.*, 27, 1–21, 2019.

40. Weir, A., Westerhoff, P., Fabricius, L., Hristovski, K., Von Goetz, N., Titanium dioxide nanoparticles in food and personal care products. *Environ. Sci. Technol.*, 46, 2242–2250, 2012.
41. Ae, K., Singh, A., Nc, S., Jp, P., Prakash, O., Tp, S., Novel eco-friendly techniques for extraction of food based lipophilic compounds from biological materials. *Nat. Prod. Chem. Res.*, 4, 5, 1000231, 2016.
42. Zhao, Y.M., de Alba, M., Sun, D.W., Tiwari, B., Principles and recent applications of novel non-thermal processing technologies for the fish industry-a review. *Crit. Rev. Food Sci. Nutr.*, 59, 728–742, 2019.
43. Hassoun, A., Ojha, S., Tiwari, B., Rustad, T., Nilsen, H., Heia, K., Cozzolino, D., Bekhit, A.E., Biancolillo, A., Wold, J.P., Monitoring thermal and non-thermal treatments during processing of muscle foods: A comprehensive review of recent technological advances. *Appl. Sci.*, 10, 6802, 2020.
44. Ali, A., Ahmad, U., Akhtar, J., Badruddeen, Khan, M.M., Engineered nano scale formulation strategies to augment efficiency of nutraceuticals. *J. Funct. Foods*, 62, 103554, 2019.
45. Dan, N., Transport and release in nano-carriers for food applications. *J. Food Eng.*, 175, 136–144, 2016.
46. Maqsoudlou, A., Assadpour, E., Mohebodini, H., Jafari, S.M., Improving the efficiency of natural antioxidant compounds via different nanocarriers. *Adv. Colloid Interface Sci.*, 278, 102122, 2020.
47. Pradhan, N., Singh, S., Ojha, N., Shrivastava, A., Barla, A., Rai, V., Bose, S., Facets of nanotechnology as seen in food processing, packaging, and preservation industry. *Biomed. Res. Int.*, 2015, 365672, 2015.
48. Thiruvengadam, M., Rajakumar, G., Chung, I.M., Nanotechnology: Current uses and future applications in the food industry. *Biotech*, 8, 1–13, 2018.
49. Lee, J.S., Hong, D.Y., Kim, E.S., Lee, H.G., Improving the water solubility and antimicrobial activity of silymarin by nanoencapsulation. *Colloids Surf. B: Biointerfaces*, 154, 171–177, 2017.
50. Wani, T.A., Shah, A.G., Wani, S.M., Wani, I.A., Masoodi, F.A., Nissar, N., Shagoo, M.A., Suitability of different food grade materials for the encapsulation of some functional foods well reported for their advantages and susceptibility. *Crit. Rev. Food Sci. Nutr.*, 56, 15, 2431–2454, 2016.
51. Akhtartavan, S., Karimi, M., Karimian, K., Azarpira, N., Khatami, M., Heli, H., Evaluation of a self-nanoemulsifying docetaxel delivery system. *Biomed. Pharmacother.*, 109, 2427–2433, 2019.
52. Ali, A., Ansari, V.A., Ahmad, U., Akhtar, J., Jahan, A., Nanoemulsion: An advanced vehicle for efficient drug delivery. *Drug Res. (Stuttg)*, 67, 617–631, 2017.
53. Kanugo, A. and Misra, A., New and novel approaches for enhancing the oral absorption and bioavailability of protein and peptides therapeutics. *Ther. Deliv.*, 11, 713–732, 2020.

54. Mehanny, M., Hathout, R.M., Geneidi, A.S., Mansour, S., Exploring the use of nanocarrier systems to deliver the magical molecule; Curcumin and its derivatives. *J. Control. Release*, 225, 1–30, 2016.
55. Ballesta, M., Izquierdo, A., Viguera, C., Perles, R., Nanoparticles and controlled delivery for bioactive compounds: Outlining challenges for new smart foods for health. *Foods*, 7, 5, 72, 2018.
56. Delfanian, M. and Sahari, M.A., Improving functionality, bioavailability, nutraceutical and sensory attributes of fortified foods using phenolics-loaded nanocarriers as natural ingredients. *Food Res. Int.*, 137, 109555, 2020.
57. Milinčić, D.D., Popović, D.A., Lević, S.M., Kostić, A.Z., Tešić, Z.L., Nedović, V.A., Pešić, M. B., Application of polyphenol-loaded nanoparticles in food industry. *Nanomaterials*, 9, 1629, 2019.
58. Zhang, S., Zhang, H., Xu, Z., Wu, M., Xia, W., Zhang, W., Chimonanthus praecox extract/cyclodextrin inclusion complexes: Selective inclusion, enhancement of antioxidant activity and thermal stability. *Ind. Crops Prod.*, 95, 60–65, 2017.
59. Engel, J.B., Heckler, C., Tondo, E.C., Daroit, D.J., da Silva Malheiros, P., Antimicrobial activity of free and liposome-encapsulated thymol and carvacrol against Salmonella and Staphylococcus aureus adhered to stainless steel. *Int. J. Food Microbiol.*, 252, 18–23, 2017.
60. Hao, J., Guo, B., Yu, S., Zhang, W., Zhang, D., Wang, J., Wang, Y., Encapsulation of the flavonoid quercetin with chitosan-coated nano-liposomes. *LWT-Food Sci. Technol., 85*, 3744, 2017.
61. Jeyakumari, A., Zynudheen, A., Parvathy, U., Binsi, P., Impact of chitosan and oregano extract on the physicochemical properties of microencapsulated fish oil stored at different temperature. *Int. J. Food Prop., 21*, 1, 943–956, 2018.
62. Oancea, A.M., Hasan, M., Vasile, A.M., Barbu, V., Enachi, E., Bahrim, G., Stănciuc, N., Functional evaluation of microencapsulated anthocyanins from sour cherries skins extracts in whey proteins isolate. *LWT-Food Sci. Technol., 95*, 129–134, 2018.
63. Pimentel-Moral, S., Ochando-Pulido, J.M., Segura-Carretero, A., Martinez-Ferez, A., Stabilization of W/O/W multiple emulsion loaded with Hibiscus sabdariffa extract through protein-polysaccharide complexes. *LWT-Food Sci. Technol., 90*, 389–395, 2018.
64. Velderrain-Rodriguez, G.R., Acevedo-Fani, A., Gonzalez-Aguilar, G.A., Martin-Belloso, O., Encapsulation and stability of a phenolic-rich extract from mango peel within water-in-oil-in-water emulsions. *J. Funct. Foods, 56*, 65–73, 2019.
65. Zhang, Y., Pu, C., Tang, W., Wang, S., Sun, Q., Gallic acid liposomes decorated with lactoferrin: Characterization, in vitro digestion and antibacterial activity. *Food Chem., 293*, 315–322, 2019.
66. Aleman, A., Marin, D., Taladrid, D., Montero, P., Gomez-Guillen, M.C., Encapsulation of antioxidant sea fennel (Crithmummaritimum) aqueous

and ethanolic extracts in freeze-dried soy phosphatidylcholine liposomes. *Food Res. Int.*, *119*, 665–674, 2019.
67. Soleimanifar, M., Jafari, S.M., Assadpour, E., Encapsulation of olive leaf phenolics within electrosprayed whey protein nanoparticles; production and characterization. *Food Hydrocoll.*, *101*, 105572, 2020.
68. FaridiEsfanjani, A. and Jafari, S.M., Biopolymer nano-particles and natural nano-carriers for nano-encapsulation of phenolic compounds. *Colloids Surf. B Biointerfaces*, 146, 532–543, 2016.
69. Liang, J., Yan, H., Yang, H.J., Kim, H.W., Wan, X., Lee, J., Ko, S., Synthesis and controlled-release properties of chitosan/beta-Lactoglobulin nanoparticles as carriers for oral administration of epigallocatechin gallate. *Food Sci. Biotechnol.*, 25, 1583–1590, 2016.
70. Dobrucka, R., Kaczmarek, M., Dlugaszewska, J., Cytotoxic and antimicrobial effect of biosynthesized silver nanoparticles using the fruit extract of Ribesnigrum. *Adv. Nat. Sci. Nanosci. Nanotechnol.*, 9, 25015, 2018.
71. Yildirim, S., Rocker, B., Pettersen, M.K., Nilsen-Nygaard, J., Ayhan, Z., Rutkaite, R., Radusin, T., Suminska, P., Marcos, B., Coma, V., Active packaging applications for food. *Compr. Rev. Food Sci. F.*, 17, 1, 165–199, 2018.
72. Jildeh, N.B. and Matouq, M., Nanotechnology in packing materials for food and drug stuff opportunities. *J. Environ. Chem. Eng.*, 13, 104338, 2020.
73. Sadeghi, R., Rodriguez, R.J., Yao, Y., Kokini, J.L., Advances in nanotechnology as they pertain to food and agriculture: Benefits and risks. *Annu. Rev. Food Sci. Technol.*, 8, 467–492, 2017.
74. Sampathkumar, K., Tan, K.X., Loo, S.C., Developing nano-delivery systems for agriculture and food applications with nature-derived polymers. *Iscience*, 12, 101055, 2020.
75. European Commission, *2009, EU Guidance to the Commission Regulation (EC)*, An official website of the European Union, 2017, https://ec.europa.eu/food/safety/chemical_safety/food_contact_materials_en. (Website Link)
76. Singh, P., Wani, A.A., Saengerlaub, S., Active packaging of food products: Recent trends. *Nutr. Food Sci.*, 41, 4, 249–260, 2011.
77. Young, E., Mirosa, M., Bremer, P., A systematic review of consumer perceptions of smart packaging technologies for food. *Front. Sustain. Food Syst.*, 4, 63, 2020.
78. Kuswandi, B., Environmental-friendly food nano-packaging. *Environ. Chem. Lett.*, 15, 2, 205–221, 2017.
79. Yam, K.L., Takhistov, P.T., Miltz, J., Intelligent packaging: Concepts and applications. *J. Food Sci.*, 70, 1, 1–10, 2005.
80. Realini, C.E. and Marcos, B., Active and intelligent packaging systems for a modern society. *Meat Sci.*, 98, 3, 404–419, 2014.
81. Goudarzi, V., Shahabi-Ghahfarrokhi, I., Babaei-Ghazvini, A., Preparation of ecofriendly UV-protective food packaging material by starch/TiO2 bio-nanocomposite: Characterization. *Int. J. Biol. Macromol.*, 95, 306–313, 2017.

82. Sanuja, S., Agalya, A., Umapathy, M.J., Synthesis and characterization of zinc oxide-neem oil–chitosan bionanocomposite for food packaging application. *Int. J. Biol. Macromol.*, 74, 76–84, 2015.
83. Wu, Z., Deng, W., Luo, J., Deng, D., Multifunctional nano-cellulose composite films with grape seed extracts and immobilized silver nanoparticles. *Carbohydr. Polym.*, 205, 447–455, 2017.
84. Tang, S., Wang, Z., Li, W., Li, M., Deng, Q., Wang, Y., Li, C., Chu, P.K., Ecofriendly and biodegradable soybean protein isolate films incorporated with ZnO nanoparticles for food packaging. *ACS Appl. Bio Mater.*, 2, 5, 2202–2207, 2019.
85. Al-Naamani, L., Dobretsov, S., Dutta, J., Chitosan-zinc oxide nanoparticle composite coating for active food packaging applications. *Innov. Food Sci. Emerg. Technol.*, 38, 231–237, 2016.
86. Svagan, A.J., Akesson, A., Cardenas, M., Bulut, S., Knudsen, J.C., Risbo, J., Plackett, D., Transparent films based on PLA and montmorillonite with tunable oxygen barrier properties. *Biomacromolecules*, 13, 2, 397–405, 2012.
87. Khan, A., Khan, R.A., Salmieri, S., Le Tien, C., Riedl, B., Bouchard, J., Chauve, G., Tan, V., Kamal, M.R., Lacroix, M., Mechanical and barrier properties of nanocrystalline cellulose reinforced chitosan-based nanocomposite films. *Carbohydr. Polym.*, 90, 4, 1601–1608, 2012.
88. George, J., High performance edible nanocomposite films containing bacterial cellulose nanocrystals. *Carbohydr. Polym.*, 87, 3, 2031–2037, 2012.
89. Liu, C., Xiong, H., Chen, X., Lin, S., Tu, Y., Effects of nano-tio2 on the performance of high-amylose starch based antibacterial films. *J. Appl. Polym. Sci.*, 132, 32, 1–7, 2015.
90. De Silva, R.T., Mantilaka, M.M., Ratnayake, S.P., Amaratunga, G.A., de Silva, K.N., Nano-MgO reinforced chitosan nanocomposites for high performance packaging applications with improved mechanical, thermal and barrier properties. *Carbohydr. Polym.*, 157, 739–747, 2017.
91. Cherpinski, A., Gozutok, M., Sasmazel, H.T., Torres-Giner, S., Lagaron, J.M., Electrospun oxygen scavenging films of poly (3-hydroxybutyrate) containing palladium nanoparticles for active packaging applications. *Nanomaterials*, 8, 7, 469, 2018.
92. Zhu, Z., Zhang, Y., Shang, Y., Wen, Y., Electrospun nanofibers containing TiO_2 for the photocatalytic degradation of ethylene and delaying postharvest ripening of bananas. *Food Bioprocess Tech.*, 12, 2, 281–287, 2019.
93. Kumar, V., Guleria, P., Mehta, S.K., Nanosensors for food quality and safety assessment. *Environ. Chem. Lett.*, 15, 2, 165–177, 2017.
94. Li, X., Wang, T., Zhang, J., Zhu, D., Zhang, X., Ning, Y., Zhang, H., Yang, B., Controlled fabrication of fluorescent barcode nanorods. *ACS Nano*, 4, 4350–4360, 2010.
95. Li, Y., Cu, Y.T., Luo, D., Multiplexed detection of pathogen DNA with DNA-based fluorescence nanobarcodes. *Nat. Biotechnol.*, 23, 7, 885–889, 2005.

96. Berekaa, M.M., Nanotechnology in food industry; advances in food processing, packaging and food safety. *Int. J. Curr. Microbiol. App. Sci.*, 4, 5, 345–357, 2015.
97. Omanovic-Miklicanina, E. and Maksimovic, M., Nanosensors applications in agriculture and food industry. *Bull. Chem. Technol. Bosnia Herzegovina*, 47, 59–70, 2016.
98. Kumar, P., Mahajan, P., Kaur, R., Gautam, S., Nanotechnology and its challenges in the food sector: A review. *Mater. Today Chem.*, 17, 100332, 2020.
99. Pal, M., Nanotechnology: A new approach in food packaging. *J. Food Microbiol. Saf. Hyg.*, 2, 02, 8–9, 2017.
100. Inbaraj, B.S. and Chen, B.H., Nanomaterial-based sensors for detection of foodborne bacterial pathogens and toxins as well as pork adulteration in meat products. *J. Food Drug Anal.*, 24, 1, 15–28, 2016.
101. Wang, J.J., Liu, B.H., Hsu, Y.T., Yu, F.Y., Sensitive competitive direct enzyme-linked immunosorbent assay and gold nanoparticle immunochromatographic strip for detecting aflatoxin M1 in milk. *Food Control*, 22, 6, 964–969, 2011.
102. Call, D.R., Borucki, M.K., Loge, F.J., Detection of bacterial pathogens in environmental samples using DNA microarrays. *J. Microbiol. Methods*, 53, 2, 235–243, 2003.
103. Duran, N. and Marcato, P.D., Nanobiotechnology perspectives. Role of nanotechnology in the food industry: A review. *Int. J. Food Sci. Technol.*, 48, 6, 1127–1134, 2013.
104. Mustafa, F. and Andreescu, S., Nanotechnology-based approaches for food sensing and packaging applications. *RSC Adv.*, 10, 33, 19309–19336, 2020.
105. Vamvakaki, V. and Chaniotakis, N.A., Pesticide detection with a liposome-based nano-biosensor. *Biosens. Bioelectron.*, 22, 12, 2848–2853, 2007.
106. McClements, D.J., Nanotechnology Approaches for improving the healthiness and sustainability of the modern food supply. *ACS Omega*, 5, 46, 29623–29630, 2020.
107. Liu, D., Chen, W., Wei, J., Li, X., Wang, Z., Jiang, X., A highly sensitive, dual-readout assay based on gold nanoparticles for organophosphorus and carbamate pesticides. *Anal. Chem.*, 84, 9, 4185–4191, 2012.
108. Wang, M. and Li, Z., Nano-composite ZrO2/Au film electrode for voltammetric detection of parathion. *Sens. Actuators B. Chem.*, 133, 2, 607–612, 2008.
109. Bagheri, H., Afkhami, A., Shirzadmehr, A., Khoshsafar, H., Khoshsafar, H., Ghaedi, H., Novel potentiometric sensor for the determination of Cd^{2+} based on a new nano-composite. *Int. J. Environ. Anal. Chem.*, 93, 5, 578–591, 2013.
110. Ji, X., Zheng, J., Xu, J., Rastogi, V.K., Cheng, T.C., DeFrank, J.J., Leblanc, R.M., (CdSe) ZnS quantum dots and organophosphorus hydrolase bioconjugate as biosensors for detection of paraoxon. *J. Phys. Chem. B*, 109, 9, 3793–3799, 2005.

111. Ensafi, A.A., Karimi-Maleh, H., Mallakpour, S., Simultaneous determination of ascorbic acid, acetaminophen, and tryptophan by square wave voltammetry using N-(3,4-Dihydroxyphenethyl)-3,5-Dinitrobenzamide-modified carbon nanotubes paste electrode. *Electroanalysis*, 24, 3, 666–675, 2012.
112. Karimi-Maleh, H., Moazampour, M., Yoosefian, M., Sanati, A.L., Tahernejad-Javazmi, F., Mahani, M., An electrochemical nanosensor for simultaneous voltammetric determination of ascorbic acid and Sudan I in food samples. *Food Anal. Methods*, 7, 10, 2169–2176, 2014.
113. Liu, S., Yu, J., Ju, H., Renewable phenol biosensor based on a tyrosinase-colloidal gold modified carbon paste electrode. *J. Electroanal. Chem.*, 540, 61–67, 2003.
114. Terry, L.A., White, S.F., Tigwell, L.J., The application of biosensors to fresh produce and the wider food industry. *J. Agric. Food Chem.*, 53, 5, 1309–1316, 2005.
115. Govindasamy, M., Mani, V., Chen, S.M., Chen, T.W., Sundramoorthy, A.K., Methyl parathion detection in vegetables and fruits using silver@ graphene nanoribbons nanocomposite modified screen printed electrode. *Sci. Rep.*, 7, 1, 1–11, 2017.
116. Shan, D., Wang, Y., Xue, H., Cosnier, S., Sensitive and selective xanthine amperometric sensors based on calcium carbonate nanoparticles. *Sens. Actuators B Chem.*, 136, 2, 510–515, 2009.
117. Yang, M., Kostov, Y., Rasooly, A., Carbon nanotubes based optical immunodetection of Staphylococcal Enterotoxin B (SEB) in food. *Int. J. Food Microbiol.*, 127, 78–83, 2008.
118. Çubukçu, M., Timur, S., Anik, U., Examination of performance of glassy carbon paste electrode modified with gold nanoparticle and xanthine oxidase for xanthine and hypoxanthine detection. *Talanta*, 74, 3, 434–439, 2007.
119. Castillo, G., Spinella, K., Poturnayova, A., Snejdarkova, M., Mosiello, L., Hianik, T., Detection of aflatoxin B1 by aptamer-based biosensor using PAMAM dendrimers as immobilization platform. *Food Control*, 52, 9–18, 2015.
120. Giannetto, M., Umiltà, E., Careri, M., New competitive dendrimer-based and highly selective immunosensor for determination of atrazine in environmental, feed and food samples: The importance of antibody selectivity for discrimination among related triazinic metabolites. *Anal. Chim. Acta*, 806, 197–203, 2014.
121. Naser-Sadrabadi, A., Zare, H.R., Benvidi, A., Photochemical deposition of palladium nanoparticles on TiO2 nanoparticles and their application for electrocatalytic measurement of nitrate ions in potato, onion and cabbage using bipolar electrochemical method. *Measurement*, 166, 108222, 2020.
122. Zou, J., Xu, M., Tang, W., Wen, L., Yang, B., Modification of structural, physicochemical and digestive properties of normal maize starch by thermal treatment. *Food Chem.*, 309, 125733, 2020.

123. Pradela-Filho, L.A., Andreotti, I.A., Carvalho, J.H., Araujo, D.A., Orzari, L.O., Gatti, A., Takeuchi, R.M., Santos, A.L., Janegitz, B.C., Glass varnish-based carbon conductive ink: A new way to produce disposable electrochemical sensors. *Sens. Actuators B Chem.*, 305, 127433, 2020.
124. Delfino, J.R., da Silva, J.L., Marques, A.L., Stradiotto, N.R., Antioxidants detection in aviation biokerosene by high-performance liquid chromatography using gold nanoparticles anchored in reduced graphene oxide. *Fuel*, 260, 116315, 2020.
125. Echegoyen, Y., Rodríguez, S., Nerín, C., Nanoclay migration from food packaging materials. *Food Addit. Contam. Part A*, 33, 3, 530–539, 2016.
126. Nguyen, N.T., Lee, J., Woo, S.M., Kim, Y.H., Min, J., The response of yeast vacuolar proteins: A novel rapid tool for Salmonella sp. screening. *Biotechnol. Appl. Biochem.*, 68, 1, 173–184, 2021.
127. Goud, K.Y., Reddy, K.K., Satyanarayana, M., Kummari, S., Gobi, K.V., A review on recent developments in optical and electrochemical aptamer-based assays for mycotoxins using advanced nanomaterials. *Microchim. Acta*, 187, 1, 1–32, 2020.
128. Wang, S., Zhang, Y., An, W., Wei, Y., Liu, N., Chen, Y., Shuang, S., Magnetic relaxation switch immunosensor for the rapid detection of the foodborne pathogen Salmonella enterica in milk samples. *Food Control*, 55, 43–48, 2015.
129. Zhao, Y., Yao, Y., Xiao, M., Chen, Y., Lee, C.C., Zhang, L., Zhang, K.X., Yang, S., Gu, M., Rapid detection of Cronobactersakazakii in dairy food by biofunctionalized magnetic nanoparticle based on nuclear magnetic resonance. *Food Control*, 34, 2, 436–443, 2013.
130. Wang, R., Ruan, C., Kanayeva, D., Lassiter, K., Li, Y., TiO_2 nanowire bundle microelectrode-based impedance immunosensor for rapid and sensitive detection of Listeria monocytogenes. *Nano Lett.*, 8, 9, 2625–2631, 2008.
131. Sahani, S. and Sharma, Y.C., Advancements in applications of nanotechnology in global food industry. *Food Chem.*, 342, 128318, 2021.
132. Song, H.S., Kwon, O.S., Lee, S.H., Park, S.J., Kim, U.K., Jang, J., Park, T.H., Human taste receptor-functionalized field effect transistor as a human-like nanobioelectronic tongue. *Nano Lett.*, 13, 1, 172–178, 2013.
133. Song, H.S., Jin, H.J., Ahn, S.R., Kim, D., Lee, S.H., Kim, U.K., Simons, C.T., Hong, S., Park, T.H., Bioelectronic tongue using heterodimeric human taste receptor for the discrimination of sweeteners with human-like performance. *ACS Nano*, 8, 10, 9781–9789, 2014.
134. Lee, M., Jung, J.W., Kim, D., Ahn, Y.J., Hong, S., Kwon, H.W., Discrimination of umami tastants using floating electrode-based bioelectronic tongue mimicking insect taste systems. *J. Food Drug Anal.*, 9, 12, 11728–11736, 2015.
135. He, X. and Hwang, H.M., Nanotechnology in food science: Functionality, applicability, and safety assessment. *J. Food Drug Anal.*, 24, 4, 671–681, 2016.

136. Sung, S.Y., Sin, L.T., Tee, T.T., Bee, S.T., Rahmat, A.R., Rahman, W.A., Tan, A.C., Vikhraman, M., Antimicrobial agents for food packaging applications. *Trends Food Sci. Technol.*, 33, 2, 110–123, 2013.
137. Appendini, P. and Hotchkiss, J.H., Review of antimicrobial food packaging. *Innov. Food Sci. Emerg. Technol.*, 3, 2, 113–126, 2002.
138. Fang, Z., Zhao, Y., Warner, R.D., Johnson, S.K., Active and intelligent packaging in meat industry. *Trends Food Sci. Technol.*, 61, 60–71, 2017.
139. Brody, A.L., Bugusu, B., Han, J.H., Sand, C.K., Mchugh, T.H., Innovative food packaging solutions. *J. Food Sci.*, 73, 8, 107–116, 2008.
140. Sridhar, A., Ponnuchamy, M., Kumar, P.S., Kapoor, A., Food preservation techniques and nanotechnology for increased shelf life of fruits, vegetables, beverages and spices: A review. *Environ. Chem. Lett.*, 19, 1715, 2020.

13

Nanointelligence in Functional Food

Muhamad Malik Mutoffar[1]*, Protity Shuvra Dey[2] and Debashrita Das[3]

[1]Department of Information Technology, Sekolah Tinggi Teknologi Bandung, West Java, Indonesia
[2]Department of Food Science & Nutrition Management, J.D. Birla Institute, Kolkata, West Bengal, India
[3]PRCS Technologies, Kolkata, West Bengal, India

Abstract

The importance of functional food is already well established as it has become an integral part in treating various ailments, as well as improving overall health and well-being. The biologically active components in these foods are considered to be responsible for the therapeutic effects. As people are getting more conscious about healthy living, the demands of functional foods are increasing as a result, the need for more research and development of new ingredients which has higher accuracy and safety is increasing as well. However, the complete utilization of these bioactive components are not achieved hence application of nanotechnology is included which helps in more bioavailability of these components due to efficient delivery of bioactive ingredients, encapsulation of nutrients and others. Moreover, the researchers have moved to a technological innovation, which is the utilization of intelligence of the machine, such as the Artificial Intelligence (AI), as assistance in creating new bioactive components. This process acts as a predictive model in innovation of new components by the use of bioinformatics. It is considered to be a useful tool as it helps in understanding the concepts of already existing nutrients, at the same time in discovery of new components from functional ingredients and their exact mechanism of action in target site to prevent diseases safely and efficiently. Hence, this chapter discusses how the synergistic use of nanointelligence in functional food has brought a development in the food, health, and nutrition sector.

Keywords: Functional foods, health, bioavailability, nanotechnology, effective nutrient delivery, artificial intelligence, food industry

**Corresponding author*: malik@sttbandung.ac.id

Tanima Bhattacharya and Shakeel Ahmed (eds.) Nanotechnology in Functional Foods, (329–358)

13.1 Functional Food

13.1.1 Introduction

In the primeval Vedic literatures originating from India as well as in the Chinese traditional remedies, the idea of functional foods had been described. The main philosophical viewpoint which exists behind the functional food advancement is the mutual origin of food along with medicine. As per the view of [1], the foods which are either enriched, whole or have been fortified with additional nutrients which are known to provide benefits for the health apart from providing the necessary vital nutrients are referred to as functional foods [2]. In the manuscript states that the functional foods can be categorized either as conventional foods which have bioactive substances present naturally or any food ingredients which have been synthesized or also modified and enriched foods with bioactive constituents. As per the study of Corbo *et al.* [3], the most recent definition given by the functional food center (FFC) considers functional food to be either normal or processed foods having bioactive substances, which are known or may be unknown and which in effective quantity can help in provision of health advantages to prevent, treat or manage the chronic disease. Luo *et al.* [4] claim that a very good example of functional food can be the fortification of margarine spread with plant-based sterols used in UK. According to the opinion of Csapó and Némethy [5], the approaches for preparing functional foods include addition of a substance having beneficial properties, which are not usually present in most of the food products, increment of concentration of a food component, which is already naturally there in the food to a certain extent, which would help in the induction of predictive impacts, replacement of macronutrient having excess intake chances with a component having beneficial effects, elimination of any substance which can cause harmful impact after consumption etc. But it must be remembered that functional food is absolutely different from that of the concept of nutraceuticals or any kind of dietary supplement.

Functional foods mainly help in the prevention of chances of any deficiencies in nutrition as they have mostly high content of vital nutrients like fiber, vitamins and minerals. In addition to this, [6] in the study states that functional foods can help in protection against chances of occurrence of various diseases as they contain high concentration of antioxidants thereby neutralizing free radicals and preventing damage of cells and tissues. Table 3.1 indicates some of the functional foods along with the major components present in them that are responsible for providing health benefits.

Table 13.1 Indicates various functional foods with the main components and benefits

Functional food	Major component	Health benefits	References
Grape juice	Phenol	Helps in improvement of heart health.	[7]
Garlic	Organic sulfur components	Helps in reduction of occurrence of heart ailments as well as cancer.	[8]
Products made of whole grains	Phytoestrogen, fiber, lignan	Helps in reduction of occurrence of heart ailments as well as cancer.	[9]
Oatmeal	Soluble fiber and beta glucan	Helps in reduction of blood cholesterol level.	[10]
Green tea	Catechins and tannins	Helps in reduction of some specific cancers.	[11, 12]
Beverages based on fruit and milk	Plant sterols	Helps in reduction of low-density lipoprotein.	[13]
Flax seed	Phenolic acids, flavonoids and lignans	Helps in reduction of hypertension and risk of cancer.	[14]
Starchy roots and tubers	Glycoalkaloids, phenols, saponins and phytic acid	Have antioxidative, antimicrobial properties and helps in reduction of hyperglycemia and high cholesterol level in blood.	[15]

Functional foods also help in considerable improvement of gastrointestinal health because the resistant starch, dietary fiber, and other similar types of complex carbohydrates act as a prebiotic, as well as help in performing different physiological purpose like that of increasing the bioavailability of micronutrients inside the body, providing relief from constipation, as well as inhibition of absorption of cholesterol in the body [6]. Irrespective of these health benefits, functional foods should be consumed along with having a healthy diet and not merely as an alternative.

13.1.2 Current Scenario

Apart from provision of plenty of health benefits, the functional foods have already emerging markets across the world and rapidly expanded its market share, which almost is supposed to be more than that of organic food products. Therefore, it is highly evident that there is new economic scope for functional foods in addition to that of health benefits. The market players who are present in the supply chain find the functional food market very attractive due to the higher profit margin for these products than that of the conventional food items. It can be studied from the manuscript of reference [16] that the size of functional food market globally is supposed to be in between as much as US$ 30 and 60 billion where Japan, Europe, and US function as the most flourishing and prospective markets. Moreover, in countries where there is prevalence of poverty like India, there are a huge number of cases of diseases related to bad dietary habits and malnutrition. In this regard, the developing countries can take the opportunity of marketing functional foods for better public health, which can also help in generation of increased employment and their income opportunities. According to [17], the manufacturers of functional food industry are currently trying their best to incorporate nutritious substances in finished food items like probiotics, prebiotics, ω-3 fatty acids, vitamins, etc., so that the content of nutrition in the food products can be increased. This current trend of functional foods is supposed to increase the demand of the functional foods and prosper the functional food market in the upcoming years. As per the opinion of [18], there is a recent trend of giving increased focus on living quality, and there has also been increment in the expectancy of life in individuals. The consumers are therefore currently more concerned about their status of health, and they are focusing more on maintaining a very good lifestyle. The market players in the functional food industry have also managed to respond to the new trend of focusing on healthy diet and lifestyle and cater to the increasing needs of the customers by developing new variants of food products, which are able to provide proper nutrition and health to the customers [19]. There should be an overall integrative, as well as innovative method, for moving forward by proper commercialization techniques of functional foods. However, even if functional foods are full of benefits and opportunities, they are not free from certain problems and challenges. First of all, in many countries, it is seen that there is no such proper category of rules and regulation for the functional foods thereby posing a severe barrier in the market development of the functional food products [20]. Institutional support, which includes education providers of food management and marketing, centers for food research and analysis, advisory service bodies for the manufacturers, etc., are of utmost importance, so that there can be proper innovation by the producers of the industry and also a healthy competition between the market players. Second, there is lack of proper generation of scientific knowledge about the functional foods. It has been seen in many cases that there is an absence of scientific evidence for the claims that are made by the functional food products, which makes it harder for the marketers to gain the confidence of the customers in the market. The most suitable research results about functional foods can be obtained by forming a partnership between the local communities and the formal science institutes, which is again a major problem for the researchers of management culture. In addition to this, it is also a severe challenge to understand the demand of the functional food products in the market. It is required that the market information can enter and pervade into the overall product value chain, which starts from the manufacturer and retailers. In this way, it would be highly possible to undertake decisions with proper strategies and collaborate and coordinate in a better way in between the players of the supply chain of functional food products. There is also lack of proper rules and regulations on protection of development of new functional food products by intellectual property rights protection, which is also a great challenge for the researchers and manufacturers of new functional food products. Above all, it must be remembered that creating proper awareness among the customers about the various public health problems and the ways in which they are connected with a proper dietary intake daily is essential so that a successful market can be enabled for the functional food products [21].

13.1.3 Problems Associated with Functional Foods

Although being a remedy for many diseases, the benefits of functional foods are not completely utilized. According to Gonçalves and colleagues, the main cause of such a scenario is the low oral bioavailability of the bioactive components. The bioavailability is the level of absorption of the nutrients by the blood streams [22]. A review by Shahidi F. and Peng H. elaborated that both internal (digestion, transportation, absorption) and external factors (interaction with other food particles, processing technique) affects the bioavailability of the phytonutrients [23]. If the amount absorbed is not enough to the required dosage, then the potential health benefits are not achieved. Hence, to maximize the benefits the strategies that are developed by the pharmacological and food industries is the use of nanotechnology in nanoformulation of these bioactive constituents to improve its performance and efficiency [24, 25].

13.2 Nanotechnology

Nanotechnology is very commonly described as a technology, in which, changes are brought in the properties of materials in the molecular level by means of controlled engineering and manipulation to achieve the requisite functions [26]. The development of nanoparticles involves two processes—the top-down (reduction of macromolecules to nanosize) and the bottom up approach (forming cluster of nanoparticles from single atoms) [27]. Nanotechnology runs on nanoscale level (1–100 nm), and when decreased to this size, a substantial change is observed in the properties of the materials which was initially absent when the molecules were in macro form. Increased surface area is considered to be the cause of such difference [28]. Although nanotechnology has become an important concept in the modern times, but its use has been from ancient time. The concept of nanotechnology was highlighted in the year 1989 by Richard Feynman in a convention organized by the American Physical Society in which he had delivered a speech on "There's Plenty of Room at the Bottom,". Later during 1974 a professor named Norio Taniguchi coined the term "Nanotechnology" [29].

13.2.1 Role of Nanotechnology in Functional Food

Nanotechnology has shown great impact is the food and health industry. Along with flavor control and increase of shelf-life, this technological advancement has also improved the functionality of certain nutrients and has shown positive effects in the overall health. In the field of functional food, the effective nutrient delivery, development of nutrients in nanoscale level, fortification and increased bioavailability have become the center of attraction of the current nanotechnology applications [30].

13.2.1.1 *Effective Nutrient Delivery*

The effective nutrient delivery involves microencapsulation of bioactive nutrients such as certain minerals, vitamins, lipids, antioxidants, carbohydrates and proteins in nanocarriers using techniques like nanostructuration, nanocomposites and nanoemulsification. This results in enhanced enzymatic, chemical and physical stability of the nutrient by obstructing unwanted reactions with other body organs, enzymes and food components especially the nutrients that are highly sensitive and easily destroyed by external environment. It is often called targeted delivery as it follows the mechanism of controlled release of the components and on specific site. The target nanodelivery of nutrients are comparatively more efficient than the normal drug delivery due to its decrease in particle size which has better absorbance and thus decreases the dosage requirement which in turn results in decreased chances of toxicity [31, 32] Phenolic compounds when encapsulated with nanoparticles by means of hydrogen bonds and hydrophobic interactions enhance the aqueous solubility of the phytochemicals [33]. Furthermore, along with the functional components, microencapsulation is also done for another important and common functional food, which is the probiotics. With the help of nanoencapsulation, probiotic bacterial strains are coated in nanocarriers which are developed to target-specific location of the gastrointestinal tract and by interacting with specific receptors and is found to improve the overall health and higher survivability of the strains in refrigerated storage [34].

13.2.1.2 *Increased Bioavailability*

Many nutrients when consumed are not completely utilized due to several factors such as low stability, interaction with other substances, lower absorption in the tissues, and others. However, when these nutrients are converted to nanoscale level, due to the alterations in the physiochemical properties, enhanced absorption and bioavailability of the nutrients is observed [28]. Moreover, nanobioactive compounds when given in the form of supplement, also known as the nanosupplements, it was found to have higher efficiency, improved delivery at target sites, better absorption and solubility and longer duration of retainment in the gastrointestinal tract [30].

13.2.1.3 Fortification

Different food processing and storage techniques often results in destruction of important nutrients due to various conditions such as heat, light, sensitivity to oxidants, and hence, restoration of them are important. On the other hand, foods may lack some specific nutrients which is essential for human health. Hence, to combat these problems, the food industries fortify the foods with essential nanonutrients [35]. Table 13.2 represents some works that have fortified foods by using nanotechnology.

The fortification process has been advanced and is also done by means of the packaging materials known as the nanopackaging. In this, the food components, which include prebiotics, flavonoids, vitamins, probiotics, minerals, and others, are encapsulated in the packaging material so that the food products can be fortified and provided with essential nutrients through controlled release under optimal conditions. [40]. However, although the packaging materials are generally regarded as safe (GRAS)-certified, there remains a risk since the physiochemical properties of these materials differ in the nanosize when compared with its macro state. It can also lead to toxicity in case of poor packaging and have detrimental effects on health [41]. The chances of bioaccumulation in different organs and tissues also increase in such cases due to its nanosize [42]. Hence, standardization and controlled examination should be done while making such packaging for safety.

Another field of technological advancement that has joined the food industry is the artificial intelligence.

13.3 Artificial Intelligence

"Intelligence" originated from the Latin Language "intelligence" means is "I understand." The basis of intelligence is an activity to be able to understand the learning process and take action. In fact, Artificial Intelligence or AI for short, originated from the presence of computers in the 1940, ancient Egypt has a very large history and contributed to the development of computers. In the current era of technology, all researchers are trying to collaborate with the field of information technology or computer science to focus computers to be able to imitate all activities carried out by humans. from this, the computer must be able to answer the ability of human intelligence and behavior [43].

Problems are solved using symbols managed by the program with a knowledge-based system, such as machine learning problems, it is

Table 13.2 Fortified foods using nanotechnology.

Fortified food	Nanocarrier	Nutrient	Effect of nanoencapsulation	References
Orange drink	Lipid nanocarriers • Nanostructured lipid carriers (NLC) • Solid Lipid Nanoparticles (SLN)	Lycopene	• Increased lycopene solubility • Decreased unpleasant smell and after taste of product	[36]
Yogurt	CaHPO4, alpha-Fe2O3 and inulin coated ZnO.	• Calcium • Iron • Zinc	• Improved digestibility • Improved overall sensory properties	[37]
Wheat biscuits	Iron oxide nanoparticle (6.19nm) with Hibiscurosa sinensis flower's extract of the petals	• Iron	• Enhanced sensory properties. • Improved total nutritional content. • Longer shelf life.	[38]
White and milk chocolate	Lyophilized colloidal nanoparticles with cinnamon extract.	• Antioxidant • Phenolic content	• Increased antioxidant and phenolic content • Flavor preservation	[39]

mandatory to deal with both input (structured or unstructured) and result knowledge in a predictable prediction scheme [44]. Madey and his friends stated that artificial intelligence is an activity to study and develop machines that are capable of having the same or even better intelligence than humans so as to produce something new in certain fields both mechanically and in the form of applications [45].

Andreas Kaplan and Heinlein revealed that artificial intelligence is applied of a system to translate data and information appropriately, learn to understand that data, and to use it in achieving certain goals and tasks through a flexible approach [46].

Machine activity in doing its job requires human intelligence, for this matter solving problems and continuous learning [47]. For example, the learning algorithm differentiates between healthy and sick people. The definition of Elaine Rich that Elter follows in his book says that "Artificial Intelligence is the study of how to make computers do things at which, at the moment, people are better" [48].

In research on intelligent systems the workings of the human brain can begin to be understood, so that models can be made and simulated on a computer. Evolving principles and ideas in the field of neural networks come from the related field of neuroscience. In the book written by Sutojo, he says that Artificial intelligence is a machine that is capable of thinking, analyzing actions to be taken and able to make decisions like those of humans [49].

In De Moraes' research previously disclosed by Fister and Piaget that AI (or computational intelligence) seeking, from nature that provides clues, so that we can become an inspiration, an intelligence system developed to imitate all human behavior, for example learning, reasoning, and evolution. AI is a bridge for algorithms in solving problems in the real world whose problems are similar to what the system solves naturally. AI can fundamentally differentiate adaptations in new forms, AI can also be based on algorithmic methodologies to adapt its environment [50].

Artificial Intelligence in which there are artificial neural networks, deep learning, genetic algorithms, expert systems, fuzzy logic, and other parts of Artificial Intelligence have been widely applied in various fields, from business, finance, industry, agriculture, medicine and pharmacy. With the large number of products produced in each of these fields, including the research that the authors conducted on the study of food nutrition, namely the synergistic use of nanointelligence in the functional fields of food, health and nutrition. Research concerns predictive models in new component innovations with the use of bio informatics, automated tasks requiring intelligent behavior.

13.3.1 Types of Artificial Intelligence

According to Hossein Hassani and friends, there are two types of AI. The first is machines that support AI, have activities to think, understand, and feel things like humans. Four subclassifications of systems based on AI [51]:

1. Reactive machines
   Hossein's example of a reactive AI engine is the IBM Deep Blue, the machine that famously beat chess Grandmaster Garry Kasparov in 1997. This machine is used automatically in response to a limited set or combination of inputs. By involving memory-based operations and the first Artificial Intelligence-based system engine, and has very limited capabilities. It can be said that this machine cannot "learn".
2. Limited memory, machine
   Machines with limited memory, machines have the ability to be reactive, it can be learn historical data to provide predictions in making decisions. Most of today's applications ranging from virtual systems, chatbots to self-driving vehicles are already part of this field.
3. Theory of mind
   Theory of mind is an advanced level of artificial intelligence system, theory of mind is currently an ongoing concept with AI systems to better understand entities interacting with each other by differentiating needs, beliefs, thought processes, and emotions.
4. Self-aware Artificial Intelligence
   Self-aware Artificial Intelligence is the final stage of AI development, all researchers have ambitions to collaborate by involving AI systems that evolve to be on par with the human brain and develop self-awareness.

Three subclassifications based on a technology-focused approach [46]:

1. Artificial Narrow Intelligence (ANI)
2. Artificial General Intelligence (AGI)
3. Artificial Superintelligence (ASI)

There are three (3) classifications, Pohl describes their definitions of that:

Artificial Narrow Intelligence: refer to as artificial intelligence that has not been optimal, it can be said that the first generation of or simply Artificial Intelligence. This is basically Computer software is currently designed and developed by programmers with input and output processes in accordance with the User Interface (UI) and user experience (UX) that are safe and friendly and can pamper and satisfy users in interacting. A model-driven architecture part of the source code is designed automatically production of human-designed code generated, running a code program is generally predictable.
Artificial General Intelligence: Artificial intelligence applied to computers has matched the level of human intelligence in almost all areas such as, how to reason, plan, make problem solving, understand difficult and complicated ideas, learn quickly, and learn from previous experiences. It can be said that the second generation of AI.
Artificial Superintelligence (ASI): Artificial intelligence applied to computers has matched or surpassed the level of human intelligence. It can be said that the third generation of AI [52].

In Figure 13.1, the ANI describes the current AI. The machine will function based on what has been programmed and entered. Hence, AI has a limited range and few abilities [51]. AGI must have an understanding like humans, computers can have understanding by processing concepts, and concepts are the action of many other concepts, from this principle, experts create knowledge acquisition systems [53]. Artificial Super Intelligence

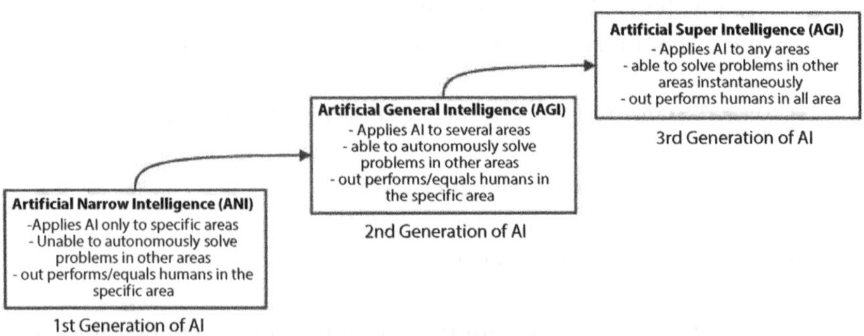

Figure 13.1 Artificial intelligence level [46].

(ASI) offers intelligence that surpasses the human mind, can be applied to hardware as it can be built and capable of handling problem faced by mankind [54].

This is in line with what Russell & Norvig expressed, artificial intelligence is divided into 4 categories: systems that mimic human behavior/acting humanly, think humanly, acting rationally, think in a rational way, rational/thinking rationally [55].

13.3.2 Function of Artificial Intelligence in Food Industry

Quality assurance in the food industry is very important because every human being needs hygienic and healthy food and all of that must ensure food safety and hygiene. To detect nutrient content, detect decay of product components, quality control, and dietary consumption are made possible through nanointelligence. AI can ensure quality by tracking the presence of microbes, toxins, and contaminants in all food processing chains by means of data retrieval to automated control functions and documentation functions automatically. Nanotechnology can make it possible its application of lowcost nanosensors in food packaging to be monitored food quality in every process in ensuring and producing quality products until consumption [56].

In the food industry, technology is usually not something new on the mind. Currently, technology is a very important part in the food industry, especially for the production and delivery process of food. We find food using the help of apps, while manufacturers produce using the help of robotics. Technology can increase the packaging optimally, increase storage time, food safety assurance [57].

Artificial Intelligence delivers Solutions and Machine Learning offers everything to maximize processes, automate processes, reduce costs, and reduce negligence and human error in the industry. AI provides benefits from restaurant, bar and cafe as well as other food businesses. Viewed from this segment, AI has a general case that can be applied in every food industry.

13.3.3 How to Introduce Artificial Intelligence in Functional Foods and Dietary Supplements

Raghuver and Tandon stated that diet is one aspect of a comprehensive lifestyle approach in obtaining good health, including regular activities, reducing stress, maintaining body weight and other positive health practices [58].

In their research, Sundaravadivel *et al.* tested and introduced the function of artificial intelligence diets in functional foods and dietary supplements [59]:

- A new 5 (five) layer deep learning model on perceptual neural network is proposed, which has tightly connect hidden layers to determine the nutritional balance in each meal.
- A new algorithm on the Bayesian network to determine the nutritional features of food ingredients and to suggest future meals or recipes. The algorithm is careful analysis of multiple Bayesian classifier and provide excellent performances.
- In the internet of things-based food monitoring, the proposed solution is food monitoring using Artificial Neural Network methods and Bayesian Algorithms.

An overview of the architecture from Sundaravadivel's research on intelligent food controlling system. It is can be regarded as a product equipped with a sensor board on a smartphone application, is also equipped with a sensor for weighing food. Under the coordination of the microcontroller integrated with the wireless module, the weight of the food is send to the cloud wirelessly via Internet connections. Therefore, system will be converted into "objects" in the IoT network. Corresponding nutrient content of food can be obtained using the camera and smartphone application. Then, the system will provide nutritional value to be consumed. Users can access the calculated nutritional value and predicted nutritional value

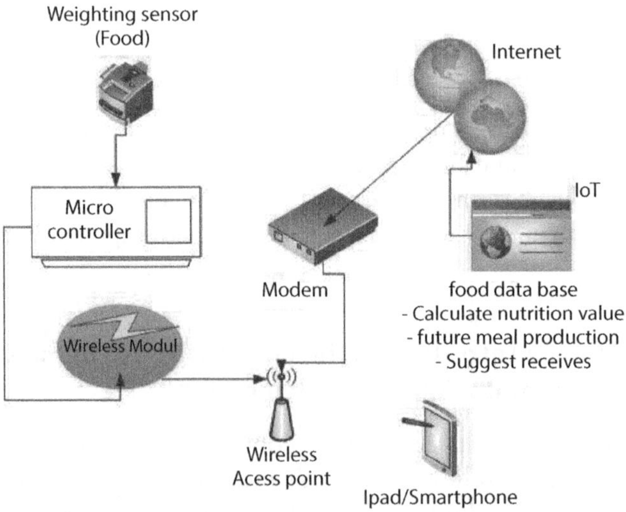

Figure 13.2 Food nutrition monitoring system [59].

using a smartphone application. Figure 13.2 shows the proposed architecture of IoT-based food monitoring system [59].

The main purpose of using food weighing sensors to measure the nutrients consumed. The output of sensor is weight of the food placed. Recording time and fluctuations in food weight can help determine what food is being consumed. Food weighing sensor solution, consisting of weighing sensor and microcontroller.

Consumption of nutrients through food has a direct effect on health and the main aspects of human activity. Recommended daily dietary nutrient intake may vary with a person's, height, age and weight. The pros, cons of diet vary with people's health status. Usually suitable for everyone to eat two meals a day. Changes in diet and the increase of people eating out have led to the need to make the right decisions for nutritional management. Not only that, food decisions will affect to physical, mental health, and dietary nutrition needs to consider health of the user [60]. Figure13.3 shows hybrid filtering-based recommendation system.

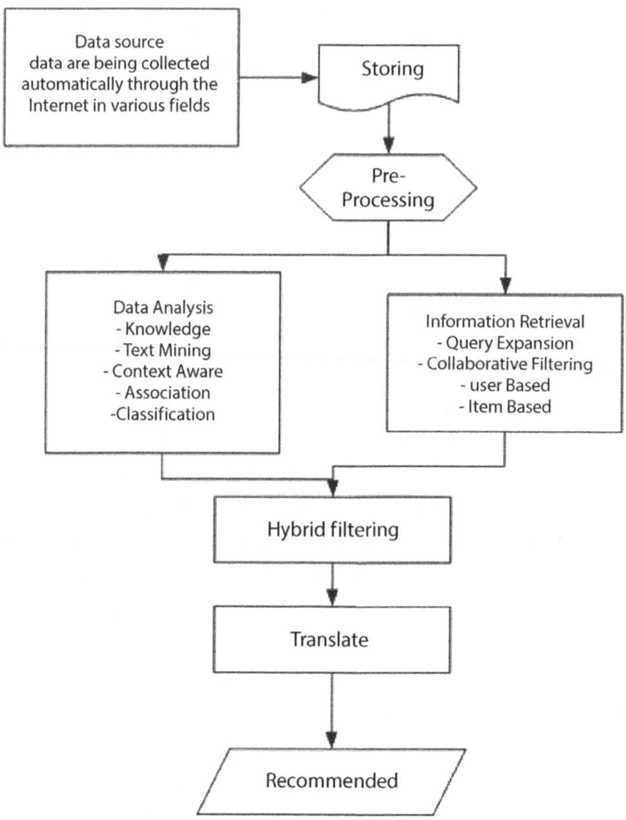

Figure 13.3 Hybrid filtering-based recommendation system [60].

Numerous examinations have been directed in the anticipation or decrease of ongoing illnesses like heftiness, hyperlipidemia, diabetes and greasy liver. Correspondingly, the clinical business has built up an assortment of items, like brilliant arm bands, individual medical care gadgets, wearable gadgets, and food sustenance applications. These items empower individuals to take part in day by day medical services utilizing different dynamic models. With the development of information technology, data will be automatically collected in various fields through the Internet. As a result, you can quickly accumulate data and use data mining to retrieve useful knowledge. Data mining finds information through preprocessing, closeness estimation and enormous information gathering. Likewise, it can likewise help clients settle on choices utilizing the information base. A suggestion framework is an innovation prescribed to clients to help clients settle on better choices. Agent strategies for creating suggestions incorporate synergistic separating and substance based sifting. Community separating is partitioned into client based sifting, and the things dependent on the objective are looked at in the proposal framework. To address the deficiencies of each separating strategy, the technique utilizes cross breed or profound AI models to build up a recommender framework with improved execution [60].

According to the amount supplied at a time, the food have different between diet and nutrition patterns. In order to analyze how many similar components are in the food, calories are applied to the normalization of 1,000 (one thousand), and cluster analysis is performed using the k-means algorithm [61]. This methods require additional surgery to recommended foods that are beneficial to the user's health and foods diet and nutrition are similar to healthy foods. In order to minimize the decomposition time, which reduces the time as much as other operations. Algorithm is repeated to find k values, where the size of each cluster is 300 (three hundred) or less. The k-means algorithm, the clustering accuracy varies according to the initialization and cluster count (k) value. In order to obtain the best clustering, cluster analysis needs to be repeated. The result of the loop is that if the value of k is 25 the size of all clusters is 300 (three hundred) or less. The 25 clustering results created the time are stored and used as a model [62].

The diet and nutrition knowledge base include weights to recommend food products tailored to the health condition of each user and same food. It has through each filter, the final will be recommended to the user. Figure 13.4 shows the structure of diet and nutrition knowledge. As filters, there are a food group filter, similarity filter, a general preference filter and a user feedback.

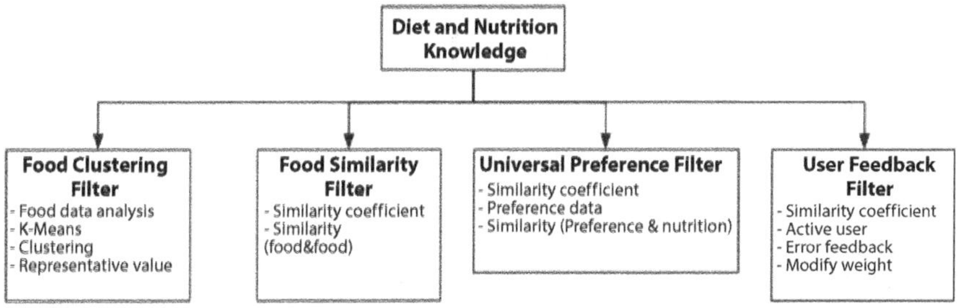

Figure 13.4 The structure of diet and nutrition knowledge [62].

In Baek *et al.*, according to the quantity supplied at a time, a food can have different between diet and nutrition patterns. To analyze the number of ingredients in the diet of 243 similar products, calories were applied for the normalization of 1000. Health decision based on mixed clustering to improve eating habits.

The food group filter refers to the representative value of the food group classified according to the diet and nutritional components shown in Table 13.3. The food similarity filter refers to the similarity coefficient filter between each food in the 25 groups. As the similarity coefficient, the Pearson correlation coefficient is used. Pearson's R is between −1 and +1, if the value is close to −1 it indicates a negative correlation; otherwise is negative. If it is close to +1, the correlation is positive. In equation (13.1), fs(a, b) represents Pearson's food a, it is b represents the nutritional value of food i and food a. i stands for carbohydrate (1), protein (2), fat (3), sugar (4), sodium (5), cholesterol (6), fatty acid (7) and trans fatty acid (8). ā is the average value of the value α. Therefore, α1 is represents the carbohydrate of α food [62].

$$fs(a,b)=\frac{\sum_{i=1}^{8}(ai-\bar{a})(bi-\bar{b})}{\sqrt{\sum_{i=1}^{8}(ai-\bar{a})^2 x \sum_{i=1}^{8}(bi-\bar{b})^2}} \tag{13.1}$$

The application of the Convolutional Neural Network (CNN) can be carried out for recognition of food image protection and evaluating its performance. Contributions can be made as follows: (1) constructing a data set use the images obtained from the food domain to conduct food recognition experiments publicly available food recording systems; (2) optimizing the CNN hyper parameter, which shows compared with

Table. 13.3 Food clusters based on diet and nutritional content [62].

Cluster	Item (1)	Carbohydrate (2)	Protein (3)	Fat (4)	Sugar (5)	Natrium (6)	Cholesterol (7)	Saturated fat (8)	Trans fat (9)
0	137	101	51.8	43.9	16.0	1713.0	38.5	20.9	0.1
1	43	150	48.2	23.6	51.9	17528.0	21.4	2.7	0.1
2	66	213	12.9	12.3	183.9	376.0	152.4	5.4	0.0
3	38	37	149.3	28.8	7.0	6292.3	869.8	6.2	0.3
4	13	41	145.1	28.4	4.3	5725.0	1935.6	6.6	0.1
5	243	102	15.6	60.7	67.0	437.6	285.8	32.7	0.5
6	94	76	74.7	46.1	3.8	2677.9	20.5	2.1	0.1
7	91	0.7	15.9	0.9	48.2	563.1	0.0	38.7	0.0
8	127	0.5	24.9	4.4	20.9	1418.2	0.0	9.1	0.0
9	129	2.1	56.0	9.9	12.7	915.1	0.0	6.8	0.0
10	266	0.2	53.2	0.6	14.1	1560.5	1.8	18357	0.0
…	…	…	…	…	…	…	…	…	…
24	171	204.2	24.8	19.3	28.9	1717.0	3060	287	0.4
Full data	3152	20422	40.6	20.5	32.6	1666.1	89.9	1596	0.1

traditional methods that use support vector machines (SVM) with hand-made functions, CNN greatly improves the accuracy of food recognition; (3) Observed by CNN, it is found that food color can dominate food recognition (4) Shows that compared with the baseline method, CNN performs significantly better on food detection tasks [63].

The previous studies in clinical nutrition research, artificial intelligence technology has been used in several projects with the aim of making tools to support diet and supplement activities, the diagnosis and prediction of disease risks [64]. Research on the application of clinical practice, it is necessary to distinguish systems that aim to develop that it can monitoring, support, and regulate the nutrition of patients with chronic diseases. Lu *et al.* By simply processing paired red, green, and blue depth images taken before and after eating food, a new AI-based system is demonstrated to accurately estimate nutrient intake [65]. Oka *et al.* A randomized controlled trial compared Artificial Intelligence supported nutritional therapy use mobile apps (n = 50) and human nutritional therapy (n = 50) [66]. Vasiloglou *et al.* used interesting technical solutions in the AI field. In clinical problems like this, control is needed for carbohydrate intake in diabetic patients [67].

13.3.4 What Algorithms are Used in Identification of Nutrients?

The food nutrition identification system using a computer system application is very important and has become an issue in recent years, because food consumption will be very influential and have an impact on health. Identifying the type of food to determine the calorie content poses its own challenges, but the process of applying this method is still being developed. This identification is very useful in fast food so that the dose and amount of consumption can be regulated, and we ourselves can decide whether food is good for health of our bodies. The type of data used in previous research is a combination of Pittsburgh Fast-food Image Dataset (PFID) and Deep Learning Neural Networks with the object classification process using SVM. The deep learning method has many data structures so that it is very easy to use to extract data features from a more complex image, SVM will classify data using a non-linear method [68]. ANN Based Modeling. To predict nutrition and functional food through neural networks, you can also use the Learning Vector Quantization (LVQ) method in Artificial Neural Networks to carry out learning on supervised layers [69]. Artificial neuron is a model to simulate the biological behavior of neurons. Neurons are represented by processing units and form part of a larger entity (neural network). Generally, neuron sends output to another neuron through axon and the axon receives information through a potential difference or force wave, depends on the neuron's potential energy. Input signal is generated by synapse, and all excitatory effects are added together. ANN simulates it with positive weight and suppresses it with negative weight. If the positive stimulus is dominant, the neuron will give a positive signal then it will send the information for other neurons through output synapse [70].

Deep learning is part of machine learning, in which the machine learning model still has to be given parameters to make accurate predictions [71]. The deep learning model has greatly improve the best results in various research fields, one of which is computer vision [72]. The deep learning model has a different concept, where this model can learn computational methods independently, with its own brain. In essence, deep learning models are designed to continuously analyze data with logic that is very similar to human logic in making decisions. To be able to make decisions according to human logic, deep learning implements a layered algorithmic structure which is currently known as an artificial neural network (ANN). ANN itself actually arises because of the development of the biological neural network that is owned by the human brain, so that with the application of this method the machine being developed will be more robust/smarter than the standard machine learning model. In its implementation, the success of implementing deep learning is very influential on the data used, because the more data is used, the more the deep learning model can understand.

There are three types of layers that make up ANN. The input layer is original data and passes it to the hidden layer. The second layer learning process is carried out. The results are collected in the output layer then output data is generated. Neural network can contain hundreds of individual units. Artificial Neural Network is a parameterize system, and weights are adjustable parameters. It is needed to estimate these parameters, ANN requires a large number of training sets. ANN acquires knowledge by scanning patterns and relationships data, acquiring knowledge through experience, rather than acquiring knowledge through programming. ANN expresses its special purpose in the need to model data sets with nonlinear correlation. Problems solving for biomedical, the original data to be literature data or experimental. The input layer scan original data then passes it to hidden layer. The second layer, learning process is carried out. The results are collected in the output layer and output data is generated. Neural network can contain many individual units. ANN is a defined framework, and its loads are flexible boundaries. Because of the need to appraise these boundaries, ANN requires countless preparing sets. ANN gets information by distinguishing examples and connections between information, that is, acquiring knowledge through experience, rather than acquiring knowledge through programming. ANN expresses its special purpose in the need to model data sets with nonlinear correlation. Problem solving biomedical, the original data to be literature data or experimental. The past two decades, among other things, artificial neural networks have used model experimental decision-making algorithms that need to be improved to evaluate the results of tests exposed to reference values, clinical data. Technique used to assess cross-contamination at the cellular level cultures based on fingerprint mass spectrometry of whole mammalian cells [64].

13.3.4.1 Convolutional Neural Network

Convolutional Neural Network (CNN) is derivative method of developing of Multilayer Perceptron (MLP) in two-dimensional data processing [73]. Figure 13.5 shows the CNN Work Concept.

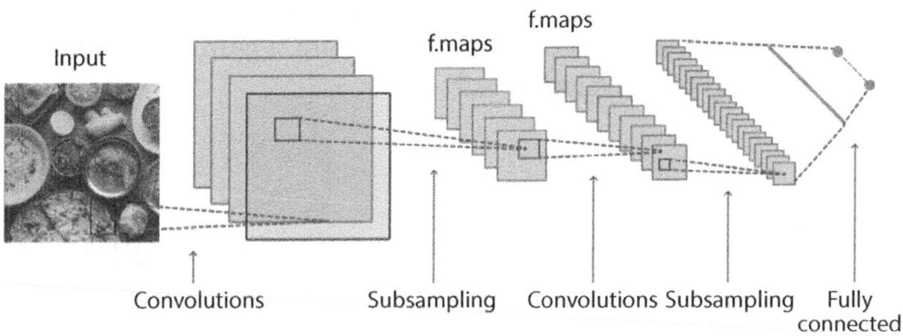

Figure 13.5 Convolution neural network concept.

CNN is included in the deep neural network family because it has a higher network and is very superior when implemented in image data.

Figure 13.6 shows that convolution is used for linear CNN operations, weights are no longer one-dimensional, but use four dimensions, and this dimension is a collection of convolution kernels. Two-dimensional data is spread over the CNN network so that it distinguishes weights and linear operating parameters [73].

The weight size of CNN = input neuron × output neuron × height × width.

CNN convolution process can only be used for data with two-dimensional structure, for example images and sound.

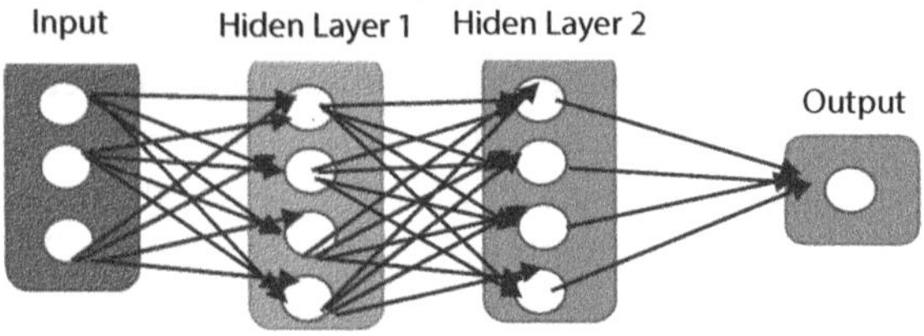

Figure 13.6 Convolution neural network process [73].

13.3.4.2 Convolutional Neural Network Architect

To build a system use the Convolutional Neural Network (CNN) method, it is recommended that you understand the structure of this algorithm. The following are the supporting components of the CNN algorithm:

1. Convolution Layer
   This stage aims to filter the food image that will be recognized. The filter will later shift throughout the image based on the stride parameter and produce an output known as the activation diagram or function diagram. Figure 13.7 shows the layer of convolution 3x3.

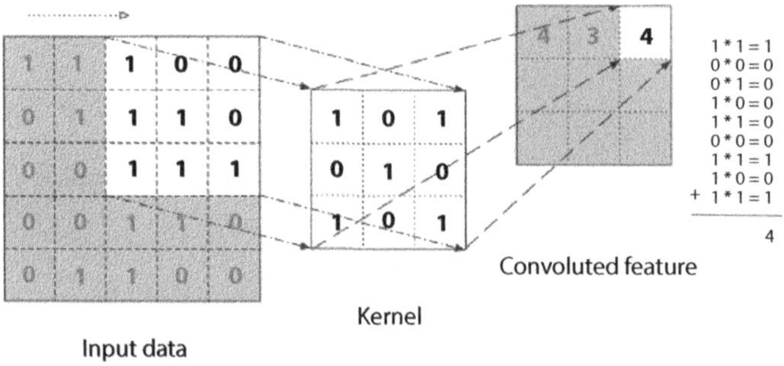

Figure 13.7 Convolution layer.

2. Subsampling Layer
   The next step is to perform a pooling operation consists filter with certain step width, which will move to the entire feature map area or activation map. The purpose of using the pooling layer is reduce the size of feature map, thereby speeding up the calculation speed, because fewer parameters need to be updated and overfitting. Figure 13.8 shows the pooling layer.

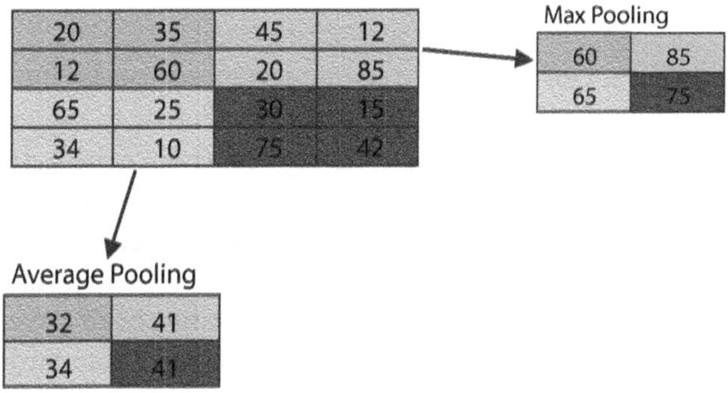

Figure 13.8 Sub sampling layer.

3. Fully Connection Layer
   This stage aims to connect all the extraction layers from the food image to predict the type of food based on the input image. Figure 13.9 shows the fully connection layer.

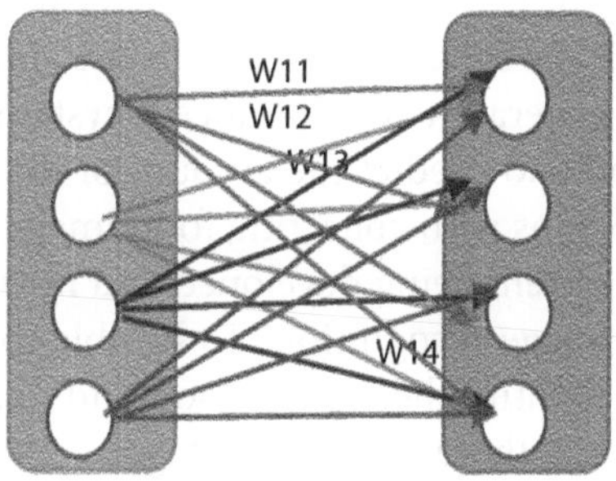

Figure 13.9 Connection layer.

4. Activation Layer
   At this stage all the possibilities that have been previously defined will be matched with the data on the types of food that already exist in the database, so that the final output of the CNN application is a prediction of the type of food based on the image input of the food that will be consumed by the user.

13.3.4.3 Testing Accuracy

Testing the accuracy of the system being developed can be interpreted as how close the measured value is to the actual value or to the value that is considered correct. If in a system accuracy test there is no actual data or values that are considered correct, then it will not be possible to determine how much the measurement accuracy of the system. Accuracy is defined as the difference or close value between the reading of the measuring instrument and the actual value. In a system test that has been developed, the true value which is never known will be replaced with a standard value that has been conventionally recognized.

$$\text{Accuracy (\%)} = \frac{correct\ Answer}{Total\ Testing} * 100$$

The value obtained by using system accuracy testing is the value in the form of a percentage (%) of system accuracy. Correct Answer is the value or number of correct results from the output produced by the system while total number of tests is the total number of data processed by the system. By testing the accuracy of the system, developers can of course evaluate the results or outputs of the system created with the aim of improving the system being built. In general, it will produce a system that can be used to identify types of food and predict the amount of nutrients. Analysis of the research problem begins with collecting food image data to be used as training data from the system being developed and how to do food recognition in the realm of image recognition. To be able to develop a system that can automatically identify the type of food and determine the nutrients contained in the food, there are several stages.

13.3.5 What Is Latest Research Application Status of AI in Functional Foods?

AI Lab is a medical data processing solution through deep learning in the form of information on the patient's disease prediction. The application will calculate, identify, and sort an object or condition of the virus present in the patient's body. This is useful as a foundation for users to make quick decisions (decision support systems) in the process of diagnosing COVID 19 [74]. This technology can be combined to detect food that may be contaminated with the virus. Food have a good effect on health, namely: Functional foods, nutritious foods, pharmaceutical foods, brand-name foods, Vita foods, phytochemical products, vinegar foods, health foods, natural foods and real foods.

According to Japanese scientists, in I Ketut's research, there are several requirements that a product must have so that it can be said to be functional food, namely: (1) It must be a food product (not in the form of capsules, tablets, or powder) derived from ingredients (ingredients) natural, (2) Can and can be eaten as part of diet or daily menu, (3) Has certain functions when digested, and can play a role in certain physical processes, such as: strengthening the body's defense mechanism, preventing certain diseases, helping to restore physical conditions after certain diseases, and maintaining physical and mental health [75], and slow down the aging process [76]. In his research, Arry uses artificial intelligence (AI) in drug discovery which has a major role in the development of new drugs from natural ingredients from previous uses [77]. Research that has been carried out with an artificial intelligence system to optimize the diet according to nutritional needs according to the user within a period of 1 month or 30 days as a dietary solution that can be produced [78].

13.3.6 Future Dimension of AI in Food and Nutrition Research

With the help of artificial intelligence (AI), specially prepared mixtures of spices, recipes and medicines with natural ingredients can be made faster and more consistent in their taste profile and composition. AI has come up with a choice of new spice blends, including Tuscan chicken, bourbon pork tenderloin and New Orleans sausage. Companies sell the mixture to the public, as well as develops new flavors for the food industry [79]. By being able to predict the need for nutrition and healthy food, the detection of disease can be predicted, even the detection of death, can be detected early as revealed in his research for the future AI [80].

13.4 Conclusion

Hence, it can be concluded that the importance of functional foods for human health has become very important as people are becoming more health conscious and aware. This is increasing the demand for development of new functional products with higher accuracy and safety. To cater to these needs, the technological advancements using nanotechnology and artificial intelligence are being explored. Nanotechnology bringing changes in the physiochemical properties of the components, resulting in new products with better delivery and bioavailability, on the other hand, by using the artificial intelligence as a predictive model the efficiency and mechanism of action are understood. Therefore, synergistic use of nanointelligence the development of functional foods has brought about the development of food, health, and nutrition sector.

References

1. Corbo, M.R. *et al.*, Functional beverages: The emerging side of functional foods: commercial trends, research, and health implications. *Compr. Rev. Food Sci. Food Saf.*, 13, 6, 1192–206, 2014. https://doi.org/10.1111/1541-4337.12109
2. Henry, C.J., Functional foods. *Eur. J. Clin. Nutr.*, 64, 657–659, 2010.
3. Martirosyan, D.M. and Singh, J., A new definition of functional food by FFC: What makes a new definition unique? *J. Funct. Food Health Dis.*, 5, 6, 209–23, 2015.
4. Luo, X., Su, P., Zhang, W., Advances in microalgae-derived phytosterols for functional food and pharmaceutical applications. *Mar. Drugs*, 13, 7, 4231–54, 2015. https://doi.org/10.3390/md13074231
5. Csapó, J. and Némethy, S., Functional, health protecting and health maintaining food products. *Ecocycles*, 4, 1, 73–82, 2018.
6. Shandilya, U.K., Functional foods and their benefits: An overview. *J. Nutr. Heal. Food Eng.*, 7, 4, 2–5, 2017.
7. Rasines-Perea, Z. and Teissedre, P.L., Grape polyphenols' effects in human cardiovascular diseases and diabetes. *Molecules*, 22, 2, 1–19, 2017. https://doi.org/10.3390/molecules22010068
8. Alali, F.Q. *et al.*, Garlic for cardiovascular disease: Prevention or treatment? *Curr. Pharm. Des.*, 23, 7, 1028–1041, 2017.
9. Benisi-Kohansal, S. *et al.*, Whole-grain intake and mortality from all causes, cardiovascular disease, and cancer: A systematic review and dose-response meta-analysis of prospective cohort studies. *Adv. Nutr.*, 7, 6, 1052–1065, 2016. https://doi.org/10.3945/an.115.011635
10. Whitehead, A. *et al.*, Cholesterol-lowering effects of oat β-glucan: A meta-analysis of randomized controlled trials. *Am. J. Clin. Nutr.*, 100, 6, 1413–1421, 2014. https://doi.org/10.3945/ajcn.114.086108
11. Kumar, N.B. *et al.*, Randomized, placebo-controlled trial of green tea catechins for prostate cancer prevention. *Cancer Prev. Res.*, 8, 10, 879–887, 2015. https://doi.org/10.1158/1940-6207.CAPR-14-0324
12. Bhattacharya, T. *et al.*, Role of Phytonutrients in nutrigenetics and nutrigenomic perspective in curing breast cancer. *Biomolecules*, 11, 8, 1176, 2021. https://doi.org/10.3390/biom11081176
13. Alemany-Costa, L. *et al.*, Sterol stability in functional fruit beverages enriched with different plant sterol sources. *Food Res. Int.*, 48, 1, 265–270, 2012.

14. Kajla, P., Sharma, A., Sood, D.R., Flaxseed—A potential functional food source. *J. Food Sci. Technol.*, 52, 4, 1857–1871, 2015.
15. Chandrasekara, A. and Josheph Kumar, T., Roots and tuber crops as functional foods: A review on phytochemical constituents and their potential health benefits. *Int. J. Food Sci.*, 20116, 15, 1–15, 2016. https://doi.org/10.1155/2016/3631647
16. Sutar N, Sutar P., Mohapatra D., New horizons in functional food sector: An Indian perspective. *J. Dairying, Foods HS.*, 29, 3/4, 166–72, 2010.
17. Khan, R.S. *et al.*, Functional food product development - Opportunities and challenges for food manufacturers. *Trends Food Sci. Technol.*, 30, 1, 27–37, 2013. https://doi.org/10.1016/j.tifs.2012.11.004
18. Mathers, C.D. *et al.*, Causes of international increases in older age life expectancy. *Lancet*, 385, 9967, 540–548, 2015.
19. Granado-Lorencio, F. and Hernández-Alvarez, E., Functional foods and health effects: A nutritional biochemistry perspective. *Curr. Med. Chem.*, 23, 26, 2929–2957, 2016.
20. Birch, C.S. and Bonwick, G.A., Ensuring the future of functional foods. *Int. J. Food Sci. Technol.*, 54, 5, 1467–1485, 2019.
21. Brown, L. *et al.*, Clinical trials using functional foods provide unique challenges. *J. Funct. Foods*, 45, 233–238, 2018.
22. Gonçalves, R.F.S. *et al.*, Advances in nutraceutical delivery systems: From formulation design for bioavailability enhancement to efficacy and safety evaluation, in: *Trends Food Sci. Technol*, vol. 78, pp. 270–291, Elsevier, USA, January 2018.
23. Shahidi, F. and Peng, H., Bioaccessibility and bioavailability of phenolic compounds. *J. Food Bioact.*, 4, 11–68, 2018.
24. Jafari, S.M. and McClements, D.J., Nanotechnology approaches for increasing nutrient bioavailability, in: *Advances in Food and Nutrition Research*, 1st, vol. 81, pp. 1–30, Elsevier Inc., USA, 2017.
25. Bhattacharya, T. *et al.*, Effect of natural leaf extracts as phytomedicine in curing geriatrics. *Exp. Gerontol*, Elsevier., 150, 111352, 2021. https://doi.org/10.1016/j.exger.2021.111352.
26. Nasrollahzadeh, M. *et al.*, An Introduction to Nanotechnology, in: *Interface Science and Technology*, 1st, vol. 28, pp. 1–27, Elsevier Ltd, India, 2019.
27. Ealias, A.M. and Saravanakumar, M.P., A review on the classification, characterisation, synthesis of nanoparticles and their application. *IOP Conf. Ser. Mater. Sci. Eng*, vol. 263, 2017.
28. Fursik, O. *et al.*, Some aspects of using the nanotechnology in food industry. *Ukr. J. Food Sci.*, 7, 2, 298–306, 2019.
29. Mohan Bhagyaraj, S. and Oluwafemi, O.S., Nanotechnology: The science of the invisible, in: *Synthesis of Inorganic Nanomaterials*, pp. 1–18, Elsevier, India, 2018.

30. Sponton, O.E. *et al.*, Nanosupplements: Improvements in nutritional functionality of food products. In: *Food Applications of Nanotechnology*, CRC Press, 99–118, 2019.
31. Magalhães, W.L.E. *et al.*, Nanotechnology applied to improve functionality in food, in: *Nutrient Delivery*, pp. 177–219, Elsevier, USA, 2017.
32. Bhattacharya, T. *et al.*, A review on natural sources derived protein nanoparticles as anticancer agents. *Curr. Top. Med. Chem.*, 21, 12, 1014–1026, 2021.
33. Punia, S. *et al.*, Nanotechnology: A successful approach to improve nutraceutical bioavailability, Springer, 119–133, 2019.
34. Mishra, S.S. *et al.*, Advances in probiotics, prebiotics and nutraceuticals. *Innov. Technol. Fermented Food Beverage Ind.*, Springer, India, 121–141, 2018.
35. Knijnenburg, J.T.N., Posavec, L., Teleki, A., Nanostructured minerals and vitamins for food fortification and food supplementation, in: *Nanomaterials for Food Applications*, pp. 63–98, Elsevier, USA, 2018.
36. Akhoond Zardini, A. *et al.*, Production and characterization of nanostructured lipid carriers and solid lipid nanoparticles containing lycopene for food fortification, in: *J. Food Sci. Technol*, vol. 55, pp. 287–298, Springer, India, 2018.
37. Santillán-Urquiza, E., Méndez-Rojas, M.Á., Vélez-Ruiz, J.F., Fortification of yogurt with nano and micro sized calcium, iron and zinc, effect on the physicochemical and rheological properties. *Lwt.*, 80, 462–469, 2017.
38. Razack, S.A. *et al.*, Green synthesis of iron oxide nanoparticles using Hibiscus rosa-sinensis for fortifying wheat biscuits, in: *SN Appl. Sci*, vol. 2, Springer, India, 2020.
39. Muhammad, D.R.A. *et al.*, Physicochemical properties and antioxidant activities of chocolates enriched with engineered cinnamon nanoparticles, in: *Eur. Food Res. Technol*, vol. 244, pp. 1185–1202, Springer, Berlin Heidelberg, 2018.
40. Thiruvengadam, M., Rajakumar, G., Chung, I.M., Nanotechnology: Current uses and future applications in the food industry, in: *3 Biotech*, vol. 8, pp. 1–13, Springer, Berlin Heidelberg, 2018.
41. Sahoo, M. *et al.*, Nanotechnology: Current applications and future scope in food. *Food Front.*, 2, 1, 3–22, 2021.
42. Singh, T. *et al.*, Application of nanotechnology in food science: Perception and overview. *Front. Microbiol.*, 8, 1–7, AUG. 2017.
43. Warwick, K., *Artificial intelligence: the basics*, Routledge, 2013.
44. Goyache, F. *et al.*, The usefulness of artificial intelligence techniques to assess subjective quality of products in the food industry. *Trends Food Sci. Technol.*, 12, 10, 370–381, 2001.
45. Sneha, B., Harave, V., Latha, R., An overview on structure, characteristics and development of an expert system. *Int. J. Comput. Sci. Trends Technol.*, 8, 2, 32–36, 2020.
46. Kaplan, A. and Haenlein, M., . Siri, Siri, in my hand: Who's the fairest in the land? On the interpretations, illustrations, and implications of artificial intelligence, in: *Bus. Horiz*, vol. 62, pp. 15–25, Kelley School of Business, Indiana University, USA, 2019.

47. Adir, O. *et al.*, Integrating artificial intelligence and nanotechnology for precision cancer medicine. *Adv. Mater.*, 32, 13, 1–15, 2020.
48. Ertel, W., *Introduction to Artificial Intelligence*, 2nd, I. Mackie (Ed.), p. 356, Springer US, Switzerland, 2017.
49. Sutojo, T. and Mulyanto, E., Vincent. Kecerdasan Buatan. Andi Offset, Yogyakarta, Indonesia, 2011.
50. de Moraes Lopes, M.H.B. *et al.*, Use of artificial intelligence in precision nutrition and fitness, in: *Artificial Intelligence in Precision Health*, pp. 465–496, 2020.
51. Hassani, H. *et al.*, Artificial Intelligence (AI) or Intelligence Augmentation (IA): What Is the Future? *Journal AI.*, 1, 2, 143–155, 2020.
52. Pohl, J., Artificial Superintelligence: Extinction or Nirvana? // Artif. Superintelligence Extinction or Nirvana? In: *International Symposuim*, 3 August, 2015.
53. Bunyamin, J.B., Agi (Artificial General Intelligence): Peluang Indonesia Melompat Jauh Kedepan. *J. Sist. Cerdas.*, 01, 02, 1–11, 2018.
54. Atreides, K., Choices in a mediated artificial super intelligence assisted world: The future before us, in: *The Transhumanism Handbook*, pp. 189–214, 2019.
55. Russell, S. and Norvig, P., *Artificial Intelligence: A modern Approach*, 3rd, Pearson Education, New Jersey, 2010.
56. Neethirajan, S. and Jayas, D.S., Nanotechnology for the food and bioprocessing industries. *Food Bioprocess Technol.*, 4, 1, 39–47, 2011.
57. Kovalenko, O., Machine learning and AI in food industry: Solutions and potential [Electronic resource] // SPD. Group, 2020. URL: https://spd.group/machine-learning/machine-learning-and-ai-in-food-industry/ (accessed: 28.02.2021).
58. Raghuveer, C., Review article consumption of functional food and our health concerns. *Pak J. Physiol.*, 5, 1, 76–83, 2009.
59. Sundaravadivel, P. *et al.*, Smart-log: A deep-learning based automated nutrition monitoring system in the IoT. *IEEE Trans. Consum. Electron. IEEE*, 64, 3, 390–398, 2018.
60. Kim, J.C. and Chung, K., Knowledge-based hybrid decision model using neural network for nutrition management, in: *Inf. Technol. Manag.*, vol. 21, pp. 29–39, Springer, US, 2020.
61. Yu, S.-S. *et al.*, Two improved k-means algorithms. *Appl. Soft Comput.*, 68, 747–755, 2018.
62. Baek, J.W. *et al.*, Hybrid clustering based health decision-making for improving dietary habits. *Technol. Heal. Care*, 27, 5, 459–472, 2019.
63. Kagaya, H., Aizawa, K., Ogawa, M., Food detection and recognition using convolutional neural network. *MM 2014 - Proc. 2014 ACM Conf. Multimed*, pp. 1085–1088, 2014.
64. Sak, J. and Suchodolska, M., Artificial intelligence in nutrients science research: A review // *Nutrients*, 13, 2, 1–17, 2021.
65. Lu, Y. *et al.*, An artificial intelligence-based system to assess nutrient intake for hospitalised patients, *IEEE Trans. Multimed.*, 23, 1136–1147, 2020.

66. Oka, R. *et al.*, Study protocol for the effects of Artificial Intelligence (AI)-supported automated nutritional intervention on glycemic control in patients with Type 2 diabetes mellitus. *Diabetes Ther. Springer Healthc.*, 10, 3, 1151–1161, 2019.
67. Vasiloglou, M.F. *et al.*, A comparative study on carbohydrate estimation: GoCARB vs. Dietitians. *Nutrients*, 10, 6, 1–11, 2018.
68. Wahyu, A.P. *et al.*, Smart canteen: Perilaku artificial intelligence: Predictive approach for foodborne bacterial pathogen inactivation Mengatur Pola Makan Dengan Membaca Nilai Nutrisi (Conventional Deep Learning Neural Network). *J. Ilm. Teknol. Infomasi Terap.*, 6, 2, 115–121, 2020.
69. Fourati, M. *et al.*, Multiresponse optimization of pomegranate peel extraction by statistical versus. *Evidence-Based Complement. Altern. Med.*, 2019, 1–18, 2019. https://doi.org/10.1155/2019/1542615.
70. Funes, E. *et al.*, A review: Artificial neural networks as tool for control food industry process. *J. Sens. Technol.*, 05, 01, 28–43, 2015.
72. Mezgec, S. and Seljak, B.K., Nutrinet: A deep learning food and drink image recognition system for dietary assessment. *Nutrients*, 9, 7, 1–19, 2017.
73. Eka Putra, W.S., Klasifikasi Citra Menggunakan Convolutional Neural Network (CNN) pada Caltech 101. *J. Tek. ITS.*, 5, 1, 65–69, 2016.
74. Setiawan, A., Gede, P., Cipta, S., Prediksi Citra Makanan Menggunakan Convolutional Neural Network Untuk Menentukan Besaran Kalori Makanan, in: *Jurnal Teknik Informatika dan Komputer,* vol. 6, pp. 30–38, 20202020, URL: https://indonesia.go.id/kategori/kuliner/1878/kecerdasan-buatan-pendeteksi-virus-covid-19 (accessed: 03.03.2021). 71.
75. Rahman, M.H. *et al.*, Resveratrol and neuroprotection: Impact and its therapeutic potential in Pendeteksi virus Covid-19 [Electronic resource] // Portal Informasi Ind Alzheimer's Disease. *Front. Pharmacol.*, 11, 61920, 2020.
76. Suter, I., Pangan Fungsional dan Prospek Pengembangannya. *In: Pentingnya Makanan Alamiah (Natural Food) Untuk Kesehatan Jangka Panjang,* Udayana University, Indonesia, 1–17, 2013.
77. Hasyim, I., Guru Besar UI Teliti Pemanfaatan AI untuk Penemuan Obat [Electronic resource]. *KOMPAS*, 2019. URL: https://tekno.tempo.co/read/1230920/guru-besar-ui-teliti-pemanfaatan-ai-untuk-penemuan-obat/full&view=ok (accessed: 05.03.2021).
78. Mustafa, W.F., Djamal, E.C., Yuniarti, R., Optimalisasi Menu Makan Diet Sehat Menggunakan Algoritma Genetika. i*n Seminar Nasional Informatika dan Aplikasinya (SNIA)*, 50-54, 2017.
79. Admin, Spice blends created by AI [Electronic resource]. *Club House For Chefs*, 2019. URL: https://www.clubhouseforchefs.ca/news/spice-blends-created-by-ai (accessed: 01.03.2021).
80. Hasan, R.A., Riset: Kecerdasan Buatan Ini Bisa Prediksi Kematian Dini pada Manusia [Electronic resource]. *Liputan6.com*, 2019. URL: https://www.liputan6.com/global/read/3930533/riset-kecerdasan-buatan-ini-bisa-prediksi-kematian-dini-pada-manusia (accessed: 28.02.2021).

14

Nanofortification: An Emerging Technique for Enhancement of the Functionality of Foods with Ample Future Scopes

Shiladitya Ghosh[1], Dolanchapa Sikdar[2] and Kakali Bandyopadhyay[2]*

[1]Amity Institute of Biotechnology, Amity University Kolkata, Rajarhat, Newtown, Kolkata, West Bengal, India
[2]Department of Food Technology, Guru Nanak Institute of Technology, Kolkata, West Bengal, India

Abstract
Fortification of common foods through addition of health-promoting and nutrition-boosting additives in the form of supplemental elements has long been a common practice to achieve the enhancement of food functionality. Nanotechnology has a very vital role to play in this field of functional foods and enhancement of functionality. Addition of competent nanoparticles alters the functionality of foods and improves the delivery and bioavailability of the fortification additives inside the host body. Present chapter focuses on the recent advancements of the "nanocarrier mediated fortification" or "nanofortification" for enhancement of the functionality of fortified foods. Remarkable varieties of nanomaterials have been synthesized and applied as carriers of different bioactive fortification agents in numerous foods, which successfully incremented the bioavailability with boosting of immunity. Effects of different nanoparticle (NP) treatment on multiple additives and other bioactive compounds and addition of the NP-treated additives in various staple foods have been presented. Application of NPs for food fortification regarding their nutrient-carrying potential, warranted biocompatibility from toxicological studies and safety assessment for enabling human consumption have been thoroughly discussed. The lingering challenges and the ample future scopes associated with this promising field of nanofortification have been vividly highlighted to pave the way for its steady progress.

Keywords: Food fortification, vitamins and minerals, bioactive compounds, nanotechnology, nanofortification, functionality enhancement, assessment of state-of-the-art

**Corresponding author*: head_ft.gnit@jisgroup.org

Tanima Bhattacharya and Shakeel Ahmed (eds.) Nanotechnology in Functional Foods, (359–394)

14.1 Introduction

Even in the 21st century a great fraction of the global population still suffer from nutrient and micronutrient deficiency and resultant health adversities [1]. Occurrence of this global health issue is more aggravated in the developing countries where options of micronutrient-rich food is limited, and food supply is not equalized among all the economic layers of the society. Micronutrients, namely vitamins and minerals, are crucially essential for development and functioning of all vital systems of human body and enables physical and mental well-being [2]. Micronutrient deficiencies may result in irreversible damage of physical and cognitive abilities and several critical illness/diseases [3]. Therefore, increasing the micronutrient content of the staple foods of different countries is a matter of primary concern and traditionally tackled through fortification. Fortification is defined as the external addition of key micronutrients like vitamins (A, D, E, C, B complex, etc.) and minerals (iron, iodine, zinc, calcium selenium, etc.) to staple foods for incrementing their micronutrient contents [4]. Depending on the reported cases of particular type of micronutrient deficiency in a country it is managed through national level fortification programs as public health strategy [5]. Another aspect of fortification is to improve the bioavailability of the micronutrients, which is known to be low for most of them [6]. Food fortification is the manual intervention, which aims at enhancing the content, as well as the bioavailability of those in staple foods for reduction of the associated mal-nutritional effects. Recently, the worldwide outbreak of the malicious coronavirus has indicated that strengthening the self-immunity of the body, and its maintenance should be considered as a priority of general life rules [7, 8]. Role of micronutrients in prevention of pathogen invasion (including the viral ones) and recovery from an infection has been observed to be crucial [7, 8]. This can only be achieved through supply of micronutrient-rich food to the communities. Hence, there is a sudden surge of focus on the food fortification, and its speedy progress has been realized as an urgent issue. Over the last decades, several techniques and methods have been devised for micronutrient fortification [9]. However, in light of the current situation, the interest is shifting toward application of the advanced technologies like nanotechnology in the area of food fortification, over the conventional ones.

Functional foods are the carriers of several bioactive ingredients that perform specific beneficial roles in our body [10, 11]. However, the definitive concept of functional food is still evolving regarding the contents and extents of the compounds responsible for their functionality. Apparently the functionality of a native food material may be enhanced by the addition of bioactive compounds to it through a specific approach [12]. Although consumption of supplements containing different micronutrients is a way of increasing their intake, several limitations like; poor solubility, low rate of absorption, lower tolerance toward the physical-chemical conditions of the digestive tract hamper their actual bioavailability [13, 14]. Hence, protected, targeted, and controlled delivery and release of these bioactive compounds has become a major concern to preserve and enable their complete bioactivity through ensuring their bioavailability. Development of encapsulated systems of food matrices carrying the specific micronutrients for targeted release to their actual site of action is practiced for this purpose [15]. Nanocarrier mediated delivery of the micronutrients has become an exciting area of research in this regard and is being considered as the most advanced fortification technique to emerge in near future. Numerous types of nanocarrier and nanodelivery systems have already been developed and used for fortification of different micronutrients [16–18]. Present chapter aims at discussing several aspects of the nanomediated fortification or nanofortification technology with respect to its application in fortification of different food products. Examples of major micronutrients, including vitamins, minerals, and other bioactive compounds, have been cited and discussed particularly focusing on their role in fortification. From the overall discussions made in the chapter, a wholesome overview of this promising technology is portrayed and the status of knowledge and its future paths have also been highlighted.

14.2 Fortification of Food and Functionality Enhancement

In a country with reported micronutrient deficiencies, population should have a well-balanced diet with improved food availability and bioavailability of micronutrients [9, 19]. Increase of food consumption with food fortification and pharmaceutical supplement is needed to be implemented and emphasized in a more in depth manner. Among all, fortification of food is a medium to offer long-term solution to meet the need of nutrient in a population toward the sufficient levels [19, 20]. Food fortification involves identification and selection of vehicles, which are actually commonly eaten foodstuffs that are processed through central metabolic routes in the body to extract and provide the micronutrients [20, 21]. Hence, the conventional food pattern does not get hampered, and no need of special consent of an individual is necessary. As staple foods in most of the developing countries, the choice of vehicles is restricted to few options and that includes milk and milk products, wheat and wheat products, corn, rice, edible fats and oils, and other beverages [9, 20, 21]. The materials, which are typically targeted and used for food fortification, are vitamins and minerals, as their availability and absorption are reported to be minimum in the staple foods of many countries.

Fortification of food is done and mainly aimed to provide sufficient amount of nutrients by normal consumption of the food vehicles. Therefore, variation of food consumption may affect the safety of the population with proper dose and efficacy [20]. Fortification of food needs careful attention toward requirement of food regulation and labeling with proper declaration. A number of basic principles were outlined by the Codex Alimentarius Commission in 1987 for adding external nutrients to foods (Codex Elimentarius, 1987), which are also obeyed by the World Health Organization (WHO) and food and drug administration (FDA) [22].

The list of general facts and principles to be followed for addition of essential nutrients to food are as follows [4, 22]:

- Considering the amount of nutrient from other sources, the essential nutrient should be added at such amount that it will not be of excessive or inadequate intake but in balanced amount.
- The essential nutrient added to food vehicles should not have any negative effect on other nutrients naturally present and their metabolism.
- During packing, storage, and distribution, the condition of the essential nutrient should be stable enough to reach and meet the customer requirements intended.
- The overall characteristics or sensory/organoleptic properties (color, flavor/aroma, texture, taste, cooking properties) and shelf-life of a food product should not be affected by the essential nutrient externally added to the food.
- The knowledge of addition of essential nutrient should be very clear to the customer as per nutritional advantage and must not be misleading.
- There should be availability of proper processing facility and efficient technology that allows the satisfactory addition of essential nutrient to food.
- There should be proper methods for measurement, monitoring and regulation at each stage of the fortification process.

14.3 Types and Methods of Food Fortification

In general, the food items of the staple diet of many countries mainly consist of processed foods observably rich in three "major or macro" nutrients, namely carbohydrates, proteins, and fats. These macronutrients are essential for development of each and every tissue of human body and basic functioning of the organs/systems formed by those as they serve as the building blocks and bulk energy provider [23]. The micronutrients, majorly the vitamins and minerals, are also deemed essential status due to their role in growth, maintenance and proper functioning of all biological systems [24, 25]. Most foods are prepared from raw food materials like cereals, grains, vegetables, fish, egg, and meat by simple traditional processing, such as cooking. Despite the sufficient supply of the three macronutrients, many recipes lack the components of raw food materials rich in micronutrients. The use of water and heat during cooking processes also lead to loss of water soluble micronutrients (iron, zinc, etc.) and the heat sensitive ones from the ingredients [26, 27]. This often results in a micronutrient-deficient food items in the diets comprising them. It is a long time realized fact that the foods deficient in micronutrients like vitamins, minerals, antioxidants and other elements important for nutrition can be enriched by external addition of these nutrients from different sources. There are several methods that are applied for the purpose of fortification of micronutrient deficient foods.

The overall purpose intended to achieve through food fortification can be generally understood by simply dividing into two classes namely:

1. **Restoration:** As mentioned in the previous section loss of micronutrients from the ingredient raw foodstuffs during processing and preparation of final form of food items is a very common phenomenon [26, 27]. In this case, the micronutrient deficiency may be overcome by restoring to the previous content through fortification with minerals and vitamins [4, 28]. This replenishment of micronutrients in a processed food is a very common form of food fortification. For this type it is important to know the natural micronutrient composition and content of the foodstuffs which is losing those during the processing. The particular minerals or vitamins lost from the foodstuff may be selectively added back to the foodstuffs through fortification.

2. **Enrichment:** The more important and major type of food fortification is the enrichment. This is defined by the addition of some specific micronutrients to a foodstuff regardless of its natural micronutrient composition or content [4, 22]. This is conducted based on the reported data obtained from the survey of nutritional status of a large population of a country. The micronutrients, which are found to be lacking in the staple food of that country, are selectively added to the food products (the vehicles for the micronutrients) at different stages of processing. This addition facilitates enrichment of the content of those particular micronutrients (minerals, vitamins or both) in the vehicle or carrier foods [4, 22]. Followed by a stringent monitoring of the stability of the micronutrients in the resulting fortified foods and satisfactory status of all aspects of consumer safety, they are made available to the population suffering from deficiency of multiple micronutrients through national policy-driven programs [29].

Survey and analysis of the literatures reported on different food fortification technologies, and in case studies, it appears that the concept fortification has a multifaceted existence. Therefore, we attempted to provide the classifications based on and considering the variation found in the literatures. Depending on the extent of application or scale of coverage, as per WHO [4], these food fortification approaches can be subclassified to the following categories:

A. **Mass fortification:** In this approach, the particular food items which are globally popular among the populations of multiple countries are specifically identified and targeted as the vehicle foods for fortification [4, 30]. The food vehicles chosen to be fortified under this approach are widely consumed by the people belonging to a broad age group (children to middle-aged adults) and large fraction of world population [9, 31]. Examples of such food vehicles include wheat, rice, milk, salt, sugar, etc. In mass fortification programs, micronutrient enrichment is intended as the major objective for the purpose of supplying multiple micronutrients collectively to all layers of a population [4, 9].
B. **Targeted fortification:** The food vehicles chosen under this category of food fortification are specifically consumed by

lower age group individuals of a population, i.e., the infants [31, 32]. The major food vehicle for this fortification is milk used in the formulation of complementary mixes for infants or young children [9, 31]. Very recently, fortification of human milk is also being attempted for targeted application on the preterm babies facing problems in growth and development post birth. In the developing countries like India, deficiency of vitamin D is a very common problem arising from the malnutrition due to inadequate and nonuniform supply of milk to the all economic layers of the population [31, 32]. Milk is predominantly used as the main fortification carrier for vitamin D in such cases. Sometimes the targeted fortification is also carried out to provide micronutrients to a displaced community, i.e., the refugees of a country [9, 31].

C. **Market driven fortification:** This is a situation oriented strategy, in which an established food manufacturing concern voluntarily fortifies a specific food product popular among a particular consumer niche of a region [4, 9, 33]. However, the fortification must comply with the governmental policies and regulations mandated for food fortification practices of that country. Several processed foods are used as the food vehicles in this approach. This is majorly adopted in the industrialized and developed countries (like European Union), where impact of processed foods on public health is well established [4]. In the industrialized countries where multiple varieties of processed foods are consumed, the demand for a particular food product in the market acts as the driving factor behind the market-driven fortification of that product. In this case the action, although voluntary, is also employed as a strategy to tackle market competition or enhance the endorsement of that product [9, 33]. Through this mode of fortification certain, selected micronutrients (e.g., calcium, iron, vitamins C and B2, etc.) are added as forticants which are not possible to add in sufficient amounts to the food vehicles chosen for mass fortification of staple foods due to economic and safety restrictions [4, 9, 33].

The techniques and methods used for food fortification have significantly advanced over the last decades. Based on the actual scientific techniques applied to achieve food fortification in conventional and advanced ways, it can be classified to the following ones:

I. Food-to-food fortification

This is the most conventionally adopted method reported for food fortification. Food-to-food fortification may be simply defined as—addition of a naturally micronutrient-rich food material of either plant or animal origin to another food material which is having the micronutrient content below the standard value marking the lower limit of the sufficient range [4, 34, 35]. As per the definition it is easy to perceive that food-to-food fortification can be achieved domestically or household basis if both types of food components namely; the micronutrient rich one and the deficient one are available locally [33, 34]. One related but distinct form of this method is to simply incorporating more micronutrient-rich foods in the overall diet, which may not necessarily require mixing of food to food and traditionally known as the dietary diversification [4, 34]. The technique of food-to-food fortification is represented in Figure 14.1.

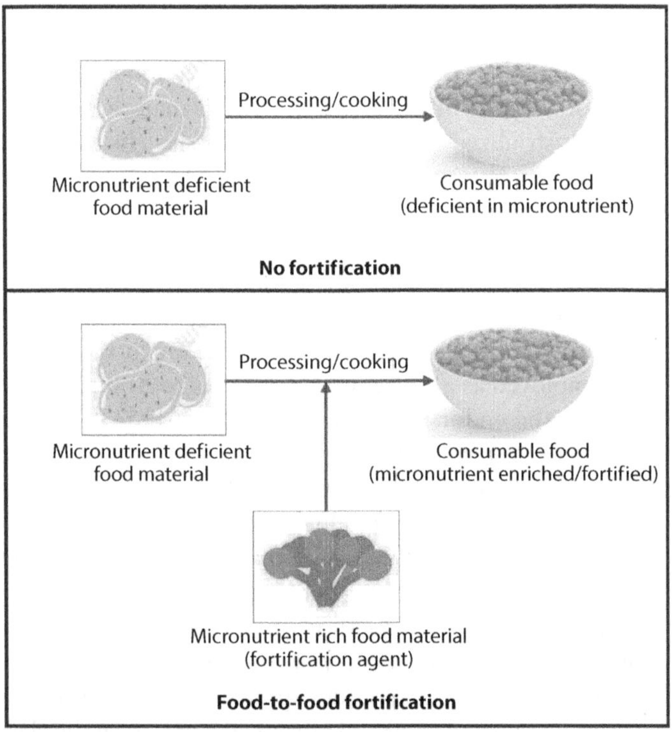

Figure 14.1 Food-to-food fortification for enrichment of micronutrient in consumable food.

However, these two should be compatible with each other for mixing and must not produce any offensive products through interreaction. This makes it necessary to practice the food-to-fortification using properly designed scientific protocols, so that the stability and safety aspects of the fortified foods can also be assessed and guaranteed [34]. Due to this reason, this mode of fortification is also considered under the advanced methods of food fortification. To achieve a fortified recipe of new food product or formulation through food-to-food fortification clear and complete knowledge of the composition and micronutrient content of both the fortifying food element as well as to-be-fortified food must be available [33, 34]. Knowledge on the bioavailability of the micronutrients and presence of any inhibitory substance is also vital regarding efficient practice of food-to-food fortification. This requires use of advanced analysis protocols and tools/equipments. With these advances, the food-to-food fortification is currently being practiced as a major food fortification technique worldwide.

II. Biofortification

As the conventional food fortification has been found unsatisfactory regarding retaining of the micronutrients in the fortified formulations and enhancement of bioavailability of those even after entering the systems of human body, more advanced strategies guaranteeing these aspects have been sought in the last decade [9, 31]. Progressively increasing reports of micronutrient deficiency driven malnutrition and diseases in many countries is another factor pushing the conventional fortification techniques to emerge into technologically advanced ones. One such fortification technology is biofortification [31, 36].

Biofortification is defined as a process of elevating the contents of essential micronutrients (vitamins, minerals, trace elements, amino/fatty acids, antioxidants, etc.) in a food crop during its production [31, 36]. Biofortification is achieved through plant breeding, agronomical practices, employing advanced biotechnological techniques, like genetic manipulation or transgenic techniques or microbial/microbiological intervention. For increased synthesis of a micronutrient within the crop/plant particularly vitamins and antioxidants, the plant breeding techniques are used conventionally

[36, 37]. However, very long timeframe required to achieve success through this technique has shifted the focus toward the advanced biotechnological techniques namely; development of transgenic or genetically modified crop variety [31, 36, 37]. These modified crops are able to synthesize the targeted micronutrients with increased amount through enhanced expression of the responsible genes, which are either modified or newly introduced. The most popular example of the biofortified transgenic crop is the "golden rice," which has the ability to produce elevated amount of the precursor provitamin A [38]. In the agronomic intervention approach particularly the mineral content (Zn, Fe, Se, etc.) of plants are targeted through the supplementation of mineral-rich fertilizers to the soil of the cultivation field and mobilizing their transport to the plant body [36, 37]. These biofortification techniques are broadly practiced in different regions of the world. Figure 14.2 depicts the techniques of

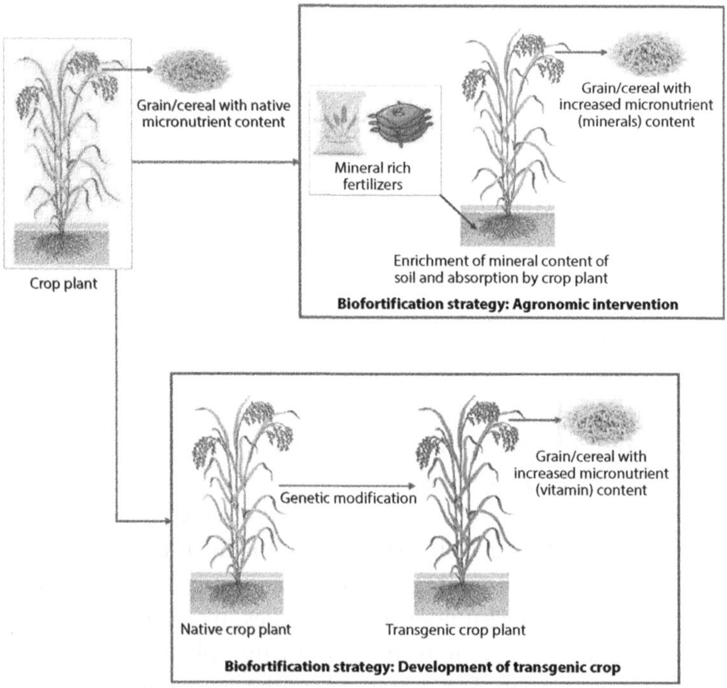

Figure 14.2 Major strategies of biofortification for enhancement of micronutrient content of raw food materials.

agronomic intervention and transgenic plant development, practiced as the major strategies of biofortification.

Biofortified food crops have been reported to exhibit measureable improvements in the health and nutrition of the consumers. This indicates the efficacy of fortification attained by this technique. As the major focus of the present chapter is not on biofortification the discussion is not extended further and interested readers are redirected to other publications detailing on biofortification [36–38].

III. Nanofortification

The abovementioned methods and techniques used for micronutrient fortification of foods have been found to suffer from many limitations. In those techniques the micronutrients are added in their native forms without any modifications as the prime fortificants to the food vehicles. Particularly, in the stage where the fortificants are entering the system of the consumers they face problems like low solubility, low stability, low permeability, and consequently lowered bioavailability in the gut and intestinal tract [13, 14]. This leads to excretory release of the unspent micronutrients despite successful delivery to the digestive system. In such scenario application of nanotechnology for the purpose of food fortification has been found to be beneficial, which has the potential to tackle these problems faced in other methods to significant extents [39, 40]. Hence, this advanced fortification technology is being noticed as the most dynamically emerging technology in the present time. The succeeding sections of this chapter dedicatedly present the principle, case studies and overall performance of this technology in the field of food fortification.

14.4 Application of Nanotechnology in Food Fortification: Nanofortification

Nanotechnology is undoubtedly one of the most swiftly progressing fields in the core areas of science and technology as a whole. Currently, its application spans from the basic applied science to complex engineered systems; concerning the food based ones as well. As per the information available in the archive of literature reporting on application of nanotechnology in food processing or food production industry, significant works have been found related to fortification of food [9, 39, 40]. Therefore, it can be viewed as one of the most progressive areas in its particular field. While on one hand, several nanostructured additives and fortificants are being incorporated within many food products for upgradation of the overall nutrient quotient, on the other hand different nanoscale materials are also used for synthesis of novel food packaging materials and in detectors [39–42]. Therefore, latest applications of nanotechnology can be found "in and around" the food products and processes.

This highlights both its importance and acceptance regarding its efficacy of application for attaining the purpose intended for, i.e., fortification of foods. In the present chapter, the concept and cases of fortification of food products through the usage of nanoscale nutrients (minerals, vitamins and bioactive agents) has been collectively termed as "nanofortication" [42]. The major aim of nanofortification, regardless of the type of nutrient is to increase the bioavailability of the particular nutrients inside the human body so as to raise the degree of absorption and its level in the body. This way, the nanofortified foods act as the carrier of several important nutrients, reports of deficiencies of which are often found due to their poor bioavailability and consequent low absorption in the nonfortified counterparts of the same food products [9, 39, 40]. Poor solubility of these nutrients in their naturally occurring state is the major hindrance factor which results in their poor bioavailability [13, 42]. Reduction of the dimension of these nutrient particles from micro scale to the nano scale ranges has been found to increase their solubility in the liquid fraction of the carrier food products and also inside the systems of the human body, in turn increasing their bioavailability [42]. This is the major engineering aspect of the synthesis of nanofortificants which are added to the food products for preparing the nanofortified foods. As the immune system of the human body is stringently dependent on the supply and availability of these nutrients for its regulation and functionality, the nanofortified foods will also enhance that. However, besides these advantageous traits there are several issues that need to be researched upon and solved properly if the full scale implementation of the nanofortifed foods is to be realized globally. Particularly, this can be a very promising and important development for the countries with scenarios of severe cases of low-nutrition containing foods and resultant malnutrition affecting all age groups.

14.5 Systems of Competent Nanoparticles and Fortification Agents

Nanofortification can be described in simple terms as a technique to deliver micronutrients in the predefined quantities to the surface or bulk of specific food vehicles [39, 40, 42]. Multiple research reports are available describing the addition of nanoscale micronutrients namely; vitamins, minerals and other bioactive compounds (antioxidants, nutraceuticals, etc.) to different food carriers as nanofortification [42, 43]. Human body is adapted to naturally occurring nanoparticles that are present in several foods but the ones specifically engineered and artificially synthesized for nanofortification may possess new morphological, physical and chemical properties that may exhibit some different response in the body. The nano-formulations comprising the nanofortificants are engineered in different applicable forms like, nanoscale powders/particles, nanoencapsulated particles, nanoemulsion, nanoliposome, nanocomposites, nanonutraceuticals, nanotubes, nanochelates, etc. The targeted micronutrients are sized-down to nanoscale materials and delivered using one of these forms through the food carrier without hampering its colour and taste [42]. In the following sections the different nanofortificant delivery systems have been discussed in a case-specific manner. The overall conceptual scheme of synthesis of nanofortificants containing different micronutrients for their use as external additives to compatible food vehicles is depicted in Figure 14.3.

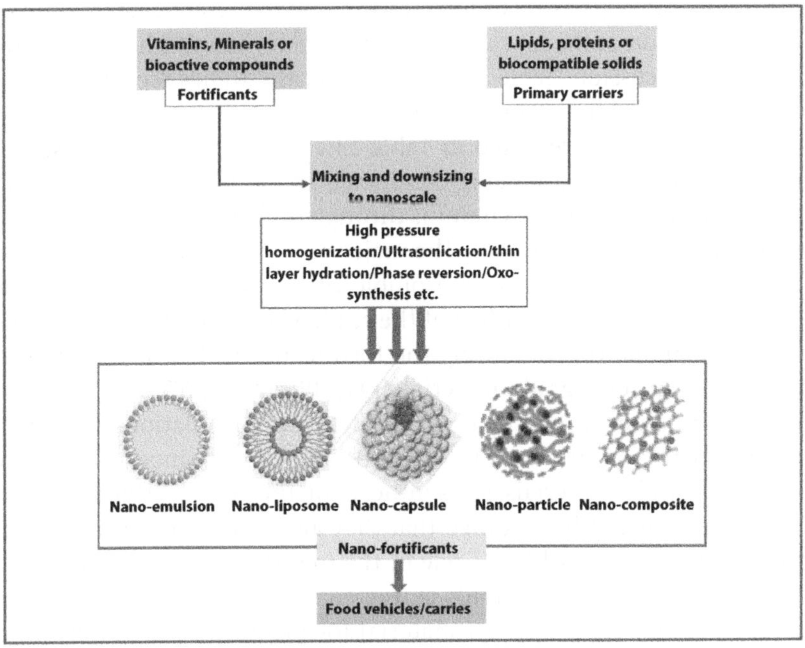

Figure 14.3 Synthesis of various types of nanofortificants containing different micronutrients.

14.5.1 Nanofortification of Vitamins

In the reported case studies of vitamin nanofortification, vitamin D and its derivative forms have been found to be the dominating fortificant that is being used as the targeted micronutrient for fortification of different food vehicles and carriers [44]. This also highlights the fact that it is the most well-known vitamin regarding its deficiency in staple foods of many countries. Being an essential micronutrient required for the proper development, bone mineralization and functioning of the skeletal system of human body the role of this vitamin is absolutely crucial. Hence, a major fraction of the research attempts have focused on nanofortification of vitamin D [44]. Vitamin K also helps in calcium absorption and metabolism in blood system and influences the development of bones along with vitamin D [45]. Vitamin A is an essential element for maintenance of health of the eyes and vision. Vitamin E is plays very important role in maintenance of external body parts like hair and skin and inside the body it serves as an immunomodulator for fighting carcinogenic diseases [45]. These vitamins are also used in nanotechnology oriented fortification studies. Although, the water soluble vitamins namely B and C are essentially required for other functional activities of human body, their water soluble nature makes them comparatively easier for fortification. In the present chapter only the nanofortification of the fat soluble or liposoluble vitamins have been considered for discussion. Table 14.1 represents case studies reporting on different liposoluble vitamins for nanofortification.

The information presented in Table 14.1 clearly indicate that significant research studies are being conducted in the recent years on nanofortification of liposoluble vitamins through different forms of nanoscale carriers and products [46–60]. The major aim of the studies on preparation of nanofortificants with liposoluble vitamins, as per Table 14.1, can be divided into two parts. Most of the studies focused on preparation of the nanoformulation containing the fortificant inside capsules/carriers prepared from food grade materials [47, 49, 52–55, 57–58, 60], while, some of the studies investigated use of those formulations for fortification of real food products [46, 48, 50, 51, 56, 59]. Since all of studies used food grade materials for synthesis of the nanofortificant loaded formulations, it can be assumed that they are fit for application in food vehicles for humans, which is one of the major concluding remarks for those studies. Lecithin or lipids extracted from soybean were found to be very popular choice of raw material for production of nanofortified vitamin formulations, Table 14.1 [54, 55, 58, 60]. One interesting fact is that in most of the cases liquid oil/lipid based carriers were used for

Table 14.1 Reported cases of nanofortification of liposoluble vitamins.

Micronutrient/ fortificant	Vehicle/carrier of fortificant	Quantity of fortificant added	Nanofortified product/ formulation	Synthesis technique/s	Response or salient findings	Ref.
Vitamin D_3, K1 and B6	Skim milk powder	D_3: 0.00055 (%w/w) K1: 0.0001 (%w/w) B_6: 0.015 (%w/w)	Nanofortified milk powder	High pressure (200-400 bar) homogenization	Increased intestinal absorption and bioavailability of calcium (tested on mice model)	[46]
Vitamin D	Chitosan coated nanostructured lipid carriers (NLCs)	NA	Nanoemulsion of fortified mixed lipids (stearic and oleic acid were found to be most compatible)	Combination of melt-emulsification and electrostatic deposition	The chitosan coating exhibited protective effect against chemical and thermal degradation of the nanofortificants present in the emulsion	[47]
Vitamin D	Whole-fat milk	0.1 to 0.5 μg/mL	Nanoemulsion (soybean oil or mixtures of the oil with cocoa butter fortified with vitamin D)	High-pressure (1000 bar) homogenization	Stable food grade nanoemulsions fortified with vitamin D was produced and used for fortification of commercial food product	[48]

(*Continued*)

Table 14.1 Reported cases of nanofortification of liposoluble vitamins. (*Continued*)

Micronutrient/ fortificant	Vehicle/carrier of fortificant	Quantity of fortificant added	Nanofortified product/ formulation	Synthesis technique/s	Response or salient findings	Ref.
Vitamin D_3	Food grade materials (vitamin E acetate and span 60)	NA	Fortified nanoniosome	Thin layer hydration and sonication	The nanoniosomes exhibited high encapsulation efficiency (>75%) of vitamin D3 with high antioxidant activity	[49]
Vitamin D_3	Low-fat milk and yogurt	1500 IU	Nanoencapsulated vitamin D	Homogenization with high tensile stress and ultrasound	Consumption of fortified milk and yogurt improved insomnia symptoms (tested on human subjects)	[50]
Vitamin D_3	Lassi (milk based Indian beverage)	2 (% w/v)	Fortified nanocapsules	Phase inversion based cold water dilution method	Fortified nanocapsules were used successfully for fortification of lassi with acceptable sensory qualities.	[51]

(*Continued*)

Table 14.1 Reported cases of nanofortification of liposoluble vitamins. (*Continued*)

Micronutrient/ fortificant	Vehicle/carrier of fortificant	Quantity of fortificant added	Nanofortified product/ formulation	Synthesis technique/s	Response or salient findings	Ref.
Vitamin E	Oil in water emulsion of food grade materials (medium chain triglyceride oil)	8 (%w/v)	Fortified nanoemulsion	Spontaneous emulsification	Vitamin E-loaded nanoemulsion may be suitably used for fortifying transparent foods and beverages	[52]
Vitamin E	Food grade natural biopolymers (whey protein isolate and gum arabic)	NA	Fortified nanoemulsion	High-pressure (12000 psi) homogenization	Both biopolymers were found to be useful for preparing fortified stable nanoemulsions for application in food	[53]
Vitamin D_3	Food grade material (Soybean phosphatidylcholine)	NA	Fortified nanoliposome	Thin film hydration-sonication technique	The fortified nanoliposomes were found to be efficient carriers of vitamin D3 with improved shelf-life and can be used for delivery of the encapsulated vitamin	[54]

(*Continued*)

Table 14.1 Reported cases of nanofortification of liposoluble vitamins. (*Continued*)

Micronutrient/ fortificant	Vehicle/carrier of fortificant	Quantity of fortificant added	Nanofortified product/ formulation	Synthesis technique/s	Response or salient findings	Ref.
Vitamin E	Food grade materials (Canola oil, soya lecithin)	0.4-1.2 (% w/w)	Fortified nanoemulsion	High-pressure (75-155 MPa) homogenization	Stability of the fortified nanoemulsion was optimized for 1 (%w/w) vitamin E.	[55]
Vitamin D	Yogurt powder (with added modified starch, maltodextrin, gum Arabic, and milk protein concentrate	NA	Fortified nanoliposomal capsules	Homogenization and sonication	Fortified yogurt powder was prepared with encapsulated vitamin D using natural food grade ingredients suitable for application on food products.	[56]
Vitamin D	Pea protein isolate	2 (%w/v)	Fortified nanoaggregates and nanoemulsion	Ultrasonication	Fortified pea protein isolate enhanced the levels of Vitamin D in blood serum and biomarkers for bone turnover in vitamin D deficient rats.	[57]

(*Continued*)

Table 14.1 Reported cases of nanofortification of liposoluble vitamins. (*Continued*)

Micronutrient/ fortificant	Vehicle/carrier of fortificant	Quantity of fortificant added	Nanofortified product/ formulation	Synthesis technique/s	Response or salient findings	Ref.
Vitamin A	Food grade nonpurified soybean lecithins	0.125-2.5 (% w/w)	Fortified nanovesicles	Ethanol injection method	Low-cost nonpurified phospholipids have been found to be efficient materials for preparation of β-carotene fortified delivery tools suitable for food products	[58]
Vitamin D_3	Dairy products (milk and yogurt)	NA	Fortified nanoemulsion	Homogenization with high tensile stress and ultrasound	The correlation of nanoencapsulated vitamin D fortified dairy products on physical and mental well-being was reassured from human subject trials	[59]
Vitamin A (palmitate form)	Food grade ingredient (lecithin-cholesterol)	NA	Fortified nanoliposome	Thin layer hydration method and ultrasound assisted homogenization	High encapsulation efficiency of vitamin A palmitate was achieved which can be added to food products for fortification	[60]

NA: Not available

preparation of the nanoformulation containing the liposoluble vitamins, probably due to their stability in soluble forms. Table 14.1 also highlights the fact that for nanofortification of the fat-soluble vitamins, nanoemulsion is the most preferred option as the nanoformulation [47, 48, 52, 53, 55, 57, 59]. As mentioned earlier these are lipid-based nanosystems comprising/encapsulating the vitamins at their bulk core. Due to their transparent appearance they are very compatible with the beverages or other liquid food products for delivering the liposoluble vitamins without affecting the sensory properties of the vehicle food itself. Those can be transferred to solid forms by spray drying or freeze drying retaining the bioactivity intact and can be added to fortify various solid food products as well.

Synthesis of the nanoemulsion from the two base phases namely; water and oil/lipid requires application of physical or physicochemical enforcements. Homogenization under a high pressurized environment of 200-1000 bar (Table 14.1) have been observed to be a very popular mechanical technique for synthesizing nanoemulsion for vitamin fortification [46, 48, 53, 55]. Sometimes it is combined with ultrasonic treatment for better dispersion of the emulsion, Table 14.1. These are known as the high energy processes. Some low energy processes like spontaneous emulsification or phase inversion emulsification are also useful for nanoemulsion preparation [51, 52]. Other than these methods like thin film/layer hydration and ethanol injection are also used for nanoemulsion preparation [49, 54, 58, 60]. Multiple analyses performed in these studies affirmed the retention of the major chemical and physicochemical properties of the vitamins in their nanoformulated forms. The encapsulation of the vitamins in the emulsion or capsules facilitated protection from the light or oxygen driven decaying and increased shelf-life during storage and delivery [54]. In some of the studies reported in Table 14.1 major focus was kept on use of low-cost waste derived food grade materials for preparation of the nanoemulsion, as cost-effective production of nanofortified products is a matter of concern regarding their large scale application [53, 57, 58]. The studies investigating animal model testing (mice or human) with fortified food products commonly reported enhancement of bioavailability of the particular vitamin, increased absorption indicated by higher level of vitamin in blood serum and better mental-physical effect on the body [46, 50, 57, 59]. Hence, nanofortification of vitamins is a promising way of fortifying the population suffering from vitamin deficiency. However, several other aspects, particularly the limiting ones, are needed to be investigated in detail before permitting and adopting its wide application in a global scale.

14.5.2 Nanofortification of Minerals

Like the vitamins, another group of essential micronutrients, the fortification of the trace metal minerals are also being attempted through nanotechnology oriented methods [9, 42]. One approach of that is to add the nanoparticles of respective minerals in the fertilizers to be used for the cultivation of certain crops or plants for their bioaccumulation in the edible part of the crop/plant [61, 62]. This nanofertilizer mediated approach is somewhat overlapped with the agronomical practices under the biofortification technologies [62]. Hence, in the present chapter we are only considering those studies which reports preparation and/or addition of the nanoparticles or nanoformulations of the particular minerals to carrier or vehicle food products for directly reaching the consumers. Among the mineral fortification studies using the nanofortification approach, iron has been observed to be most prominent one.

Various types of food products have been chosen as the carrier or vehicles for the iron nanofortification. Very recently, in a study amyloid fibrils were synthesized from soy derived proteins like; soy β-conglycinin, soy glycinin and soy protein isolate, and their ability to reduce iron from Fe (III) to Fe (II) has been used for nanofortification of iron [63]. The iron reduction ability facilitated formation of fibril-Fe (II) nanoparticle composites. Since the bioavailability of the reduced form Fe (II) is more than Fe (III) state these fibril-Fe (II) nanoparticle composites has been used as fortificants indifferent liquid foods [63]. The fortified foods did not undergo any distinct color change, hence, are suitable for application as nanoiron-fortified foods [63]. In another study, stable nanoaggregates of whey protein isolate, gum arabic and ferric ions have been formed through self-assembly under heating (85°C) and acidic pH (4) condition [64]. The nanoaggregated iron exhibited better targeted release of iron with a good adsorptivity on oil-water interface. Based on these observations this nanoaggregated iron has been recommended as a promising fortificant for nanofortification of food emulsions [64]. Food-borne nanoparticles (FNs) with high water solubility were synthesized from roasting sturgeon as carriers for bioactive Fe (II) [65]. Significant amount (1 $\mu g\ mm^{-2}$) of the stable FN-Fe (II)) complex was found in the digestive tract which exhibited a high retention rate of almost 95.18% after completion of the gastrointestinal digestion [65]. Hence, its high potential for iron delivery through nanofortification was confirmed as a biocompatible, efficient and stable iron carrier [65]. Whey protein derived peptides obtained through ultrafiltration and hydrolysation treatments have been observed to have iron-chelating properties and used for synthesis of chelated peptide-iron nanocomplexes (size range 422.9 ±

3.41 nm) with iron content of 36.42 μg Fe/mg protein [66]. These high iron loaded peptide-iron nanocomplexes can be used as a promising alternative for chemical iron fortificants for multiple food products, as confirmed by these recent studies. Natural amyloid fibrils from the inexpensive milk protein β-lactoglobulin have been reported to possess natural reducing effects [67]. It was used as an anti-oxidizing nanocarrier and a colloidal stabilizer for reductive synthesis of iron nanoparticles. The hybrid material resulting from the interaction of β-lactoglobulin with $FeCl_3$, $6H_2O$ is a stable protein–iron colloidal dispersion of nanoparticle [67]. This nanocomposite released Fe ions during acidic and enzymatic digestion under *in vitro* mode. In mice model testing this nanocomposite showed high *in vivo* iron bioavailability and hence, recommended as potential nanodelivery systems for iron nanofortification for various solid and liquid food matrices [67]. Altogether these studies indicate that nanofortification of iron is a promising route for obtaining sufficient fortification of iron in deficient food products and to iron deficient populations.

To combat the deficiency of iron and development of iron deficiency anemia in infants and young children fortification of formula milk is attempted. In one study ^{57}Fe(III) oxo-hydroxide nanoparticles has been synthesized and used as the additive iron-fortificant to the formula milk powder [68]. The effect of nanofortification of the milk was tested using rat model systems and the absorption of Fe in the intestine was also monitored. It was observed that the iron was successfully delivered through the nanoparticles and absorbed in the body [68]. The iron level of the body was found to be increasing and reaching sufficient values, particularly in red blood cells, comparison to the nonfortified rat groups [68]. Nanoiron composites of ferric oxyhydroxide were synthesized based on chondroitin sulfate and protamine sulfate or chitosan in aqueous solution under neutral and ambient conditions [69]. The iron delivery efficiency was evaluated by using cell culture (polarized human intestinal epithelial (Caco-2) cells) which exhibited positive results [69]. This study indicated the feasibility of synthesizing bioavailable nanosized iron through a green route which is desirable for the application of nanofortification. The confirmation of efficient delivery and incorporation of the iron in intracellular environment was indicated experimentally by the use of two different animal cell lines namely; HT-29 and Caco-2 in one study [70]. Iron uptake by both cells through sugar (sucrose) coated iron oxide/hydroxide nanoparticles was found to be much higher than that achieved through the use of inorganic iron (from salt) and ferritin nanoparticles [70]. Interestingly, the cellular storage of iron after delivery was found to be higher in case of ferritin NPs than the sucrose coated iron oxide/hydroxide NPs. No apparent cell toxicity was observed, however, significant DNA damage was detected when the iron (Fe) concentration exceeded 500 μmol/L [70]. Hence, these studies confirmed that Fe(III) oxo-hydroxide nanoparticles are suitable carriers of iron for the purpose of nanofortification and should the adopted for fortification of humans after passing suitable trial tests using compatible vehicle foods.

Zinc is identified as the second most important mineral micronutrient after iron. Hence, the nanofortification of zinc is also being attempted in the recent years on different food vehicles for tackling zinc deficiency. A study used Zn nanoparticles for replacing the conventional bulk form of Zn (zinc oxide) using biscuits as the food vehicle [71]. The nano-zinc-fortified biscuits (13.5 ppm, 27 ppm and 54 ppm) were fed to rats suffering from Zn deficiency diseases. Observations revealed that, compared to the group fed with bulk Zn the ones fed with nanofortified Zn containing biscuits grew rapidly and overall healthy status was obtained [71]. No apparent toxicity was observed. The study recommended the 13.5 ppm dose of the Zn-NPs for tackling mild Zn deficiency and 27 ppm for the more severe cases of Zn deficiency [71]. Owing to the anti-microbial ability of Zn NPs, one study used them as the preservative for prevention of microbial degradation of minced fish meat (Tilapia) during storage [72]. Besides acting as the anti-microbial agent the Zn NPs will also act as the nanofortificants for delivering Zn to the consumers [72].

Possibilities of Zn nanofortification of the dairy based products have been researched by using whey protein isolate (WPI) nanocomposites as the encapsulating material, based on the fact that peptide groups possess the ability to bind mineral ions [73]. The results indicated that WPI nanocomposites were able to efficiently encapsulate significant amount of Zn (5-10 mM $ZnCl_2$) [73]. In a more recent study whey protein nanoparticles (WP-NPs) were used for encapsulating zinc citrate for combating oxidative stress complications in rats [74]. Use of Zn-WP-NPs exhibited better hepatoprotective effect and histological changes in rats. Hence, whey protein isolates can be used for Zn nanofortification of the compatible dairy products [74].

During search and survey of literatures on minerals fortification it was observed that other than iron (Fe), the technique commonly used for fortification of zinc (Zn) and selenium (Se) is agronomic practice [75, 76]. Even in nanotechnology oriented methods preparation and use of nanofertilizers containing these minerals in nanoscale form for direct fortification of soil and consequent translocation and accumulation in the plant body is a more preferred method than nanofortification of the vehicle foods. It is expected that in near future more studies will focus on the direct nanofortification of minerals using food carriers/vehicles.

14.5.3 Nanofortification of Bioactives

Nowadays the completeness of functional food is not bound to the conventionally known macro- and micro- nutrients anymore, rather, with progress and advancement of analytical research several compounds have also been identified as bioactive ones [77]. These bioactive compounds are also having specific beneficial effects on the modulation of the overall health and immunity of human body [77–79]. As many of these are not present together naturally in each and every food, their supply can only be possible through extraction from a rich natural source and fortification to particular food vehicle. In this regard nanofortification has gained notable popularity for delivery of several bioactive compounds through nanoscale carriers. A list of important bioactive compounds reported for nanofortification purpose is presented in Table 14.2.

The tabulated information primarily presents the applicability of nanofortification in case of several different types of bioactive compounds like precursor of vitamin (β-carotene), phenolic compounds, pigments, polyunsaturated fatty acids, flavonol, etc. [80–90]. In all of these studies regardless of the bioactive compound being used and the type of nanoformulation is prepared the nanofortificant was found to be stable. Hence, these can be used for fortification of specific food vehicles.

14.6 Overall Assessment of Nanofortified Foods: Performance and Limitations

From the analysis of the reported literatures on nanofortification of functional or bioactive compounds namely; vitamin, minerals and other bioactives it is highlighted that this field is observing swift progress in the recent years. Regardless of the fortificant being used for synthesis of nanostructured carriers/formulations, the studies reported successful results. The studies investigating the synthesis of nanocarrier embedded fortificants have monitored the physico-chemical characteristics like loading, stability, retention, release and shelf-life of the nanofortificants through different analytical techniques [46–60, 63–74, 80–90]. Some studies have furthered the investigation toward the assessment of compatibility of the synthesized nanofortificants for fortifying the particular food vehicles. The nanofortified food vehicles have been applied in different *in vivo* model systems like animal cell line/culture, rat/mice and/or human subjects in some of the studies [46, 50, 57, 59, 69, 70, 74]. Based on these facts, it can be inferred that the nanofortification technology is currently at its primary

Table 14.2 Nanofortification efforts of bioactive compounds and their function.

Bioactive compound	Function/importance in body	Nanocarriers/vehicles (components)	Ref.
β-carotene	• Used as natural colour/pigment in the food industry • Acts as vitamin A precursor • Acts as natural antioxidant for enhancement of immune response and facilitates protection against cancer and cardiovascular disorders	Nanostructured lipid (Glyceryl tristearate and high oleic sunflower oil)	[80]
		Nanostructured lipid carriers (glycerol palmito stearate, octyl octanoate, Poloxamer 407)	[81]
β-sitosterol	• Anti-inflammatory and anti-tumor effect against many diseases • Reduction of total cholesterol, LDL-cholesterol levels in blood • Prevention of arteriosclerosis and cardiovascular diseases • Exhibit antidiabetic and antioxidant activities • Reported to have gallstone and prostatitis reducing activities	Nanostructured lipid carriers (Glyceryl palmtostearate, Polyoxyethylene sorbitan monooleate, Poly ethylene Glycol (PEG), Medium chain triglycerides)	[82]
		Nanostructured lipid carriers (Propolis wax, glyceryl behenate and pomegranate seed oil)	[83]

(Continued)

Table 14.2 Nanofortification efforts of bioactive compounds and their function. (*Continued*)

Bioactive compound	Function/importance in body	Nanocarriers/vehicles (components)	Ref.
Rutin (glycoside)	• Identified as a form of quercetin belonging to the group of flavonols which are important bioactive compounds due to its significant radical scavenging activity • Known to have nutritional and health promoting effects • Provides antioxidant effects against several oxidizing radicals (hydroxyl, superoxide and peroxyl) through radical-scavenging activity	Nanostructured lipid carriers (Cacao butter, oleic acid, citric acid, Tween 80)	[84]
Resveratrol (polyphenol)	• Identified as a natural antioxidant which prevents free radical activities causing oxidative stress triggered cardiovascular diseases, atherosclerosis cancer, etc. • It has hepato-protective, anti-inflammatory, anti-diabetic, anti-obesity, anti-ageing and neuro-protective effects and activities	Pro-niosome or nanovescicles (maltodextrin, lactose monohydrate, pullulan, Span 60, cholesterol, ethanol)	[85]
		Nanoencapsulation (starch from water chestnut, lotus stem and horse chestnut)	[86]
Phenolic compounds	• Identified as important micronutrients for functional foods • Acts as natural antioxidants for prevention of certain diseases i.e., cancer, internal inflammation and neurodegenerative disorders	Nanoemulsion (soybean oil, span 80, olive leaf extract, pectin)	[87]

(*Continued*)

Table 14.2 Nanofortification efforts of bioactive compounds and their function. (*Continued*)

Bioactive compound	Function/importance in body	Nanocarriers/vehicles (components)	Ref.
Eicosapentaenoic acid (EPA), Docosahexaenoic acid (DHA)	• The long-chain polyunsaturated fatty acids are beneficial for health • Alleviates cardiovascular disease, nervous system disorder, autoimmune disorders, regulates blood pressure and anti-inflammatory • Aids in proper development and functioning of the cells of brain and retina	Nanoliposomes (fish oil, sunflower oil and lecithin)	[88, 89]
Carotenoids	• Used as natural food colorants and fragrances • Used as feed additives for combating various chronic diseases like skin cancer and cardiovascular disorders	Nanoemulsion (limonene, glycerol, water and nonionic surfactants)	[90]

phase considering overall development. For some specific micronutrients, like vitamin D or iron, the progress can be considered as early stage of advanced phase, as the nanofortification studies of these have been tried in real animal or human systems and positive results have been obtained. It is expected that, in the near future, nanofortification will gain more popularity as the major route of fortification for all micronutrients under focus.

However, safe use and administration of the nanofortificants is a primary concern of this field. Although human system is adapted to natural nanoparticles, toxicity of engineered nanoparticles on microorganisms and mammalian cells is reported in many cases, hence, it must be considered as a safety concern before full adoption of this method can be realized [91, 92]. The major physicochemical factors responsible for the toxicity of NPs may be listed as: 1) particle size and high surface area; 2) surface charge, groups, or coating; 3) shape and morphology; 4) high penetrance and translocation; 5) high persistence and stability; and 6) composition and stability [93, 94]. Multiple review articles are available which discussed about this issue in detail. Based on the results of toxicology studies of NPs, some guidelines and recommendations have been set for application of NPs in foods regarding the monitoring of toxicity, exposure, detection and safe handling practices [95, 96]. One of which is strict use of generally recognized as safe (GRAS) materials for synthesis of the nanomaterials for fortification. Other than assessing the primary and potent toxicity characteristics of the NPs the dosage permissible for safe administration has to be optimized in case of specific nanofortificants. Most importantly, the biochemical and molecular basis of interaction of the nanofortificants with that of the cells of the biosystem in which it is being administered has to be studied and understood clearly before approving its real system application. However, gaining complete knowledge of the longer-term effects of the engineered nanofortificants is going to take more time, which is the major limiting factor for the implementation of nanofortification of many micronutrients.

14.7 Conclusion and Recommendations

Latest research on nanofortification has revealed that this technology is fully compatible and widely applicable to an array of essential micronutrient species, including vitamins, minerals, and multiple varieties of bioactive compounds. The compatibility of the nanofortificants for addition to specific food carrier/vehicles has also been affirmed by multiple *in vitro* studies and found to be suitable for targeting some staple foods.

Nanofortificants developed for each of these nutrients have been found to exhibit positive effects regarding their delivery, bioavailability, absorption, and elevation of levels in the body toward the sufficient values. More importantly, the concept and application of functional foods has now reached new dimensions due to the current spread of the severe global pandemic. In this regard, nanofortification has a major role to play in near future as it is conceptualized on direct delivery of the fortificants to the body through food vehicles. However, specific human trials are not yet conducted for all types of nanofortificants due to the safety concerns associated with the NPs. Understanding of the molecular mechanisms of interaction of nanofortificants with the components of the gastrointestinal system (enzymes, ions, etc.) is crucial for its safe and successful implementation. Combined conduction of the *in vivo* and *in vitro* studies and advanced monitoring and analysis of the results is required to achieve this objective. Global adoption of this promising technology will require conjugated attention from the scientists, policy makers, food manufacturers, consumer educators, and consumers and must be implemented through strategically planned fortification programs. Therefore, ample future scope of research, innovation, and development is due in this field.

Acknowledgments

The authors express their gratitude toward the editors of this book for inviting and giving the opportunity to contribute this chapter to be a part of the book.

References

1. Ritchie, H. and Roser, M., *Micronutrient deficiency*, Our World in data, 2017. https://ourworldindata.org/micronutrient-deficiency
2. Lukaski, H.C., Vitamin and mineral status: Effects on physical performance. *Nutrition*, 20, 632–644, 2004.
3. Whatham, A., Bartlett, H., Eperjesi, F., Blumenthal, C., Allen, J., Suttle, C., Gaskin, K., Vitamin and mineral deficiencies in the developed world and their effect on the eye and vision. *Ophthalmic Physiol. Opt.*, 28, 1–12, 2008.
4. Dary, O. and Hurrell, R., *Guidelines on food fortification with micronutrients*, pp. 1–376, World Health Organization, Food and Agricultural Organization of the United Nations, Geneva, Switzerland, 2006.

5. Palacios, C., Cormick, G., Hofmeyr, G.J., Garcia-Casal, M.N., Peña-Rosas, J.P., Betrán, A.P., Calcium-fortified foods in public health programs: Considerations for implementation. *Ann. N. Y. Acad. Sci.*, 1485, 1, 3, 2021.
6. Platel, K. and Srinivasan, K., Bioavailability of micronutrients from plant foods: An update. *Crit. Rev. Food Sci. Nutr.*, 56, 1608–1619, 2016.
7. Gorji, A. and Ghadiri, M.K., The potential roles of micronutrient deficiency and immune system dysfunction in COVID-19 pandemic. *Nutrition*, 82, 111047, 2020.
8. Gasmi, A., Tippairote, T., Mujawdiya, P.K., Peana, M., Menzel, A., Dadar, M., Benahmed, A.G., Bjørklund, G., Micronutrients as immunomodulatory tools for COVID-19 management. *Clin. Immunol.*, 220, 108545, 2020.
9. Ohanenye, I.C., Emenike, C.U., Mensi, A., Medina-Godoy, S., Jin, J., Ahmed, T., Sun, X., Udenigwe, C.C., Food fortification technologies: Influence on iron, zinc and vitamin A bioavailability and potential implications on micronutrient deficiency in sub-Saharan Africa. *Sci. Afr.*, 11, e00667, 2020.
10. Milner, J.A., Functional foods: The US perspective. *Am. J. Clin. Nutr.*, 71, 1654S–1659S, 2000.
11. Hasler, C.M., The changing face of functional foods. *J. Am. Coll. Nutr.*, 19, 499S–506S, 2000.
12. Lu, X., Chen, J., Guo, Z., Zheng, Y., Rea, M.C., Su, H., Zheng, X., Zheng, B., Miao, S., Using polysaccharides for the enhancement of functionality of foods: A review. *Trends Food Sci. Technol.*, 86, 311–327, 2019.
13. Rein, M.J., Renouf, M., Cruz-Hernandez, C., Actis-Goretta, L., Thakkar, S.K., da Silva Pinto, M., Bioavailability of bioactive food compounds: A challenging journey to bioefficacy. *Br. J. Clin. Pharmacol.*, 75, 588–602, 2013.
14. Neilson, A.P., Goodrich, K.M., Ferruzzi, M.G., Bioavailability and metabolism of bioactive compounds from foods, in: *Nutrition in the Prevention and Treatment of Disease*, pp. 301–319, Academic Press, Elsevier, Cambridge, Massachusetts, United States, 2017.
15. Pérez-Esteve, É., Ruiz-Rico, M., Martínez-Máñez, R., Barat, J.M., Mesoporous silica-based supports for the controlled and targeted release of bioactive molecules in the gastrointestinal tract. *J. Food Sci.*, 80, E2504–E2516, 2015.
16. Ezhilarasi, P.N., Karthik, P., Chhanwal, N., Anandharamakrishnan, C., Nanoencapsulation techniques for food bioactive components: A review. *Food Bioprocess Technol.*, 6, 628–647, 2013.
17. Salvia-Trujillo, L., Martín-Belloso, O., McClements, D.J., Excipient nanoemulsions for improving oral bioavailability of bioactives. *Nanomaterials*, 6, 17, 2016.
18. Hu, B., Liu, X., Zhang, C., Zeng, X., Food macromolecule based nanodelivery systems for enhancing the bioavailability of polyphenols. *J. Food Drug Anal.*, 25, 3–15, 2017.
19. Lockyer, S., White, A., Buttriss, J.L., Biofortified crops for tackling micronutrient deficiencies–what impact are these having in developing countries

and could they be of relevance within Europe? *Nutr. Bull.*, 43, 319–357, 2018.

20. Dary, O. and Hurrell, R., *Guidelines on food fortification with micronutrients*, pp. 1–376, World Health Organization, Food and Agricultural Organization of the United Nations, Geneva, Switzerland, 2006.
21. Dwyer, J.T., Wiemer, K.L., Dary, O., Keen, C.L., King, J.C., Miller, K.B., Philbert, M.A., Tarasuk, V., Taylor, C.L., Gaine, P.C., Jarvis, A.B., Bailey, R.L., Fortification and health: Challenges and opportunities. *Adv. Nutr.*, 6, 124–131, 2015.
22. Kwak, N.S. and Jukes, D.J., Functional foods. Part 2: The impact on current regulatory terminology. *Food Control*, 12, 109–117, 2001.
23. Kerksick, C.M. and Kulovitz, M., Requirements of energy, carbohydrates, proteins and fats for athletes, in: *Nutrition and enhanced sports performance*, pp. 355–366, Academic Press, Elsevier, Cambridge, Massachusetts, United States, 2013.
24. Singh, M., Role of micronutrients for physical growth and mental development. *Indian J. Pediatr.*, 71, 59–62, 2004.
25. Tolonen, M., *Vitamins and minerals in health and nutrition*, Elsevier, Amsterdam, Netherlands, 1990.
26. Suri, D.J. and Tanumihardjo, S.A., Effects of different processing methods on the micronutrient and phytochemical contents of maize: From A to Z. *Compr. Rev. Food Sci. Food Saf.*, Amsterdam, Netherlands, 15, 912–926, 2016.
27. Lee, S., Choi, Y., Jeong, H.S., Lee, J., Sung, J., Effect of different cooking methods on the content of vitamins and true retention in selected vegetables. *Food Sci. Biotechnol.*, 27, 333–342, 2018.
28. Caballero, B., *Encyclopedia of human nutrition*, Elsevier, Amsterdam, Netherlands, 2005.
29. Babu, U.S. and Calvo, M.S., Modern India and the vitamin D dilemma: Evidence for the need of a national food fortification program. *Mol. Nutr. Food Res.*, 54, 1134–1147, 2010.
30. Lynch, S.R., The impact of iron fortification on nutritional anaemia. *Best Pract. Res. Clin. Haematol.*, 18, 333–346, 2005.
31. Mannar, M.V. and Hurrell, R.F., Food fortification: Past experience, current status, and potential for globalization, in: *Food fortification in a globalized world*, pp. 3–11, Academic Press, Elsevier, Cambridge, Massachusetts, United States, 2018.
32. Morlacchi, L., Mallardi, D., Giannì, M.L., Roggero, P., Amato, O., Piemontese, P., Consonni, D., Mosca, F., Is targeted fortification of human breast milk an optimal nutrition strategy for preterm infants? An interventional study. *J. Transl. Med.*, 14, 1–8, 2016.
33. Klassen-Wigger, P. and Barclay, D.V., Market-Driven Fortification, in: *Food Fortification in a Globalized World*, pp. 63–67, Academic Press, Elsevier, Cambridge, Massachusetts, United States, 2018.

34. Chadare, F.J., Idohou, R., Nago, E., Affonfere, M., Agossadou, J., Fassinou, T.K., Kénou, C., Honfo, S., Azokpota, P., Linnemann, A.R., Hounhouigan, D.J., Conventional and food-to-food fortification: An appraisal of past practices and lessons learned. *Food Sci. Nutr.*, 7, 2781–2795, 2019.
35. Kruger, J., Taylor, J.R., Ferruzzi, M.G., Debelo, H., What is food-to-food fortification? A working definition and framework for evaluation of efficiency and implementation of best practices. *Compr. Rev. Food Sci. Food Saf.*, 19, 3618–3658, 2020.
36. Saltzman, A., Birol, E., Bouis, H.E., Boy, E., De Moura, F.F., Islam, Y., Pfeiffer, W.H., Biofortification: Progress toward a more nourishing future. *Global Food Secur.*, 2, 9–17, 2013.
37. Bouis, H.E. and Saltzman, A., Improving nutrition through biofortification: A review of evidence from HarvestPlus, 2003 through 2016. *Global Food Secur.*, 12, 49–58, 2017.
38. Khush, G.S., Lee, S., Cho, J.I., Jeon, J.S., Biofortification of crops for reducing malnutrition. *Plant Biotechnol. Rep.*, 6, 195–202, 2012.
39. Jampilek, J., Kos, J., Kralova, K., Potential of nanomaterial applications in dietary supplements and foods for special medical purposes. *Nanomaterials*, 9, 296, 2019.
40. Katouzian, I. and Jafari, S.M., Nano-encapsulation as a promising approach for targeted delivery and controlled release of vitamins. *Trends Food Sci. Technol.*, 53, 34–48, 2016.
41. Bumbudsanpharoke, N. and Ko, S., Nano-food packaging: An overview of market, migration research, and safety regulations. *J. Food Sci.*, 80, R910–R923, 2015.
42. Fathima, S.J., Nallamuthu, I., Khanum, F., Vitamins and minerals fortification using nanotechnology: Bioavailability and Recommended Daily Allowances, in: *Nutrient Delivery*, pp. 457–496, Academic Press, Elsevier, Cambridge, Massachusetts, United States, 2017.
43. Jampilek, J. and Kralova, K., Potential of nanonutraceuticals in increasing immunity. *Nanomaterials*, 10, 2224, 2020.
44. Maurya, V.K. and Aggarwal, M., Enhancing bio-availability of vitamin D by nano-engineered based delivery systems-An overview. *Int. J. Curr. Microbiol. Appl. Sci.*, 6, 340–53, 2017.
45. Wasserman, R.H. and Taylor, A.N., Metabolic roles of fat-soluble vitamins D, E, and K. *Annu. Rev. Biochem.*, 41, 179–202, 1972.
46. Erfanian, A., Mirhosseini, H., Rasti, B., Hair-Bejo, M., Mustafa, S.B., Manap, M.Y.A., Absorption and bioavailability of nano-size reduced calcium citrate fortified milk powder in ovariectomized and ovariectomized-osteoporosis rats. *J. Agric. Food Chem.*, 63, 5795–5804, 2015.
47. Rabelo, R.S., Oliveira, I.F., da Silva, V.M., Prata, A.S., Hubinger, M.D., Chitosan coated nanostructured lipid carriers (NLCs) for loading Vitamin D: A physical stability study. *Int. J. Biol. Macromol.*, 119, 902–912, 2018.

48. Golfomitsou, I., Mitsou, E., Xenakis, A., Papadimitriou, V., Development of food grade O/W nanoemulsions as carriers of vitamin D for the fortification of emulsion based food matrices: A structural and activity study. *J. Mol. Liq.*, 268, 734–742, 2018.
49. Talebi, V., Ghanbarzadeh, B., Hamishehkar, H., Pezeshki, A., Ostadrahimi, A., Effects of different stabilizers on colloidal properties and encapsulation efficiency of vitamin D3 loaded nano-niosomes. *J. Drug Delivery Sci. Technol.*, 61, 101284, 2019.
50. Sharifan, P., Khoshakhlagh, M., Khorasanchi, Z., Darroudi, S., Rezaie, M., Safarian, M., Vatanparast, H., Afshari, A., Ferns, G., Ghazizadeh, H., Ghayour Mobarhan, M., Efficacy of low-fat milk and yogurt fortified with encapsulated vitamin D3 on improvement in symptoms of insomnia and quality of life: Evidence from the SUVINA trial. *Food Sci. Nutr.*, 8, 4484–4490, 2020.
51. Maurya, V.K. and Aggarwal, M., Fabrication of nano-structured lipid carrier for encapsulation of vitamin D3 for fortification of 'Lassi'; A milk based beverage. *J. Steroid Biochem. Mol. Biol.*, 193, 105429, 2019.
52. Saberi, A.H., Fang, Y., McClements, D.J., Fabrication of vitamin E-enriched nanoemulsions: Factors affecting particle size using spontaneous emulsification. *J. Colloid Interface Sci.*, 391, 95–102, 2013.
53. Ozturk, B., Argin, S., Ozilgen, M., McClements, D.J., Formation and stabilization of nanoemulsion-based vitamin E delivery systems using natural biopolymers: Whey protein isolate and gum arabic. *Food Chem.*, 188, 256–263, 2015.
54. Mohammadi, M., Ghanbarzadeh, B., Hamishehkar, H., Formulation of nanoliposomal vitamin D3 for potential application in beverage fortification. *Adv. Pharm. Bull.*, 4, 569, 2014.
55. Mehmood, T., Optimization of the canola oil based vitamin E nanoemulsions stabilized by food grade mixed surfactants using response surface methodology. *Food Chem.*, 183, 1 7, 2015.
56. Jafari, S.M., Vakili, S., Dehnad, D., Production of a functional yogurt powder fortified with nanoliposomal vitamin D through spray drying. *Food Bioprocess Technol.*, 12, 1220–1231, 2019.
57. Almajwal, A.M., Abulmeaty, M., Feng, H., Alruwaili, N.W., Dominguez-Uscanga, A., Andrade, J.E., Razak, S., ElSadek, M.F., Stabilization of vitamin D in pea protein isolate nanoemulsions increases its bioefficacy in rats. *Nutrients*, 11, 75, 2019.
58. Michelon, M., Mantovani, R.A., Sinigaglia-Coimbra, R., de la Torre, L.G., Cunha, R.L., Structural characterization of β-carotene-incorporated nanovesicles produced with non-purified phospholipids. *Food Res. Int.*, 79, 95–105, 2016.
59. Sharifan, P., Bagherniya, M., Bajgiran, M.M., Safarian, M., Vatanparast, H., Eslami, S., Tayefi, M., Khadem-Rezaiyan, M., Baygan, A., Khoshakhlagh, M., Ghayour-Mobarhan, M., The efficacy of dairy products fortified with nano-encapsulated vitamin D3 on physical and mental aspects of the health

in obese subjects; the protocol of the SUVINA trial. *Transl. Metab. Syndr. Res.*, 4, 1–9, 2021.

60. Pezeshky, A., Ghanbarzadeh, B., Hamishehkar, H., Moghadam, M., Babazadeh, A., Vitamin A palmitate-bearing nanoliposomes: Preparation and characterization. *Food Biosci.*, 13, 49–55, 2016.
61. Elemike, E.E., Uzoh, I.M., Onwudiwe, D.C., Babalola, O.O., The role of nanotechnology in the fortification of plant nutrients and improvement of crop production. *Appl. Sci.*, 9, 499, 2019.
62. Raliya, R., Saharan, V., Dimkpa, C., Biswas, P., Nanofertilizer for precision and sustainable agriculture: Current state and future perspectives. *J. Agric. Food Chem.*, 66, 6487–6503, 2017.
63. Xiang, N., Wu, S., Wei, Z., Shao, P., Sun, P., Characterization of iron reducibility of soy protein amyloid fibrils and their applications in iron fortification. *Food Chem.*, 353, 129420, 2021.
64. Yao, X., Xu, K., Shu, M., Liu, N., Li, N., Chen, X., Nishinari, K., Phillips, G.O., Jiang, F., Fabrication of iron loaded whey protein isolate/gum Arabic nanoparticles and its adsorption activity on oil-water interface. *Food Hydrocolloids*, 115, 106610, 2021.
65. Song, Y., Liu, K., Su, W., Hou, S., Che, T., Tan, M., Construction and evaluation of an iron delivery system by ultra-small nanoparticles from roast sturgeon (*Acipenser schrenckiid*). *Food Funct.*, 12, 1147–1155, 2021.
66. Athira, S., Mann, B., Sharma, R., Pothuraju, R., Bajaj, R.K., Preparation and characterization of iron-chelating peptides from whey protein: An alternative approach for chemical iron fortification. *Food Res. Int.*, 141, 110133, 2021.
67. Shen, Y., Posavec, L., Bolisetty, S., Hilty, F.M., Nyström, G., Kohlbrecher, J., Hilbe, M., Rossi, A., Baumgartner, J., Zimmermann, M.B., Mezzenga, R., Amyloid fibril systems reduce, stabilize and deliver bioavailable nanosized iron. *Nat. Nanotechnol.*, 12, 642, 2017.
68. Fernández-Menéndez, S., Peixoto, R.R.A., Fernández-Colomer, B., Romero, M.C., Sanz-Medel, A., Fernández-Sánchez, M.L., Searching for enhanced iron fortification of formula milk via nanoparticles and Isotope Pattern Deconvolution. *Spectrochim. Acta Part B: At. Spectrosc.*, 148, 165–171, 2018.
69. Feng, G., Tang, S., Gang, Y., Zeng, M., Wu, H., Chondroitin sulfate and its nanocomposites with protamine or chitosan stabilize and deliver available nanosized iron. *Int. J. Biol. Macromol.*, 150, 501–508, 2020.
70. Turiel-Fernández, D., Bettmer, J., Montes-Bayón, M., Evaluation of the uptake, storage and cell effects of nano-iron in enterocyte-like cell models. *J. Trace Elem. Med. Biol.*, 49, 98–104, 2018.
71. Raya, S.D.H.A., Hassan, M.I., Farroh, K.Y., Hashim, S.A., Salaheldin, T.A., Zinc oxide nanoparticles fortified biscuits as a nutritional supplement for zinc deficient rats. *J. Nanomed. Res.*, 4, 10–15406, 2016.

72. Pati, K., Chowdhury, S., Nath, S., Murmu, P., Rahman, F.H., Extending the shelf life of tilapia mince by zinc oxide nanoparticle–A precursor of value-added fishery product. *Curr. J. Appl. Sci. Technol.*, 39, 73–82, 2020.
73. Gülseren, İ., Fang, Y., Corredig, M., Zinc incorporation capacity of whey protein nanoparticles prepared with desolvation with ethanol. *Food Chem.*, 135, 770–774, 2012.
74. Hassan, M.A., El-Nekeety, A.A., Abdel-Aziem, S.H., Hassan, N.S., Abdel-Wahhab, M.A., Zinc citrate incorporation with whey protein nanoparticles alleviate the oxidative stress complication and modulate gene expression in the liver of rats. *Food Chem. Toxicol.*, 125, 439–451, 2019.
75. Bala, R., Kalia, A., Dhaliwal, S.S., Evaluation of efficacy of ZnO nanoparticles as remedial zinc nanofertilizer for rice. *J. Soil Sci. Plant Nutr.*, 19, 379–389, 2019.
76. El-Ramady, H., Abdalla, N., Taha, H.S., Alshaal, T., El-Henawy, A., Salah, E.D.F., Shams, M.S., Youssef, S.M., Shalaby, T., Bayoumi, Y., Schnug, E., Selenium and nano-selenium in plant nutrition. *Environ. Chem. Lett.*, 14, 123–147, 2016.
77. Bernhoft, A., A brief review on bioactive compounds in plants. *Bioactive compounds in plants-benefits and risks for man and animals*, vol. 50, pp. 11–17, 2010.
78. Liu, R.H., Dietary bioactive compounds and their health implications. *J. Food Sci.*, 78, A18–A25, 2013.
79. Patil, B.S., Jayaprakasha, G.K., Chidambara Murthy, K.N., Vikram, A., Bioactive compounds: Historical perspectives, opportunities, and challenges. *J. Agric. Food Chem.*, 57, 8142–8160, 2009.
80. Oliveira, D.R.B., Michelon, M., de Figueiredo Furtado, G., Sinigaglia-Coimbra, R., Cunha, R.L., β-Carotene-loaded nanostructured lipid carriers produced by solvent displacement method. *Food Res. Int.*, 90, 139–146, 2016.
81. Pezeshki, A., Hamishehkar, H., Ghanbarzadeh, B., Fathollahy, I., Nahr, F.K., Heshmati, M.K., Mohammadi, M., Nanostructured lipid carriers as a favorable delivery system for β-carotene. *Food Biosci.*, 27, 11–17, 2019.
82. Bagherpour, S., Alizadeh, A., Ghanbarzadeh, S., Mohammadi, M., Hamishehkar, H., Preparation and characterization of Betasitosterol-loaded nanostructured lipid carriers for butter enrichment. *Food Biosci.*, 20, 51–55, 2017.
83. Soleimanian, Y., Goli, S.A.H., Varshosaz, J., Maestrelli, F., Propolis wax nanostructured lipid carrier for delivery of β sitosterol: Effect of formulation variables on physicochemical properties. *Food Chem.*, 260, 97–105, 2018.
84. Babazadeh, A., Ghanbarzadeh, B., Hamishehkar, H., Novel nanostructured lipid carriers as a promising food grade delivery system for rutin. *J. Funct. Foods*, 26, 167–175, 2016.
85. Shruthi, P.A., Pushpadass, H.A., Franklin, M.E.E., Battula, S.N., Naik, N.L., Resveratrol-loaded proniosomes: Formulation, characterization and fortification. *LWT*, 134, 110127, 2020.

86. Ahmad, M. and Gani, A., Development of novel functional snacks containing nano-encapsulated resveratrol with anti-diabetic, anti-obesity and antioxidant properties. *Food Chem.*, 352, 129323, 2021.
87. Mohammadi, A., Jafari, S.M., Assadpour, E., Esfanjani, A.F., Nanoencapsulation of olive leaf phenolic compounds through WPC–pectin complexes and evaluating their release rate. *Int. J.Biol. Macromol.*, 82, 816–822, 2016.
88. Ojagh, S.M. and Hasani, S., Characteristics and oxidative stability of fish oil nano-liposomes and its application in functional bread. *J. Food Meas. Charact.*, 12, 1084–1092, 2018.
89. Ghorbanzade, T., Jafari, S.M., Akhavan, S., Hadavi, R., Nano-encapsulation of fish oil in nano-liposomes and its application in fortification of yogurt. *Food Chem.*, 216, 146–152, 2017.
90. Chaari, M., Theochari, I., Papadimitriou, V., Xenakis, A., Ammar, E., Encapsulation of carotenoids extracted from halophilic Archaea in oil-in-water (O/W) micro- and nano-emulsions. *Colloids Surf. B: Biointerfaces*, 161, 219–227, 2018.
91. Card, J.W., Zeldin, D.C., Bonner, J.C., Nestmann, E.R., Pulmonary applications and toxicity of engineered nanoparticles. *Am. J. Physiol.-Lung Cell. Mol. Physiol.*, 295, L400–L411, 2018.
92. Lojk, J., Repas, J., Veranič, P., Bregar, V.B., Pavlin, M., Toxicity mechanisms of selected engineered nanoparticles on human neural cells *in vitro*. *Toxicology*, 432, 152364, 2020.
93. Ganguly, P., Breen, A., Pillai, S.C., Toxicity of nanomaterials: Exposure, pathways, assessment, and recent advances. *ACS Biomater. Sci. Eng.*, 4, 2237–2275, 2018.
94. Sukhanova, A., Bozrova, S., Sokolov, P., Berestovoy, M., Karaulov, A., Nabiev, I., Dependence of nanoparticle toxicity on their physical and chemical properties. *Nanoscale Res. Lett.*, 13, 1–21, 2018.
95. McClements, D.J. and Xiao, H., Is nano safe in foods? Establishing the factors impacting the gastrointestinal fate and toxicity of organic and inorganic food-grade nanoparticles. *NPJ Sci. Food*, 1, 1–13, 2017.
96. Chau, C.F., Wu, S.H., Yen, G.C., The development of regulations for food nanotechnology. *Trends Food Sci. Technol.*, 18, 269–280, 2007.

Index

Printed and bound by CPI Group (UK) Ltd, Croydon, CR0 4YY
17/09/2026
14964275-0004